Nutraceuticals and Functional Foods as Immunomodulators

Rajesh K. Kesharwani • Raj K. Keservani •
Anil K. Sharma
Editors

Nutraceuticals and Functional Foods as Immunomodulators

Springer

Editors
Rajesh K. Kesharwani
Department of Computer Application
Nehru Gram Bharati (Deemed to be University)
Prayagraj, Uttar Pradesh, India

Raj K. Keservani
Faculty of B. Pharmacy
CSM Group of Institutions
Prayagraj, Uttar Pradesh, India

Anil K. Sharma
All India Council for Technical Education
New Delhi, India

ISBN 978-981-19-2506-1 ISBN 978-981-19-2507-8 (eBook)
https://doi.org/10.1007/978-981-19-2507-8

This Springer imprint is published by the registered company Springer Nature Singapore Pte Ltd.
The registered company address is: 152 Beach Road, #21-01/04 Gateway East, Singapore 189721, Singapore

Preface

The food ingredients have been considered essential for maintenance of routine body functionalities from the beginning of human civilization. On several instances, the food has demonstrated to have effects such as cure, treatment, mitigation, or alleviation of manifestations of ailments. The same spirit could be witnessed in famous quote of Hippocrates *Let food be thy medicine and medicine be thy food.*

The market for nutraceuticals and functional foods attained peaks following the health-related claims; this triggered the regulators across the globe to regulate the whole domain of nutraceuticals. Now only safe, pure, and efficacious products are given market authorizations. The emergence of nutraceuticals and functional foods as alternative to synthetic drugs has been proved to be beneficial in numerous clinical conditions. The need for immunomodulators as a prophylactic has witnessed an exponential demand during the disastrous pandemic Covid-19 which of course has led to an increased awareness and consumption of nutraceuticals and functional foods possessing immunity boosting abilities.

In view of manifold merits associated with nutraceuticals and functional foods, this book has strived to present its readers a holistic picture of these products as immunomodulators. The chapters are contributed by experts from academia as well as industry so as to provide the nuances of domain in an elaborated fashion.

This book has 14 chapters which have been divided into two parts.

Part I: Nutraceuticals and Functional Foods

The role of nanotechnology-assisted immunotherapy has been described in Chap. 1 entitled "Nanotechnology-based bacterial immunotherapy" written by Rajeshwar Kamal Kant Arya and colleagues. The authors have highlighted the role of immunity in cancer, and role of bacteria in cancer genesis, how the bacterial immunotherapy used in combating with cancer, and how the nanotechnology-based bacterial immunotherapy works on cancer regression.

The basics of immune response and overview of different nutrients as immunomodulators have been presented in Chap. 2 entitled "Immunomodulatory properties of nutraceuticals and functional foods" written by Adriana García-Gurrola and associates. The chapter begins with introduction to immune system followed by relationship between immunity and nutrition. Then many food ingredients having a role in immunomodulation are discussed.

The structure activity relationship of phytoconstituents in immunomodulation is discussed in Chap. 3 entitled "Immunoregulatory bioactive phytoconstituents: recent trends and future challenges" written by Sreeharsha Nagaraja and group. The chapter embraces introduction to immunomodulators, examples, and applications of isolated phytoconstituents. The primary focus of this chapter is to provide a deep insight into scientific data on immunomodulatory properties of plant-derived nutraceuticals, mechanism of action, and challenges and clinical restraints.

The pharmacological immunity modifying action of plant-derived substances has been given in Chap. 4 entitled "Effect of plant-derived immunomodulators on the immune system" drafted by Divya Vani Koraganji and associates. This chapter starts with introduction to immunomodulators and naturally existing plant-derived compounds and types. It also discusses the quercetin and its derived compounds with physicochemical properties and bioavailability. Further, it highlights the mechanism of action of quercetin on immune system with future challenges.

The way polyphenols facilitate in modifying immunity is described in Chap. 5 entitled "Polyphenols and its effect on immune system" written by Kanchan Gairola and colleagues. The chapter begins with a general introduction to immunomodulators with specific description of polyphenols covering mechanism of action and clinical applications.

The mineral nutraceuticals and their role in immunity is reviewed in Chap. 6 entitled "Mineral nutraceuticals and immunity enhancement" written by Manoj Kumar Mishra and coworkers. The chapter emphasizes on probiotics, essential amino acids, vitamins, dietary antioxidants, β-glucans, and other immunomodulators. In the global marketplace, nutraceuticals and functional foods have become a multi-billion-dollar industry, but the optimum intake level and recommended amounts of functional food have not yet been established which can be overcome by both experimental animals and in humans.

Chapter 7 entitled "Nutraceuticals as disease preventive food and immunity boosters" written by Bhushan R. Rane and group. The chapter covers introduction, classification, and applications of a variety of nature-borne substances having immense potential in management of ailments involving compromised immune status.

Part II: Micronutrients and Macronutrients

In Chap. 8 entitled "Current and future prospects of flavonoids for human immune system" authored by Sippy Singh and Durgesh Singh, polyphenolic compounds that promote human immune system and protect body against several health ailments are discussed. Flavonoids are potent substances which are widely used at present because of their antioxidative, anti-inflammatory, antiviral/bacterial, cardioprotective, antidiabetic, antiaging, anticarcinogenic properties and still have more to be explored and used in future. This chapter provides view of current and futuristic application of flavonoids.

The significance of Resveratrol in immunomodulation is described in Chap. 9 entitled "Resveratrol and immunomodulation" drafted by Mayela Govea Salas and associates. The chapter depicts a wholesome picture of this phytoconstituent covering its chemistry, mechanism of action, and applications in human health.

The healthy benefits offered by household condiment turmeric are quite known to mankind for a long time. The immunomodulatory function of its key constituent curcumin is discussed in Chap. 10 entitled "Immunomodulation impact of curcumin and its derivative as a natural ingredient" by Eknath D. Ahire and his coauthors. The chapter begins with chemical structure of curcumin and covers applications in immunomodulation. The authors express the need for exhaustive clinical studies so as to establish safety and efficacy in human.

The applications of colchicine and andrographolide in management of malignancies and infectious diseases is presented in Chap. 11 entitled "Colchicine and andrographolide as natural immunomodulators" written by S. Yasri and V. Wiwanitkit. In this chapter, the authors have summarized and discussed some important natural immunomodulators. The summary of colchicine and andrographolide as a natural immunomodulator is provided. At present, both colchicine and andrographolide are well-known biochemicals that are widely used in clinical practice. The immunomodulating role of both biosubstances seems interesting.

Immuno-booster activity of catechin and epigallocatechin-3-gallate is discussed in Chap. 12 entitled "Immune booster property of epigallocatechin-3-gallate and catechin" written by Leidy Johana Valencia-Hernández and coworkers. The chapter is aimed at describing the important biological activity of both compounds as booster of the immune system.

A general discussion on nutraceuticals and functional foods is presented in Chap. 13 entitled "Recent advances of nutraceutical and functional foods in immune health" drafted by Saumya Das and associates. The authors have listed dietary substances and their benefits in human health in particular immune status.

An overview of introduction and immunomodulatory role of *Withania somnifera* is depicted in Chap. 14 entitled "Therapeutic effects of *Withania somnifera*: An overview with special focus on Alzheimer's disease and infertility among youth" written by Deepika Saini and colleagues. The chapter has special focus on the mechanistic insights into the therapeutic effect of *Withania somnifera* in treating Alzheimer's and infertility problem. This chapter also describes the latest attempts in biotechnological production of withanolides.

Prayagraj, Uttar Pradesh, India Rajesh K. Kesharwani
Prayagraj, Uttar Pradesh, India Raj K. Keservani
New Delhi, India Anil K. Sharma

Contents

Part II Micronutrients and Macronutrients

About the Editors

Rajesh K. Kesharwani, PhD, MTech has more than 12 years of research and 9 years of teaching experience from various institutes of India in bioinformatics and biotechnology education. He has received several awards, including the NASI Swarna Jayanti Puruskar by the National Academy of Sciences of India. He has supervised 1 PhD and more than 20 UG and PG students for their research work, has authored over 52 peer-reviewed articles, 40 book chapters, and 30 edited books with international publishers. Dr. Kesharwani has been a member of many scientific communities as well as a reviewer for many international journals. He has presented many papers at various national and international conferences. He has received his MTech and PhD from the Indian Institute of Information Technology, Allahabad, and worked at NIT Warangal for a couple of semesters. He has received GATE fellowship from the Ministry of Human Resource Development, India, and Senior Research Fellowship from the Indian Council of Medical Research, India. His research fields of interest are medical informatics, protein structure and function prediction, computer-aided drug designing, structural biology, drug delivery, cancer biology, nano-biotechnology, and biomedical sciences.

Raj K. Keservani, PhD, MPharm is a faculty at the CSM Group of Institutions, Allahabad, India. He has more than 14 years of academic experience from various institutes of India in pharmaceutical education. He has published 30 peer-reviewed papers in the field of pharmaceutical sciences in national and international journals, 64 book chapters, 2 coauthored books, and 35 edited books. He is also active as a reviewer for several international scientific journals. Mr. Keservani graduated with a pharmacy degree from the Department of Pharmacy, Kumaun University, Nainital (UA), India. He received his Master of Pharmacy (MPharm) (specialization in pharmaceutics) from the School of Pharmaceutical Sciences, Rajiv Gandhi Proudyogiki Vishwavidyalaya, Bhopal, India. His research interests include nutraceutical and functional foods, novel drug delivery systems (NDDS), transdermal drug delivery/drug delivery, health science, cancer biology, and neurobiology.

Anil K. Sharma, PhD, MPharm is associated with All India Council for Technical Education, New Delhi, India. He has expertise in the area of pharmaceutics with a background in drug delivery. He has taught these subjects for nearly 13 years at institutions such as the Delhi Institute of Pharmaceutical Sciences and Research, University of Delhi, and School of Medical and Allied Sciences, G D Goenka University. Dr. Sharma has served in academic position such as a lecturer (pharmaceutics), Delhi Institute of Pharmaceutical Sciences and Research, University of Delhi. Anil Kumar Sharma holds a PhD (Pharmaceutical Sciences) from the University of Delhi, MPharm (Pharmaceutics) from Rajiv Gandhi Proudyogiki Vishwavidyalaya, and BPharm from the University of Rajasthan.

Part I
Nutraceuticals and Functional Foods

Chapter 1
Nanotechnology-Based Bacterial Immunotherapy

Rajeshwar Kamal Kant Arya, Deepak Sati, Dheeraj Bisht, and Raj K. Keservani

Abstract Recently, the nanotechnology-based bacterial immunotherapy emerged as a new combinatory therapeutic approach for the effective treatment of cancer, which combines the bacterial immunotherapy with nanotechnology for treating cancer. Although both bacterial immunotherapy and nanotechnology are very effective and advantageous solely, single treatment system is insufficient for complete eradication of cancer. Combining nanotechnology with bacterial immunotherapy opens new avenues for treating various diseases, abates the complication of bacterial immunotherapy, and overcomes the deficiency of both systems. Nanotechnology is helpful in targeting deep into the cancer cell due to its small size, enhanced permeability and retention (EPR) effect, and immunomodulatory activity. It also plays an important role in thermal and radio immunotherapy and cancer diagnostic. In this chapter, we highlighted the role of immunity in cancer and the role of bacteria in cancerogenesis, how bacterial immunotherapy is used in combating cancer, and how nanotechnology-based bacterial immunotherapy works on cancer regression.

Keywords Nanotechnology · Immunotherapy · Cancerogenesis · Bacterial immunotherapy · Immunosuppressant

1.1 Introduction

Immunotherapy is a therapy system used to combat cancer by stimulating the inherent defense mechanism of the body. Immunostimulants are used to boost the defense system to fight with cancer cells and destroy them. Immunotherapy is a kind of natural treatment; it is comprised of white platelets, organs, and tissues of the lymph framework. Biological therapy is a sort of treatment that utilizes substances

R. K. K. Arya (✉) · D. Sati · D. Bisht
Department of Pharmaceutical Sciences, Bhimtal Campus, Kumaun University Nainital, Bhimtal, Uttarakhand, India

R. K. Keservani
Faculty of B. Pharmacy, CSM Group of Institutions, Prayagraj, Uttar Pradesh, India

R. K. Kesharwani et al. (eds.), *Nutraceuticals and Functional Foods as Immunomodulators*, https://doi.org/10.1007/978-981-19-2507-8_1

produced using living life-forms to treat malignancy. Immunotherapy therapy attracted attention as an emergent therapy system in cancer treatment due to its potential to shatter immune tolerance and induce immune effects on selected cancer cell without any side effect (Hu et al. 2015). Immunotherapy can slow down the progress of cancer or arrest the spread of cancer. Bacteria-based immunotherapy for cancer was initiated way back in the nineteenth century (McCarthy 2006). William Coley in 1983 prepared bacterial endotoxin called "Coley toxin," found its regressing effect on carcinoma, and then successfully treated various types of carcinomas (Kaimala et al. 2018). For the treatment of cancer using bacterial immunotherapy, a thorough knowledge of bacterial immunity against cancer cells is to be explored in a more aggressive way. In the last 10 years, various researches have been performed for exploring the relationship between bacteria and cancer (Linnebacher et al. 2011). The bacterial and microbial flora system maintains a balance; the imbalance between microbial floras induces the carcinogenesis by various ways, e.g., bacteria induced inflammation or immunomodulation (Song et al. 2019). Bacteria are causative agents for various communicable diseases, but they also have contribution in causing cancer, e.g., *H. pylori*-associated gastric cancer; mutagenic bacterial metabolite is also thought to be a main factor for cancer due to the inflammatory potential of the *H. pylori* that is associated with gastric cancer (Parsonnet 1995). The chronicity of the infection is directly proportional to the cancer; as long as the inflammation occurs, the chances of cancer development also enhanced (Singh et al. 2019). The intestinal microflora contains microbes that show symbiotic relationship and support the health system by enhancing the immunity against microbial pathogenesis (Zheng et al. 2020). The microflora interactions maintain the balance by protecting from detrimental microbes and provide a biological fencing on the membrane for infiltrating infectious or immunogenic molecule into the blood and also enhance forbearance toward the antigens available on the mucosa (Linnebacher et al. 2011). This process impedes pathogenic colonization, influences the mucosal barrier, and induces the immune response.

The basic concept of immunotherapy was to exploit the patient's immune system as an effective tool for cancer treatment by stimulating the immune system to attack cancer cells. Some old reports demonstrated that an inflammation-related cancer was reduced by accidental infection by erysipelas (Patyar et al. 2010). The anaerobic bacteria can grow in cancer cell, but being pathogenic, they are not good for cancer treatment. Further, studies show that obligate anaerobic bacteria, e.g., *Clostridia*, can proliferate in necrotic or anaerobic region of solid tumor and regress the cancer (Barbé et al. 2006). Various in vivo studies show that *Clostridium*, *Salmonella*, *Bifidobacterium*, and *Escherichia* can be accumulated in cancer cell than the normal cell and can germinate spores in the cancer cell (Oelschlaeger 2010).Various studies suggest that microbes influence the oncogenesis and immunotherapy; they can have the inhibitory or stimulatory effect on the initiation of cancer and progression (Inamura 2020). The bacteria generate toxin and carcinogenic substance that stimulate inflammation or immunosuppression and promote oncogenesis (Nath et al. 2010). *E. coli* causes inflammation and stays in inflammatory cells that influence the microflora composition that promote the carcinogenesis. Beside these carcinogenic effects, several corroborations suggest bacteria can influence the chemotherapeutic and immunotherapeutic potential (Song et al. 2019).

Bacterial infection stimulates the immune system by modifying various cellular components, e.g., CD4, CD8 T cell, myeloid-derived suppressor cell (MDSC), regulator T cell (Treg), and tumor-associated macrophages (TMA). They also influence inherent immune receptors, e.g., Toll-like receptor (TLR), that are responsible for secretion of pro-inflammatory cytokines (Kaimala et al. 2018). Some bacterial exotoxins also stimulate immune response against the cancer cell and directly destroy cancer cell (Zahaf and Schmidt 2017). Immunotherapy is an impressive strategy for cancer treatment, but the effectiveness is very limited due to their inadequate assemblage at the cancer cell and various unwanted effects. Recently, nanotechnology has gained popularity for conquering these technical shortcomings due to their physicochemical property and versatility. The use of nanotechnology in bacterial immunotherapy can enhance the immunotherapy potential multifolds with their target specificity and intensify the delivery of tumor vaccine, immunomodulator checkpoint inhibitors, etc. (Gupta et al. 2010; Li et al. 2019).

1.2 Role of the Immune System in Cancer

The presence of inflammatory cell in cancer tissue has given the idea of understanding the relation between inflammation and cancer (Singh et al. 2019). Now, the capability of the immune system for initiating and arresting the cancer is established. The imbalance between intrinsic and acquired immunities show immunosuppressant action that instigates the cancerogenesis and proliferation of cancer. Intrinsic immunity generates macrophages against the infection and repairs the tissue (Gonzalez et al. 2018). Macrophages release IFN-γ which destroy the cancer cells and IL-4 and IL-13. TME induces protumorigenic tumor-associated macrophages (TAM) that stimulate angiogenesis, lymphogenesis, and cancer growth (Dhabekar 2011), releases immunosuppressant (IL-10 and TGF-β), prevents maturing of DC cell, and diminishes effector T-cell activity (Taylor et al. 2006). It releases EGF and VEGF that help in cell development, angiogenesis, and restructuring of the ECM by secreting metalloproteinases (MMPs) (Anteby et al. 2004). Damaged and inflammatory tissues contain neutrophils that release the neutrophil extracellular trap (NET), kill microbes, do phagocytosis (Riyapa et al. 2012), and also suppress inflammation. The neutrophils show both protumor and antitumor activities (Brandau et al. 2013). Cancer and stromal cells release chemokines that recruit tumor-associated neutrophils (TANs) to TME. The natural killer (NK) and effector T cells release the TNFα that stimulates the anticancer anti-metastatic hepatocyte growth factor (c-MET). TANs generate new inflammation during cancerogenesis and progression and induce immunosuppression in T cells by PD-L1 expression induced by tumor-derived granulocyte-macrophage CSF (GM-CSF) (Masucci et al. 2019). PMN-MDSCs induce the chronic inflammation and antigen-specific tolerance by T cells (Dorhoi and Du Plessis 2018). NK cells destroy infected cells and induce the apoptosis. The antigen-presenting cells (APCs) or dendritic cells (DCs) interplay between intrinsic and acquired immunities (Steinman 2006) and present internal and foreign antigens to T cells in the context of MHC molecules (Alberts et al. 2003) and DCs found in all tissues throughout the body. During cancer progression, the DCs prime naïve, memory T cells, and antigen presentation cells develop resistance for antigens and evoke effector T-cell response. Various types of cancer cell carry the cancer-infiltrating DCs (Hubert et al. 2019). At the initiatory phase of cancer, T cell is generated, and naïve T cell invades lymph nodes, on enticement migrated to cancer cell environment, exerts immune response, and destroys cancer cells. T effector cell exhibits antigen-dependent anticancer activity. The low immunogenic cancer cells escaped from T cell, and cancer cells develop a system to prevent themselves from destruction from effector T cell (TAMs, NK cells, and TANs). The effector cells regulate immune checkpoints on CTLS and CD4+T and prevent tissue damage. The checkpoint molecules CTLA-4 and PD-1 inhibit T-cell function (Takeuchi and Saito 2017). The programmed cell death protein (PD1) and programmed cell death protein ligand (PDL-1) are expressed by immune cells and cancer cells, inhibit T-cell activity, and suppress the anticancer function such as T-cell migration, proliferation, secretion of cytotoxic mediators, and restriction of cell killing (Han et al. 2020). The cancer cell environment seize the immune checkpoints and suppress anticancer activities by recruiting regulatory $CD4^+$ T cells (Tregs) that suppress cell destroying activity of Th1 CD4 T cells, CTLs, macrophages, NK cells, and neutrophils (Lee et al. 2020). The expression of PDL-1 LAG-3, CD39/73, or PD1 regulates the Treg-derived immunosuppressive function, and LAG-3 and CD39/73 also exhibit immunosuppressant function by contact-independent mechanisms, which sequester IL-2 and produce immunosuppressants IL-10, TGF-β, prostaglandin E2, adenosine, and galectin-1 (Cai et al. 2019). Invariant NK T (iNKT) cells, subset of T cells, identify the NK cell like lipid antigen CD1d; upon activation, it secretes effector cytokines such as IFN-γ, IL-4, and IL-17 (Fujii and Shimizu 2017). B cell exhibits humoral immunity for microbes by extracting specific antibodies that converted into protein and B cells. B cell assists cancer proliferation by releasing IL-35 and enhances immunosuppressive function by secreting IL-10 and TGF-β (Klinker and Lundy 2012). B cell also induces angiogenesis and chronic inflammation by activating myeloid cells via FcRγ (Rivera and Bergers 2015).

1.3 Role of Bacteria in Immunotherapy

Bacteria functioned as a dual-edge blade for cancer. In one hand, it instigates the cancer, and in other hand, it induces the immune system and blocks cancerogenesis (Linnebacher et al. 2011). In the last 10 years, the killed or attenuated microbes have gained popularity as an important weapon for cancer immunotherapy (Kaimala et al. 2018). The explicit mechanism of bacterial immunotherapy is not so clear till date, but it is believed that weak antigenic cancer with facultative anaerobic bacteria

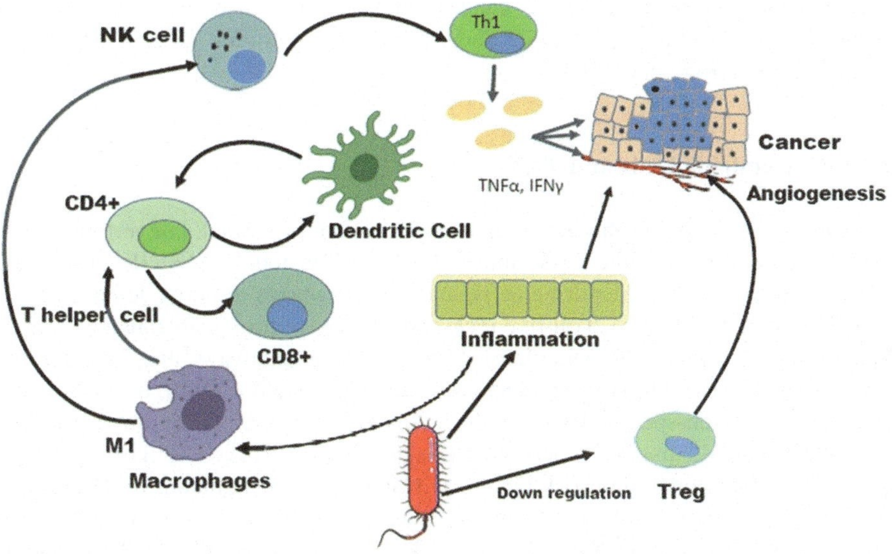

Fig. 1.1 Bacterial immunotherapy

promotes immunogenicity for cancer cell. There are several examples of bacterial immunotherapy using the underlying mechanism (Fig. 1.1). The attenuated bacterial infection harmonizes the activity of various cellular components (CD4+, CD8+ T cells, Treg) and TAM of the immune system to wield their antitumoral immunity (Nelson et al. 2015).

1.3.1 Effect of Bacteria on Myeloid Cell

1.3.1.1 Tumor-Associated Macrophages

A study on mice reveals *S. typhimurium* helps in reversing the cancer by helping in the maturation of myeloid cell (MDCs), reduces the suppressive function, and increases anticancer immune response (Kaimala et al. 2014). Another study revealed that *Salmonella* shows the immunomodulatory effect by elevating the level of TILs, CD11+myeloid cell, CD4+, and CD8 + T in mice and induces the inhibitory effect on cancer cell. The *Salmonella* treatment enhances the level of TAM which is responsible for the activation IFN-γ-dependent Sca-1 and MHC class II proteins. Leschner and colleague have shown that *S. typhimurium* SL7207 on i.v. administration *activated the* TNF-*α and* invades *the tumor cell by disrupting the vasculature* (Leschner et al. 2009). Recombinant attenuated *Salmonella enterica* serovar *typhimurium* vaccine arrests cancer development by inhibiting the HER-2/neu expression. The vaccine shows anticancer effect by diminishing the immunosuppressive function in cancer environment, increases the level of $CD11b^{+}Gr\text{-}1^{+}$ myeloid cell MDSC and TNF-α, and modifies Treg, $CD4^{+}CD25^{+}$ $Foxp3^{+}$ Tregs (Hong et al. 2013).

1.3.1.2 Dendritic Cells and TANs

L. monocytogenes infect the melanoma cell in human or animal, convert them into antigen-presenting cell like DCs, and express CD11c, F4/80, MHCII, CD40, and CD83 similar to mature dendritic cells, and APC manages the host defense mechanism (Gorvel et al. 2014). Toll-like receptors (TLR) are responsible against microbial infection. *Salmonella* stimulates TLR signaling pathways that increase the expression of co-stimulatory molecules (e.g., CD86 and CD80) on APCs and activate the antigen-specific CD8 T cells and natural killer cells that are responsible for killing and relapse of cancer cell (Hajam et al. 2017). The TLR signal, a cytoplasmic adaptor protein (MyD88a), is responsible for inducing the TLRs and IL-1/IL-18 receptors and activation of downstream of IL-1 receptor-associated kinases (IRAKs) and NF-κB (Cohen 2014). A study show the TLR-MyD88 is necessary for bacterial immunotherapy and mice with MyD88 deficiency were found to be unable to reverse the cancer after *Salmonella* treatment (Kaimala et al. 2014). *Salmonella typhi* vaccine also activates and recruits neutrophils at cancer cell that releases the TNFα producing additive cytotoxic effect in the presence IFN-γ (Vendrell et al. 2011).

1.3.1.3 Effect of Bacteria on Tumor-Associated Lymphoid Cells

Natural Killer Cell

The cancer microenvironment nurtures the lymphoid cells, e.g., NK cells, B cells, CD4+ T cells, CD8+ T cells, and Tregs (Kaimala et al. 2018). The lymphoid cells have both inhibitory and stimulatory effects on the cancer (Wagner and Koyasu 2019). The microenvironment can affect the function of lymphoid cells in cancer cells, and the NK cells can identify and remove cancer cell without any previous antigenic exposure (Kaimala et al. 2018). *Salmonella* infection *enhances* IFN-γ and NK level in mice model (Harrington et al. 2007). The cancer depresses the host immunity that disturbs the NK cell response for cancer cell (Guerrouahen et al. 2020). Recombinant *Salmonella* activates NK cell and IL-2 expression. This strain exhibits excellent anticancer activity in vivo against metastatic cancer and also enhances IFN-γ or TNF-α expression that stimulates the anticancer activity (Lin et al. 2021).

1.3.2 CD4+ and CD8+ T Cells

CD4 T cells exert anticancer activity due the helper cells (Th1, Th2, or Th17) that secrete cytokines (IFN-γ and IL-12, IL-4, IL-5 and IL-13). Th1 cell shows anticancer activity by stimulating the CD8 cell or kills the cancer cell by releasing TNF-α and/or IFN-γ (Bascuas et al. 2018). Th1 ceases the progression of cancer by allowing DCs and macrophages and enhances the antigen-presenting activity that allows TCD8+ cell for strong anticancer activity. Th2 has insignificant effect in cancer treatment, but T17 enhances the cancer progression by secreting IL-17 (Lee et al. 2019). The attenuated *Salmonella* increases the infiltration of CD4+ and CD8+ T cell about threefold into cancer cell in mice model. M1 secretes macrophage inflammatory protein 1-alpha that recruited stimulated T cell in cancer cell; *Salmonella* follows this mechanism and increases the T-cell penetration into cancer cell. *Salmonella typhi* increases the leukocyte count and decreases the cancer progression and mitotic index and increases the life span of mice (Vendrell et al. 2011).

1.3.3 Gamma-Delta (γδ) T Cells

Gamma-delta T cells belong to T-lymphocyte cell, having a surface antigen recognition complex type 2. During stimulation, it exhibits excellent anticancer potential due to the release of ample amount of IFN-γ and TNF-α (Silva-Santos et al. 2015). *Mycobacterium vaccae*, *M. obuense*, and BCG stimulate the activation of γδ T cells that result in upregulation of granzyme expression and synthesis of Th1 cytokines (IFN-γ and TNF-α) (Fowler et al. 2012).

1.3.4 Effect of Bacteria on Immunosuppressor Cells

1.3.4.1 Myeloid-Derived Suppressor Cells

Cancer or autoimmune diseases can modify the myeloid-derived suppressor cells (MDSCs), found in the bone marrow, peripheral blood, and spleen, that help cancer cell in evading from the host's immune system (Gabrilovich and Nagaraj 2009). MDSCs do nitration of T cell that represses the CD4 and CD8 T-cell-induced anticancer effect (Ostroumov et al. 2018). MDSCs can be a good target for cancer treatment. *Salmonella* triggers the differentiation of intratumoral MDSCs and improves immunomodulatory properties. *L. monocytogenes* LLO reduces the immunosuppressant activities of MDSCs and Tregs in surroundings of cancer tissues. MDSCs reduce Arg1 expression and IL-10 by Tregs. It both causes the above action (Wallecha et al. 2013) and also depresses the expression of T-cell receptor and immunosuppressive cytokines (Lindau et al. 2013).

1.4 Regulatory T Cells (Tregs)

Tregs are involved in autoimmunity and cancer diseases. Generally, $CD4^{+}CD25^{+}Foxp3^{+}$ has the preventive immunogenic reaction for autoantigen. These cells inhibit the tumor antigen-specific CLTs and decrease the anticancer immunity (Manzo et al. 2015). Study shows attenuated *S. typhi* reduces Tregs cells in cancer-draining lymph node that diminish lung metastasis and cancer progression, improve animal life (Vendrell et al. 2013), and also decrease the proportion of $CD25^{+}FoxP3^{+}$ cells in the spleen and CD4+ T cells in a colon cancer model. It also elicits the conversion of immunosuppressive MDSCs into TNF-α-secreting neutrophils and reduces the generation of Treg cells, especially in the presence of tumor-specific CTLs (Hong et al. 2013). *Salmonella* vaccine also acts on the downregulation of CD44, a key cell surface molecule on Tregs as well as tumor cells, and contributes to angiogenesis and proliferative potential (Liu and Chopra 2010).

1.5 Nanotechnology in Immunotherapy

The cancer immunotherapy shows promising results, but developing a low toxic dosage form with high targeting efficiency and excellent efficacy is a difficult task. Nanotechnology has the revolutionary potential to reform the cancer therapy. Nanoparticles have already established its significance in the modern drug delivery system; it ameliorates the immunostimulation, cancer cell targeting, bio-distribution, and release kinetics. All the nanoparticulated system (nanosized vesicular system, e.g., nanoparticles, liposomes, etc.) could be utilized for passive and active immunotherapy (Sharma et al. 2017; Keservani et al. 2017a, b). Nanotechnology improves the cancer immunotherapy by protecting the burden during circulation, targeting the immunotherapeutics to the cancer tissues, inducing the immune system, and modulating the immunosuppressant action for cancer cell (Qian et al. 2018; Keservani et al. 2017a, b). The nanosized immunotherapeutic system can improve the therapeutic techniques and conquer the difficulties related to the poor therapeutic responses, for example, insufficient stimulation, side effects, and suppressed pharmacological activity. Researchers are continuously evolving the nanomaterials with improved structural properties and surface modification ability that can be exploited for guiding nanotechnology-based immunotherapeutic into the vasculature, avoidance of opsonization, and assemblage of nanoparticles at the cancer cell due to enhanced permeability and retention (EPR) properties. A controlled drug release can be used to design nanoparticle systems specific for microbiome intervention in cancer. However, this represents a generally unexplored area (Song et al. 2019). The first cancer vaccine, Sipuleucel-T was approved by the US Food and Drug Administration (FDA) in the year 2010 for prostate cancer treatment (Cheever and Higano 2011).

1.5.1 Nanoparticle-Based Bacterial Immunotherapy

Recently, new combinatory therapeutic approaches emerged for the effective treatment of cancer because single-treatment method is insufficient for complete eradication of cancer. The advantages of bacterial immunotherapy and nanoparticle-based immunotherapy are combined together to overcome the deficiency of both systems and also exploit its advantages. Both systems together work effectively with enhanced cancer targeting and excellent efficacy; recently, the researcher has prepared various vaccine and outer membrane vesicles (OVM) and nanobodies for the therapeutic and diagnostic purpose as well; some example are given here. Figure 1.2 shows the nanotechnology-based immunotherapy.)

1.5.2 Nano-vaccine

Live or attenuated bacterial oral vaccine is unable to penetrate into the cancer vasculature. To enhance the penetrability, Hu et al. (2015) prepared oral nano-vaccine containing attenuated *Salmonella* bacteria; the cationic nanosized enwrapped bacterial vectors efficiently deliver the vaccine into the cancer cell. The nano-vaccine protects bacteria from phagocytosis and enhances the circulatory time and acid tolerability in GIT. The nano-vaccine stimulates the T cell, increases the

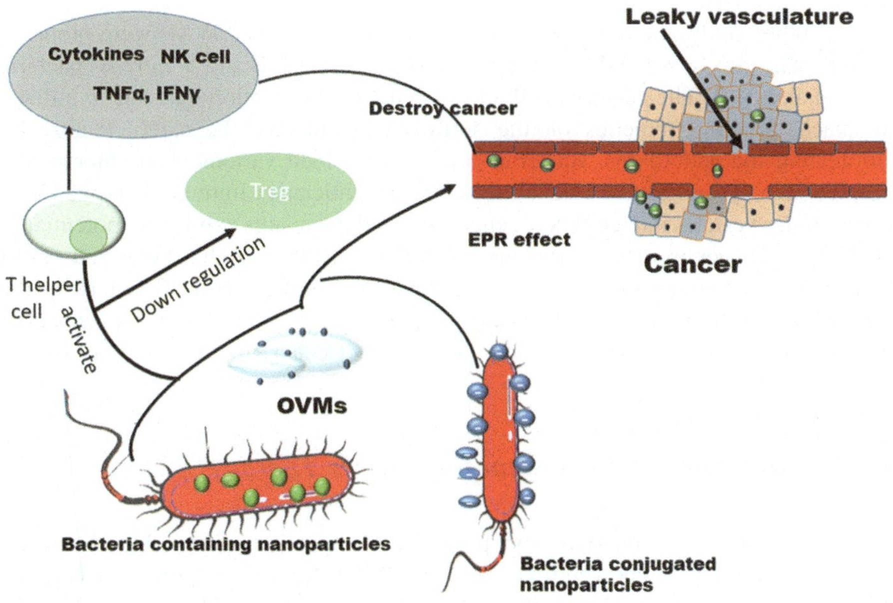

Fig. 1.2 Nanotechnology-based bacterial immunotherapy

secretion of cytokine, and inhibits the cancer growth by suppressing the angiogenesis. In 2019, Terán-Navarro and team synthesized gold nanoparticles containing a peptide (listeriolysin O) derived from bacteria for treating skin cancer. The vaccine works on dendritic cell, modifies the cancer cell sensitivity toward the immune system, and activates the cell killing function that ameliorates the cell viability. The anti-PD-1 or anti-CTLA-4 checkpoint inhibitors when incorporated into the nano-vaccine enhance the cancer eradicating function manyfold. The study shows the nano-vaccine is a safe therapeutic system and enhances the immunity against the cancer cell.

The outer membrane vesicles (OMVs) derived from Gram-negative bacteria are effective vaccine platforms with precedence for safe use in humans (Davitt et al. 2016). Unlike most subunit antigens and synthetic nanoparticles, OMVs contain endogenous immunostimulatory ligands and deliver the antigens in their native orientation. OMVs could induce antigen-presenting cells by engaging the intrinsic receptors and CMI by activating the APC, co-stimulation, and cytokine production. After the uptake, the DCs enhance the surface MHC classes I and II, CD80 and CD86 expressions. OMV-induced DCs produce the T-cell polarizing cytokines IL-1β, IL-6, IL-18, and IL-12 and their expression contributed by TLR4, whereas the NLRP3 inflammasome was required for OMV-induced production of IL-1β and IL-18. Following vaccination of mice, Th1- and Th17-type T-cell responses were observed after ex vivo restimulation with OMVs and HK bacteria. In addition, OMV vaccination produced functional $CD8^+$ T cells capable of killing bacteria-infected cells. Collectively, these results support the utility of the OMV platform as a nonliving vaccine that can lead to protective CMI against bacterial pathogens (Davitt et al. 2016).

In another study, the nanosized outer membrane vesicles (OVM) were prepared from the Gram-positive and Gram-negative (*E. coli* Δmsb B and *S. aureus*) bacteria (Kim et al. 2017). The bacterial cell allows genetic modification that can be utilized to attach targeting moieties on the surface and enhance the safety feature by eliminating the endotoxins. Bacterial cell can also load various chemotherapeutic agents. Such system can be utilized for effective anticancer immunotherapy. These extracellular vesicles induce IFN-γ generation, and the results exhibit the potential of OVM as a cancer immunotherapeutic without side effects. The nanosized OVM assemblage in the cancer cells produces the IFN-γ and cytokine CXCL10 in the cancer microenvironment and elicits the anticancer effect. This system also enhances the memory cell production for longer anticancer effects (Kim et al. 2017).

1.5.3 Nanoparticulate System for Cancer Therapy

In 2016, Lizotte and his team member prepared self-aggregating nano-based inhalant system containing cowpea mosaic virus (CPMV) for treating lung cancer; the inhalant system produces effective systemic cancer immunotherapy against the B16F10 in the skin. The nano-based inhalant system stimulates cytokines and enhances the potency of the immune system against lung cancer, ovarian cancer, metastatic breast cancer, and colon cancer.

As we know, phagocytosis minimizes the circulation time of nanoparticles in the body. To overcome this problem, the attenuated *Salmonella* bacterial membrane vesicles were combined with nanoparticles, and the nanoparticle surface was modified with polyethylene glycol (PEG) and also attached to the Arg-Gly-Asp (RGD) peptide in which the Asp (RGD) peptide enhances the circulatory time and cancer targetability and stability (Chen et al. 2020). Tegafur is loaded to this system that removes MDSc and synergistically enhanced the anticancer effect and prevents metastasis (Chen et al. 2020). This system also induces the intrinsic immunity and accumulated into the cancer tissue due to the enhanced permeability and retention (EPR) and active targeting effects.

Probiotic bacteria were utilized to deliver the nanoparticulated checkpoint blocker to the affected area. The probiotic bacterial nanoparticulated system used a synchronized lysing integrated circuit (SLIC) using *E. coli* Nissle 1917 strain for releasing the target program cell death ligand and cytotoxic T-lymphocyte-associated protein 4 (CTLA-4) after intratumoral injection. The computerized SLIC model was used for validating the lysis mechanism for higher therapeutic efficiency. The formulation exhibited higher therapeutic efficiency over the other conventional system and suppressed the cancer in mouse by potentiating the immunity; the formulation enhanced the activation of T cells, an abscopal effect, and enhanced T memory cell in mice. The formulation enhances the cytokine granulocyte-macrophage colony-stimulating factor.

The swimming ability of *Salmonella* could be utilized for effective targeting of cancer cell by incorporating the nanomedicine; it freed the drug by degrading itself at cancer environment; for this purpose, doxorubicin incorporated nanosized liposome into bacterial cell (Zoaby et al. 2017); and the bacterial motility is pH- and glucose-dependent, which favors the cancer environment. The nanoparticles are incorporated into the bacterial cell by incubation or electroporation. Electroporation builds ephemeral aperture on the bacterial cell wall that provides entry to the nanosized liposomes inside the bacteria. The metabolic process causes the drug release from the liposome; mainly, ammonia causes the osmotic imbalance in the liposome and releases the drug that destroys the bacterial carrier and kills the cancer cell. This process enhances in situ therapeutic efficacy (Zoaby et al. 2017).

1.5.4 Thermotherapy

Several cancer therapies are available, but cancer monotherapy can result in reappearance of cancer due to partial destruction of cancer tissue. To overcome these problems, Park et al. (2020) developed nanoemulsion containing anaerobic *Clostridium novyi*-NT spores for synergistic image-guided combinational cancer therapy. The nanoemulsion also carries scintillators for guiding MRI and for an image-guided X-ray photodynamic therapy (PDT) of the normoxic peripheral cancer. The nanoemulsion was used for the treatment of the hypoxic central tumor tissues. Photosensitizer-coated NSs (PS-NSs) and *C. novyi*-NT spores are emulsified with clinically available ethiodized oil (Lipiodol) to be the nanoemulsion and injected into the tumor with computed tomography image guidance. Following the image-guided X-ray PDT and anaerobic *C. novyi*-NT combination treatment, apoptotic cell death in cancer tissues, including both peripheral and central tumor regions, is significantly higher than in the control groups. This nanoemulsion system can overcome the limitations of conventional cancer therapy, resulting in increased cancer therapeutic efficacy.

1.5.5 Radiotherapy and Diagnostic Purpose

Nanoparticles coated with bacterial wall was prepared that also contains the immunostimulatory PC7A/CpG Polyplex core and imide groups. The bacterial nanoparticles with various attachments can ameliorate the antigen salvage, improve the affinity of antigen toward the MHC-1, and also augment the innate immunity by proliferation of the T cells. Nano-bacterial system on combining with radiotherapy arrests the neoantigens and increases their absorption in DCs and from there delivered to MHC-presenting cell and activated the effector T cell; PC7A and CpG assist in modulating the antigen uptake in DCs, raise the expression of MHC-I in TME, and increase the effector T cell and type I IFN. Treating mice with this system exhibits 100% relapse of primary cancer with high survival rate and low metastasis rate. For improving the immunotherapeutic efficiency of the nano-bacterial system, the immune checkpoint can also be combined with the nano-system that possibly impedes the immunosuppressive signals. Such nano-bacterial system can ameliorate cancer immunity and long-term immune response (Patel et al. 2019). In another study, melanin-containing OMVs were prepared from *E. coli* for therapeutic and diagnostic purpose (Gujrati et al. 2019). OMVs show good biocompatibility, biodegradability, and low-cost manufacturing. The strong acoustic waves are converted into optical image; OMVs show superiority over fluorescence method due to high resonating efficiency. OMVs can be utilized for monitoring and imaging of cancer in vivo. The nanosized bacterial cell escapes out from the phagocytosis and the EPR effect; both the two events enhance the residence time in the cancer cell. OMVs could penetrate into deep tissue due to high absorption coefficient. NIR irradiation killed the cancer cell. OMVs also induce the production of anticancer cytokines, including IFN-γ. Dosing for longer time induces IFN-γ production in a sustained manner (Gujrati et al. 2019).

1.6 Challenges and Opportunities for Nanotechnology-Based Immunotherapy

Designing the nanotechnology-based bacterial immunotherapy faces several challenges, e.g., understanding cancer site, microenvironment, microbial flora, and inflammation-derived cancerogenesis and bacterial efficacy against cancer with the toxicological effect or side effects of microflora, and their relationships with cancerogenesis, progression, and treatment of cancer (Song et al. 2019), in the last 10 years of research, several preclinical and clinical studies have been performed on nanotechnology-based immunotherapy, but just a few approaches are so far clinically approved. For example, attenuated herpes simplex virus (genetically modified)-based cancer treatment successfully passed phase III (Shukla and Steinmetz 2016). There is a lacuna between the understanding of cancerogenesis and immunology and in vivo behaviors of nanoparticles that include long-time assemblage at the nontarget site and their safety concern. They can cause toxicity, e.g., hollow silica nanoparticles induce the release of pro-inflammatory cytokines that damages the hepatic tissues. Nanoparticles can influence the constitution of microflora; a depth understanding of nanoparticles and microflora is required for developing new strategy (Song et al. 2019). The intersubject variability should also be considered. The microflora is patient-specific that changes with the food intake, drugs, and other environmental factors, and the nanoparticles can modify the composition of microflora that cannot be predicted in vivo or by clinical investigation. Other challenges may include the production, targeting efficiency, and distribution. Targeting the nanoparticles to the specific organ or tissues is affected by surface morphology, surface properties, and physicochemical properties of nanoparticles (Song et al. 2019).

1.7 Advantage and Disadvantage

The distinct characteristics of nanoparticulated system and bacterial membrane make them an excellent drug delivery system; nanoparticle-based bacterial immunotherapy has excellent immunomodulatory properties, penetrability, biocompatibility, and targeting ability to specific cells and sustained release (Li et al. 2013). The nanosized bacterial system enhances the circulatory time and targeting into deep vasculature because of EPR and also augments the efficacy of thermotherapy and radiotherapy (Park et al. 2020; Patel et al. 2019). The nanoparticle surface also allows surface modification and attachment of ligands that improve the targeting (Li et al. 2013). The nanoparticle-based bacterial system not only improves cancer targeting but also helps in cancer diagnosis. Nanoparticles that may cause immunogenic response have short residence time due to rapid blood clearance.

1.8 Conclusion

The understanding of the role of immunity in cancer pushes scientist to make new system that plays an important role in cancer. Cancer immunotherapy holds great promise in cancer therapy. Collaboration between cancer immunologists and engineers will further the understanding of the complex and underlying immunology, therefore driving technological development. With emerging nanotechnological interventions, the efficacy of many such immunotherapies could be drastically improved, and a merger of these two rapidly growing fields of science could facilitate clinical translation of many cancer immunotherapies.

References

Alberts B, Johnson A, Lewis J, Raff M, Roberts K, Walter P (2003) Molecular biology of the cell. 4th edn. Ann Bot 91:401

Anteby EY, Greenfield C, Natanson-Yaron S, Goldman-Wohl D, Hamani Y, Khudyak V, Ariel I, Yagel S (2004) Vascular endothelial growth factor, epidermal growth factor and fibroblast growth factor-4 and -10 stimulate trophoblast plasminogen activator system and metalloproteinase-9. Mol Hum Reprod 10:229–235

Barbé S, Mellaert LV, Anné J (2006) The use of clostridial spores for cancer treatment. J Appl Microbiol 101:571–578

Bascuas T, Moreno M, Grille S, Chabalgoity JA (2018) Salmonella immunotherapy improves the outcome of CHOP chemotherapy in non-Hodgkin lymphoma-bearing mice. Front Immunol 9:7

Brandau S, Dumitru CA, Lang S (2013) Protumor and antitumor functions of neutrophil granulocytes. Semin Immunopathol 35:163–176

Cai J, Wang D, Zhang G, Guo X (2019) The role of PD-1/PD-L1 axis in Treg development and function: implications for cancer immunotherapy. Onco Targets Ther 12:8437–8445

Cheever MA, Higano CS (2011) Provenge (Sipuleucel-T) in prostate cancer: the first FDA-approved therapeutic cancer vaccine. Clin Cancer Res 17:3520–3526

Chen Q, Bai H, Wu W, Huang G, Li Y, Wu M, Tang G, Ping Y (2020) Bioengineering bacterial vesicle-coated polymeric nanomedicine for enhanced cancer immunotherapy and metastasis prevention. Nano Lett 20:11–21

Cohen P (2014) The TLR and IL-1 signalling network at a glance. J Cell Sci 127:2383–2390

Davitt CJ, Petersen HE, Kikendall NL, Lavelle EC, Morici LA (2016) Naturally-derived bacterial nano-particles engage diverse innate receptors, driving the activation of dendritic cells and leading to the establishment of potent adaptive immune responses. J Immunol 196:76.11

Dhabekar GS (2011) Role of macrophages in malignancy. Ann Maxillofac Surg 1:150

Dorhoi A, Du Plessis N (2018) Monocytic myeloid-derived suppressor cells in chronic infections. Front Immunol 8:1895

Fowler DW, Copier J, Wilson N, Dalgleish AG, Bodman-Smith MD (2012) Mycobacteria activate γδ T-cell anti-tumour responses via cytokines from type 1 myeloid dendritic cells: a mechanism of action for cancer immunotherapy. Cancer Immunol Immunother 61:535–547

Fujii S-I, Shimizu K (2017) Exploiting antitumor immunotherapeutic novel strategies by deciphering the cross talk between invariant NKT cells and dendritic cells. Front Immunol 8:886

Gabrilovich DI, Nagaraj S (2009) Myeloid-derived-suppressor cells as regulators of the immune system. Nat Rev Immunol 9:162–174

Gonzalez H, Hagerling C, Werb Z (2018) Roles of the immune system in cancer: from tumor initiation to metastatic progression. Genes Dev 32:1267–1284

Gorvel L, Textoris J, Banchereau R, Ben Amara A, Tantibhedhyangkul W, von Bargen K, Ka MB, Capo C, Ghigo E, Gorvel J-P, Mege J-L (2014) Intracellular bacteria interfere with dendritic cell functions: role of the type I interferon pathway. PLOS One 9:e99420

Guerrouahen BS, Maccalli C, Cugno C, Rutella S, Akporiaye ET (2020) Reverting immune suppression to enhance cancer immunotherapy. Front Oncol 9:1554

Gujrati V, Prakash J, Malekzadeh-Najafabadi J, Stiel A, Klemm U, Mettenleiter G, Aichler M, Walch A, Ntziachristos V (2019) Bioengineered bacterial vesicles as biological nano-heaters for optoacoustic imaging. Nat Commun 10:1114

Gupta S, Yadav BS, Kesharwani RK, Mishra KP, Singh NK (2010) The role of nanodrugs for targeted drug delivery in cancer treatment. Arch Appl Sci Res 2(1):37–51

Hajam IA, Dar PA, Shahnawaz I, Jaume JC, Lee JH (2017) Bacterial flagellin—a potent immunomodulatory agent. Exp Mol Med 49:e373–e373

Han Y, Liu D, Li L (2020) PD-1/PD-L1 pathway: current researches in cancer. Am J Cancer Res 10: 727–742

Harrington L, Srikanth CV, Antony R, Shi HN, Cherayil BJ (2007) A role for natural killer cells in intestinal inflammation caused by infection with *Salmonella enterica* serovar Typhimurium. FEMS Immunol Med Microbiol 51:372–380

Hong E-H, Chang S-Y, Lee B-R, Pyun A-R, Kim J-W, Kweon M-N, Ko H-J (2013) Intratumoral injection of attenuated Salmonella vaccine can induce tumor microenvironmental shift from immune suppressive to immunogenic. Vaccine 31:1377–1384

Hu Q, Wu M, Fang C, Cheng C, Zhao M, Fang W, Chu PK, Ping Y, Tang G (2015) Engineering nanoparticle-coated bacteria as oral DNA vaccines for cancer immunotherapy. Nano Lett 15: 2732–2739

Hubert M, Gobbini E, Bendriss-Vermare N, Caux C, Valladeau-Guilemond J (2019) Human tumor-infiltrating dendritic cells: from in situ visualization to high-dimensional analyses. Cancers 11: 1082

Inamura K (2020) Roles of microbiota in response to cancer immunotherapy. Semin Cancer Biol 65:164–175

Kaimala S, Mohamed YA, Nader N, Issac J, Elkord E, Chouaib S, Fernandez-Cabezudo MJ, Al-Ramadi BK (2014) Salmonella-mediated tumor regression involves targeting of tumor myeloid suppressor cells causing a shift to M1-like phenotype and reduction in suppressive capacity. Cancer Immunol Immunother 63:587–599

Kaimala S, Al-Sbiei A, Cabral-Marques O, Fernandez-Cabezudo MJ, Al-Ramadi BK (2018) Attenuated bacteria as immunotherapeutic tools for cancer treatment. Front Oncol 8:136

Keservani RK, Sharma AK, Kesharwani RK (2017a) Drug delivery approaches and nanosystems, Drug targeting aspects of nanotechnology, vol 2. Apple Academic Press, Toronto. isbn:9781771885843

Keservani RK, Sharma AK, Kesharwani RK (2017b) Drug delivery approaches and nanosystems, Novel drug carriers, vol 1. Apple Academic Press, Toronto. isbn:9781771885836, E-Book isbn:978131522537

Kim OY, Park HT, Dinh NTH, Choi SJ, Lee J, Kim JH, Lee S-W, Gho YS (2017) Bacterial outer membrane vesicles suppress tumor by interferon-γ-mediated antitumor response. Nat Commun 8:626

Klinker MW, Lundy SK (2012) Multiple mechanisms of immune suppression by B lymphocytes. Mol Med 18:123–137

Lee HL, Jang JW, Lee SW, Yoo SH, Kwon JH, Nam SW, Bae SH, Choi JY, Han NI, Yoon SK (2019) Inflammatory cytokines and change of Th1/Th2 balance as prognostic indicators for hepatocellular carcinoma in patients treated with transarterial chemoembolization. Sci Rep 9: 3260

Lee WS, Yang H, Chon HJ, Kim C (2020) Combination of anti-angiogenic therapy and immune checkpoint blockade normalizes vascular-immune crosstalk to potentiate cancer immunity. Exp Mol Med 52(9):1475–1485

Leschner S, Westphal K, Dietrich N, Viegas N, Jablonska J, Lyszkiewicz M, Lienenklaus S, Falk W, Gekara N, Loessner H, Weiss S (2009) Tumor invasion of *Salmonella enterica* serovar typhimurium is accompanied by strong hemorrhage promoted by TNF-α. PLOS One 4:e6692

Li XD, Gao JY, Yang Y, Fang HY, Han YJ, Wang XM, Ge W (2013) Nanomaterials in the application of tumor vaccines: advantages and disadvantages. Onco Targets Ther 6:629

Li Y, Ayala-Orozco C, Rauta PR, Krishnan S (2019) The application of nanotechnology in enhancing immunotherapy for cancer treatment: current effects and perspective. Nanoscale 11:17157–17178

Lin Q, Rong L, Jia X, Li R, Yu B, Hu J, Luo X, Badea SR, Xu C, Fu G, Lai K, Lee M, Zhang B, Gong H, Zhou N, Chen XL, Lin S, Fu G, Huang J-D (2021) IFN-γ-dependent NK cell activation is essential to metastasis suppression by engineered Salmonella nature communications. Nat Commun 12:2537

Lindau D, Gielen P, Kroesen M, Wesseling P, Adema GJ (2013) The immunosuppressive tumour network: myeloid-derived suppressor cells, regulatory T cells and natural killer T cells. Immunology 138:105–115

Linnebacher M, Maletzki C, Klier U, Klar E (2011) Bacterial immunotherapy of gastrointestinal tumors. Langenbecks Arch Surg/Dtsch Gesell Chirurg 397:557–568

Liu T, Chopra AK (2010) An enteric pathogen Salmonella enterica serovar Typhimurium suppresses tumor growth by downregulating CD44high and CD4T regulatory (Treg) cell expression in mice: the critical role of lipopolysaccharide and Braun lipoprotein in modulating tumor growth. Cancer Gene Ther 17:97–108

Lizotte PH, Wen AM, Sheen MR, Fields J, Rojanasopondist P, Steinmetz NF, Fiering S (2016) In situ vaccination with cowpea mosaic virus nanoparticles suppresses metastatic cancer. Nat Nanotechnol 11:295

Manzo T, Heslop HE, Rooney CM (2015) Antigen-specific T cell therapies for cancer. Hum Mol Genet 24:R67–R73

Masucci MT, Minopoli M, Carriero MV (2019) Tumor associated neutrophils. Their role in tumorigenesis, metastasis, prognosis and therapy. Front Oncol 9:1146

McCarthy EF (2006) The toxins of William B. coley and the treatment of bone and soft-tissue sarcomas. Iowa Orthop J 26:154

Nath G, Gulati AK, Shukla VK (2010) Role of bacteria in carcinogenesis, with special reference to carcinoma of the gallbladder. World J Gastroenterol 16:5395–5404

Nelson MH, Diven MA, Huff LW, Paulos CM (2015) Harnessing the microbiome to enhance cancer immunotherapy. J Immunol Res 2015:e368736

Oelschlaeger TA (2010) Bacteria as tumor therapeutics? Bioeng Bugs 1:146–147

Ostroumov D, Fekete-Drimusz N, Saborowski M, Kühnel F, Woller N (2018) CD4 and CD8 T lymphocyte interplay in controlling tumor growth. Cell Mol Life Sci 75:689–713

Park W, Cho S, Kang D, Han J-H, Park J-H, Lee B, Lee J, Kim D-H (2020) Tumor microenvironment targeting nano-bio emulsion for synergistic combinational X-ray PDT with oncolytic bacteria therapy. Adv Healthcare Mater 9:e1901812

Parsonnet J (1995) Bacterial infection as a cause of cancer. Environ Health Perspect 103(Suppl 8): 263–268

Patel RB, Ye M, Carlson PM, Jaquish A, Zangl L, Ma B, Wang Y, Arthur I, Xie R, Brown RJ, Wang X, Sriramaneni R, Kim K, Gong S, Morris ZS (2019) Development of an in situ cancer vaccine via combinational radiation and bacterial-membrane-coated nanoparticles. Adv Mater 31:e1902626

Patyar S, Joshi R, Byrav DP, Prakash A, Medhi B, Das B (2010) Bacteria in cancer therapy: a novel experimental strategy. J Biomed Sci 17:21

Qian H, Liu B, Jiang X (2018) Application of nanomaterials in cancer immunotherapy. Mater Today Chem 7:53–64

Rivera LB, Bergers G (2015) Intertwined regulation of angiogenesis and immunity by myeloid cells. Trends Immunol 36:240–249

Riyapa D, Buddhisa S, Korbsrisate S, Cuccui J, Wren BW, Stevens MP, Ato M, Lertmemongkolchai G (2012) Neutrophil extracellular traps exhibit antibacterial activity against *Burkholderia pseudomallei* and are influenced by bacterial and host factors. Infect Immun 80: 3921–3929

Sharma AK, Keservani RK, Kesharwani RK (2017) Nanobiomaterials, Applications in drug delivery. Apple Academic Press, Toronto. isbn:9781771885911

Shukla S, Steinmetz NF (2016) Emerging nanotechnologies for cancer immunotherapy. Exp Biol Med (Maywood) 241:1116–1126

Silva-Santos B, Serre K, Norell H (2015) γδ T cells in cancer. Nat Rev Immunol 15:683–691

Singh N, Baby D, Rajguru JP, Patil PB, Thakkannavar SS, Pujari VB (2019) Inflammation and cancer. Ann Afr Med 18:121–126

Song W, Anselmo AC, Huang L (2019) Nanotechnology intervention of the microbiome for cancer therapy. Nat Nanotechnol 14:1093–1103

Steinman RM (2006) Linking innate to adaptive immunity through dendritic cells. Novartis Found Symp 279:101–109. discussion 109-113, 216–219

Takeuchi A, Saito T (2017) CD4 CTL, a cytotoxic subset of CD4+ T cells, their differentiation and function. Front Immunol 8:194

Taylor A, Verhagen J, Blaser K, Akdis M, Akdis CA (2006) Mechanisms of immune suppression by interleukin-10 and transforming growth factor-β: the role of T regulatory cells. Immunology 117:433–442

Vendrell A, Gravisaco MJ, Pasetti MF, Croci M, Colombo L, Rodríguez C, Mongini C, Waldner CI (2011) A novel *Salmonella typhi*-based immunotherapy promotes tumor killing via an antitumor Th1-type cellular immune response and neutrophil activation in a mouse model of breast cancer. Vaccine 29:728–736

Vendrell A, Gravisaco MJ, Goin JC, Pasetti MF, Herschllik L, De Toro J, Rodríguez C, Larotonda G, Mongini C, Waldner CI (2013) Therapeutic effects of *Salmonella typhi* in a mouse model of T-cell lymphoma. J Immunother 36:171–180

Wagner M, Koyasu S (2019) Cancer immunoediting by innate lymphoid cells. Trends Immunol 40: 415–430

Wallecha A, Singh R, Malinina I (2013) Listeria monocytogenes (Lm)-LLO immunotherapies reduce the immunosuppressive activity of myeloid-derived suppressor cells and regulatory T cells in the tumor microenvironment. J Immunother 36:468–476

Zheng D, Liwinski T, Elinav E (2020) Interaction between microbiota and immunity in health and disease. Cell Res 30:492–506

Zoaby N, Shainsky-Roitman J, Badarneh S, Abumanhal H, Leshansky A, Yaron S, Schroeder A (2017) Autonomous bacterial nanoswimmers target cancer. J Control Release 257:68–75

Chapter 2
Immunomodulatory Properties of Nutraceuticals and Functional Foods

Adriana García-Gurrola, Abraham Wall-Medrano, Miguel A. Olivas-Aguirre, Francisco J. Olivas-Aguirre, and Alberto A. Escobar-Puentes

Abstract The immune system is a complex assembly of coordinated cells to maintain human health. Nowadays, there is a growing interest in maintaining the immune system in proper conditions to counteract noncommunicable chronic diseases, harmful entities (toxic xenobiotics, allergens, and microbial pathogens), cancer, and emerging infectious and autoimmune diseases that currently affect health systems worldwide. Adequate human nutrition is essential for the maintenance and modulation of the immune system. This is due to the different functional ingredients that can be found in food matrices such as fruits, vegetables, dairy products, meat, cereals, and oils. Among these, polyphenolic antioxidants, vitamins, probiotics, prebiotics, and peptides can be highlighted. All the above dietary functional ingredients act as immunomodulators since they have a direct or indirect immunostimulatory response by interacting with immune cells and triggering stimulating responses for antibodies and anti-inflammatory cytokine and chemokine secretions but also for suppressing pro-inflammatory cytokines. Therefore, these immunomodulator compounds have been considered as adjuvants in the prevention and treatment of some infectious diseases and some others related to autoimmunity.

Keywords Functional foods · Nutraceuticals · Immunostimulating response · Nutrition and immune system

---

A. García-Gurrola · A. A. Escobar-Puentes (✉)
Facultad de Medicina y Psicología, Universidad Autónoma de Baja California, Tijuana, Mexico
e-mail: escobar.alberto@uabc.edu.mx

A. Wall-Medrano
Instituto de Ciencias Biomédicas, Universidad Autónoma de Ciudad Juarez, Anillo Envolvente del Pronaf y Estocolmo s/n, CP, Chihuahua, Mexico

M. A. Olivas-Aguirre
Laboratorio de Inmunobiología y Regulación del Transporte Iónico, Centro Universitario de Investigaciones Biomédicas, Universidad de Colima, Colima, Mexico

F. J. Olivas-Aguirre
Departamento de Ciencias de la Salud, Universidad de Sonora (Campus Cajeme), Hermosillo, Mexico

R. K. Kesharwani et al. (eds.), *Nutraceuticals and Functional Foods as Immunomodulators*, https://doi.org/10.1007/978-981-19-2507-8_2

2.1 Introduction

It is well known that diet provides essential nutrients for energy production and to sustain diverse body functions. However, growing evidence supports the idea that nutrients and other food-derived xenobiotics have a greater impact on human health including their immunomodulatory action. The influence of nutrition on the immune system was initially delineated by the fact that malnutrition is a leading cause of immunosuppression (Alwarawrah et al. 2018; Chandra 1974). Strong evidence supports the notion that malnutrition leads to several immune abnormalities including the impairment of physical barriers (e.g., mucosal layers) to protect tissues, lymphoid organ atrophy, lymphopenia, subsequent recurrent infections, and reduced, delayed, or null response to vaccination (Rytter et al. 2014). Immunomodulators are compounds that regulate the immune system in terms by either stimulating or suppressing the action of key immune cell populations or their response against a determined stimulus. The importance of nutraceuticals and functional foods as immunomodulators relies on their potential to prevent or even treat immunocompromised illnesses such as infectious diseases or chronic metabolic perturbations (Chalamaiah et al. 2018; Keservani et al. 2020a, b). Such observations suggested that food possesses functional ingredients with immunomodulatory effects. In this context, a great effort has been made to elicit the identity and mechanisms of the compounds modifying the immune response. Between them, phenolic antioxidants, probiotics, prebiotics, vitamins, and peptides have been identified as primary modulators of immune cells, cytokine release, and antibody production (Cunningham-Rundles et al. 2005). This chapter aims to provide a general overview of the immune system (innate + acquired), the pleiotropic effects of certain functional foods and nutraceuticals with demonstrated immunomodulatory properties (Fig. 2.1), and their role in preventing or treating immunocompromised disorders.

2.2 The Immune System

2.2.1 Hematopoiesis and the Immune System

Most of the immune cells arise from a process of proliferation, differentiation, and maturation of hematopoietic stem cells (HSC) called hematopoiesis (Dutra 2017). Such process requires the coordinated interplay between diverse cells in primary organs, where HSC reside and develop all the immune cell subsets. An HSC residing in the bone marrow has the potential to divide into common myeloid progenitors that eventually will differentiate into leukocytes with specific functions such as monocytes, macrophages, dendritic cells (DC), eosinophils, neutrophils, etc. (Fig. 2.2). HSC can differentiate into lymphoid progenitors that coordinate with other lymphoid organs to produce B lymphocytes, T lymphocytes, and natural killer

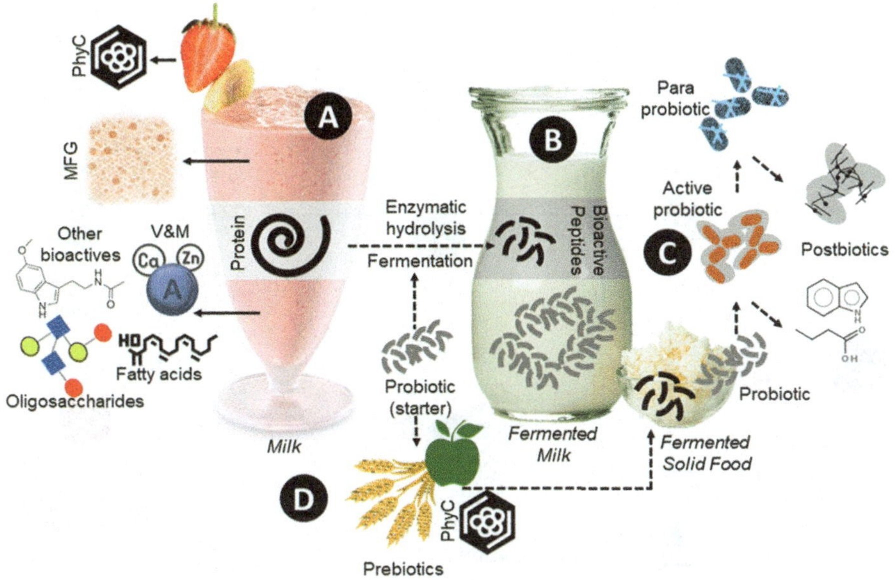

Fig. 2.1 Selected dietary components with immunomodulatory action. Common foods and new designed foods may contain several bioactive components with immunostimulatory activity. For example, (**a**) whole milk is rich in oligosaccharides, vitamins and minerals (V&M), free fatty acids, and many other molecules that make up milk fat globule (MFG). If fruit is added to whole milk to turn it into a milkshake, other phytochemicals (PhyC) with immunostimulatory action (e.g., phenolic compounds) while enhancing, even more, this biological action. (**b**) If whole milk is fermented by the action of probiotic bacteria (starter) or it is subject to protease hydrolysis, milk proteins are transformed into bioactive peptides, some of which will enhance the cellular and humoral immune response. (**c**) The hydrolysis of milk oligosaccharides/sugars (lactose) and proteins will increase the amount, metabolic maturity, and activity of probiotic bacteria as monitored by a sharp increase of several fermentation metabolites (postbiotics). Lastly, under colonic conditions, the senescent probiotic bacteria will release cellular components also with immunostimulatory potential (paraprobiotics). (**d**) A similar "probiotic" pathway may follow cereal and vegetable sources rich in immunoreacting phytochemicals such as prebiotics

(NK) cells, which represent ~30% of the total blood leukocyte count. An adequate hematopoiesis is necessary to generate a competitive immune system, and interestingly some dietary compounds can alter the hematopoietic system and promote either immunosuppression or boost of the immune system, as it will be discussed later.

Evolution has led the organism to the development of the *immune system* as an integral strategy to protect the body from harmful or unknown entities, although it is as complex as its protective mechanisms. The innate immune system (IIS, aka natural immunity, Fig. 2.3) is the first nonspecific line of defense and recruits many cell lines, namely, neutrophils, monocytes, macrophages, NK, and DC, that help the host to eliminate toxic xenobiotics, allergens, and microbial pathogens that enter the body through mucosal surfaces, discriminating self from nonself (Dutra

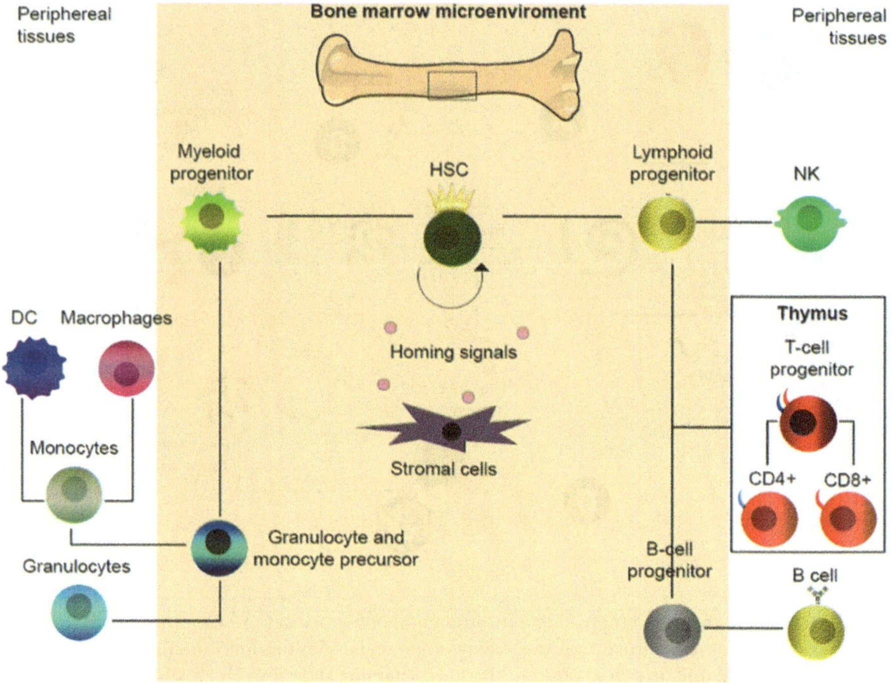

Fig. 2.2 Overview of the hematopoietic system

2017). IIS also includes several physical barriers like mucous layers, membrane-bound receptors such as toll-like receptors (TLRs), germline-encoded pattern recognition receptors (PRRs, e.g., Dectin-1, Nod12), host defense peptides (HDPs, e.g., defensins and cathelicidins), and other small molecules that are either constitutively present or timely released from activated cells (e.g., cytokines, chemokines). PRRs link the IIS to the adaptive immune system (AIS, aka acquired immunity) which, in turn, is divided into two types: cell-mediated and antibody-mediated (humoral) immunity (Palm and Medzhitov 2009). Cell-mediated AIS consists of a plethora of highly specific immune antigen receptors located on the surface of T lymphocytes [mainly CD4+ (helper T cells, Th) and CD8+ (cytotoxic T cells, Tc)] which are responsible for antigen recognition, while CD4+ bearing T-regulatory cells (Treg) are responsible for the tolerance to nonself but innocuous components such as food (Childs et al. 2019).

The major histocompatibility complex (MHC) I (MHCI) or II (MCHII) presents antigens to CD8+ or CD4+ bearing T lymphocytes by simulating T-cell receptors (TCRs) that mediate intracellular calcium influx and favoring the expression and production of cytokines necessary for the activation and function of effector cells, and it is noteworthy that vertebrates use *somatic diversification* to generate a highly diverse repertoire of antigen receptors (Boehm 2011). Th cells secrete several cytokines such as interferon γ (IFN-γ) and interleukins (IL)-2, IL-4, IL-5, IL-6,

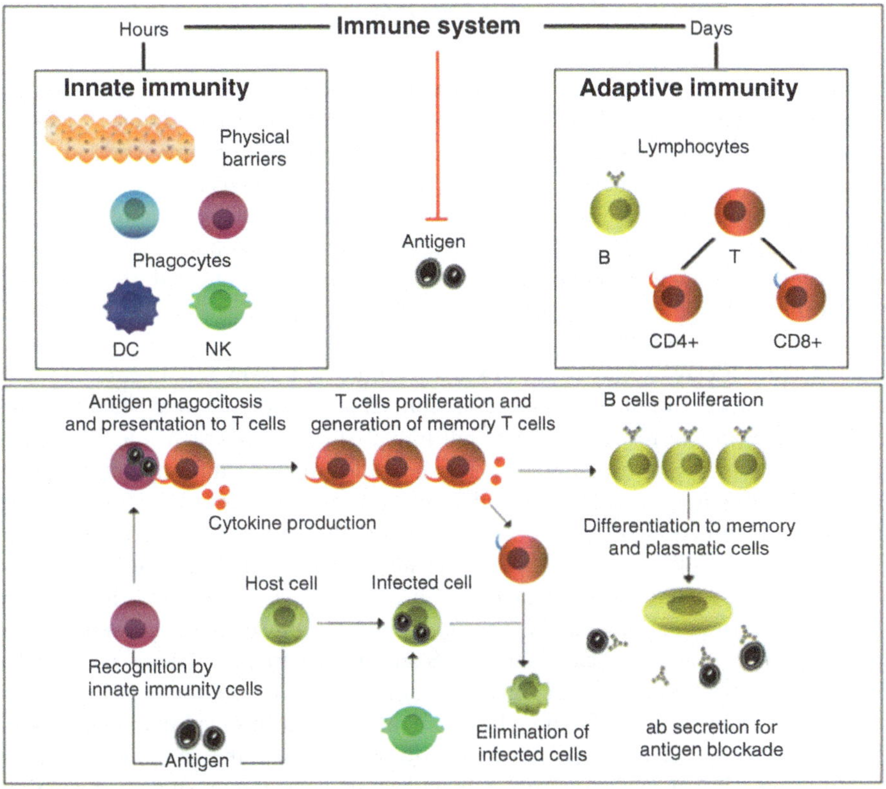

Fig. 2.3 Overview of the immune system. Composition and classification of the immune system (upper panel). Interplay of antigens with both the IIS and the AIS (lower panel). Dendritic cell (DC), antibody (ab)

IL-10, IL-13, and IL-25, while the expression of IL-37 (a formerly IL-1 member) in macrophages and epithelial cells abrogates pro-inflammatory cytokines (Dinarello et al. 2016). These and other cellular/molecular events link inflammation (systemic/local) with immunity to such an extent that the systematic evaluation of immune cell proliferation/activation and cytokine profiling are commonly assayed when validating the immunomodulator effect of nutraceuticals and functional foods such as dietary peptides. Lastly, B lymphocytes produce antibodies (ab) upon Th activation, initiating the *humoral response*. ab (namely, IgA [immunoglobulin A], IgD [immunoglobulin D], IgE [immunoglobulin E], IgG [immunoglobulin G], IgM [immunoglobulin M]) can be found in several body fluids and within the bloodstream, and their function is to recognize and block antigens of very diverse nature (Simon et al. 2015; Keservani et al. 2016).

2.2.2 The Immune System in Disease

Many noncommunicable chronic diseases (e.g., obesity, type II diabetes, cancer), malnutrition, and emerging infectious diseases [severe acute respiratory syndrome coronavirus 2 (SARS-CoV2)], which currently affect health systems worldwide, are immunocompromised illnesses. For example, it has been observed that regardless of the subtype, patients with *diabetes mellitus* exhibit a lesser peripheral blood mononuclear cell count (PBMC, T and B cells, NK cells, and monocytes) and a lower responsiveness to activating stimuli (Table 2.1), while uncontrolled diabetes often results in polyneuropathy that negatively impacts the physical barriers of IIS promoting the susceptibility to infections.

Unhealthy lifestyles may result in kidney failure and uremia, both coursing with alterations on the immune system function, and considering that such patients usually have protein-restrictive diets, nutritional support approaches to burst their immune system must be explored. It has been documented that malnutrition also leads to abnormalities in neutrophil and T-cell subsets and their response to mitogens (activator stimulus) and further production of cytokines (Takele et al. 2016; Fernández et al. 2005); such panorama was attributed to a decrease on K^+ currents, necessary for sustained Ca^{+2} intake and lymphocyte activation for the immune response against infections. Lastly, obesity has a great impact on the immune system associated with a leptin enhancement of IIS and AIS (Fernández-Riejos et al. 2010). However, in obese patients with great number of adipocytes, leptin levels generate a sustained activation of the IIS and AIS which resembles a chronic inflammatory state (Fernández-Riejos et al. 2010). Another disease altering the immune system is hypertension. It has been described that through autonomic system regulation, hypertension can lead to the immunosuppression of different immune cell subsets; however, some dietary components such as antihypertensive biopeptides can also modulate AIS by promoting the cytokine expression and lymphocyte proliferation (Singh et al. 2014).

Lastly, chronic inflammation is also a characteristic of most of cancers. Alterations of IIS and AIS have been observed and include the augmentation and recruitment of immune cells to the tumor of metastatic sites. Interestingly, the role of the immune system infiltrates in tumor development has been discussed in terms of antitumor and pro-tumor effects, depending on cancer type, stage, and immune cell type (Gonzalez et al. 2018). However, in nonsolid cancers (such as leukemia, lymphoma, or myeloma), normal hematopoiesis is deregulated as cancer cells compete and modify the bone marrow microenvironment, the main lymphopoietic site; such modifications displace the production of healthy B an T cells, resulting in nonfunctional lymphoblast, therefore augmenting the susceptibility to infections (Enciso et al. 2015). In conclusion, during sickness, the IIS/AIS repertoire is deregulated with a disease-specific signature (see Table 2.1), and so, a greater effort to reestablish immunocompetence not only by using drugs but with nutrition-based approaches remains as an open scientific field.

Table 2.1 Effect of principal diseases on the immune system

Condition	Effect on the immune system	Immune derangement	Ref.
Diabetes mellitus	↓ PBMC count ↓ Response to activating stimuli	↓ Antigen phagocytosis ↑ Susceptibility to infection Delayed or null response to vaccination	Daoud et al. (2009)
Renal failure	↓ Monocyte count ↓ Function of DC Impaired maturation ↓ Endocytosis	↓ Antigen phagocytosis ↓ Antigen presentation ↓ T-cell activation ↑ Susceptibility to infection	Lim et al. (2007)
Severe malnutrition	↓ K+ currents in T cells ↓ T-cell activation ↓ Cytokine production	↓ B-cell activation ↓ Effector function ↑ Susceptibility to infection	Fernández et al. (2005)
Obesity	Chronic inflammation ↑ Innate immunity ↑ Phagocytosis ↑ Adaptive immunity ↑ T-cell activation	↑ Monocyte phagocytosis ↑ Cytokine production ↑ T-cell activation ↑ Potentiation of mitogens	Fernández-Riejos et al. (2010)
Solid tumors	Chronic inflammation Recruitment and activation of innate and adaptive immune cells	↑ Pro-inflammatory macrophages ↑ Neutrophils ↑ T-cell infiltration	Gonzalez et al. (2018)
Nonsolid tumors	Immunosuppression ↓ Adaptive immunity	↓ Lymphopoiesis ↓ Functional B cells ↓ Functional T cells ↑ Susceptibility to infection	Enciso et al. (2015)
Hypertension	↑ Adaptive immunity ↑ Innate immunity	↑ Pro-inflammatory cytokines ↑ Lymphocyte activation	Singh et al. (2014)

DC dendritic cells; *PBMC* peripheral blood mononuclear cell count

2.2.3 Nutrition and Immunity

The role of nutrition over the immune system was initially highlighted by the fact that malnutrition is the main cause of immunodeficiency (Childs et al. 2019; Chinen and Shearer 2010); interestingly, diet intake patterns can significantly determine the systemic activity of immune components and the hematopoietic niche (Hastreiter et al. 2020). Particularly, balanced diets containing high amount of vegetables, fish, and oils significantly reduce serum inflammatory markers (Ahola et al. 2017) and hematopoiesis (Xia et al. 2015). Unhealthy dietary habits promote inflammatory states associated with a higher production of antigenic molecules of microbial origin such as lipopolysaccharide (LPS) that possesses immunogenic activity by activating patter recognition receptors (Ahola et al. 2017). Also, many cytokines such as tumor necrosis factor α (TNF-α) and chemokines are deregulated in hyperglycemic states and prolonged intake of high-fat diets, altering NK cell activity and favoring inflammation which, in turn, stimulate the proliferation and recruitment of macrophages to adipose tissue, alter the migration of immune cells, etc. (Alwarawrah et al. 2018). Therefore, fat- and glucose-lowering diets must be implemented to reach immunocompetence. Also, protein-/amino acid-deficient diets can promote the susceptibility to infections, and many studies have demonstrated that certain amino acids such as glutamine are necessary for immunoglobulin synthesis and cytokines, while its restriction can impair the activation of lymphocytes, macrophages, and NK cells (Li et al. 2007). Amino acids can also alter humoral and cellular immunity at different levels, favoring gene expression, lymphocyte metabolism, and production of ab and enhancing the proliferation, between others. The lipids consumed in diet can also affect immune cell function. Fatty acids (FA) have demonstrated the ability to reduce T-lymphocyte proliferation. It has been attributed to a reduction on cytokine production, such as IL-2, necessary for T-cell proliferation, or TNF-α (Xia et al. 2015). Additionally, it has been proposed that lipids can alter lymphocyte metabolism and migration. As for dietary fiber, its daily intake seems to be determinant for improving the gut microbiota and short-chain fatty acids (SCFAs) which in turn improves local lymphocyte CD4+ and CD8 count and function and Ig (immunoglobulin) production (Yang et al. 2020). Lastly, not only macronutrients but also micronutrients can also impact on the immune system: certain vitamins (VTs) are essential for epithelial maintenance, and the lack of them increases the odd for infections; particularly, liposoluble VTs are involved in immune cell proliferation, differentiation, homing of lymphocytes, antibody production, and inflammation (Aslam et al. 2017). All these dietary components are just few examples of nutraceuticals and functional foods with immunomodulatory properties, as it will be discussed in further sections.

2.2.4 Immunosenescence and Nutrition

Aging leads to the gradual deterioration of diverse systems, and the immune system is not the exception. Immunosenescence is a term commonly used to describe the differences observed between younger and older individuals in a broad range of immune parameters (Pawelec 2018). It is known that during elderly, the susceptibility to chronic diseases, cancer, and infections is augmented. Correspondingly, it has been observed that age has a significant effect on HSC, reducing the homing in the bone marrow, the response to macroenvironmental signals, and the differentiation to lymphoid progenitors. Therefore, a lower humoral immunity in aged persons is observed, as the Ig diversity and affinity are severely compromised. Also, aging strongly impacts on thymus function. It has been observed that the thymus suffers atrophy and together with the bone marrow the adipose tissue increases; consequently, lymphocyte maturation is decreased, and an adipocyte-mediated chronic inflammation is observed (Larbi et al. 2018). Thus, recently, novel interventions potentiating the immune system and limiting the effects of age in the immune system are being studied, and nutrition represents an attractive strategy. Indeed, several compounds with immunomodulatory activity have been described and recommended to reestablish the immune system during elderly, as some nutrients as VTs and probiotics can boost the immune system; these and other dietary senescence immunomodulators will be extensively discussed in this chapter.

2.3 Immunomodulatory Response of Polyphenolic Compounds

Polyphenolic compounds (PC) (Fig. 2.4) are the most abundant naturally produced phytochemicals by vegetal organisms, and consequently, they have been present in human diet since a long time ago. By its abundant distribution in plants and fruits, it is not difficult to find important sources of these compounds (tea, coffee, berries, spices, pomegranate, and all types of vegetables are abundant in polyphenols) as well as beneficial claims associated with their consumption (Rothwell et al. 2013). Each of the molecules that conforms this variated family can reach to different target cells, interact with specific receptors, and elicit signaling pathways on multiple processes. The antioxidant capacity of PC family is the main mechanism described

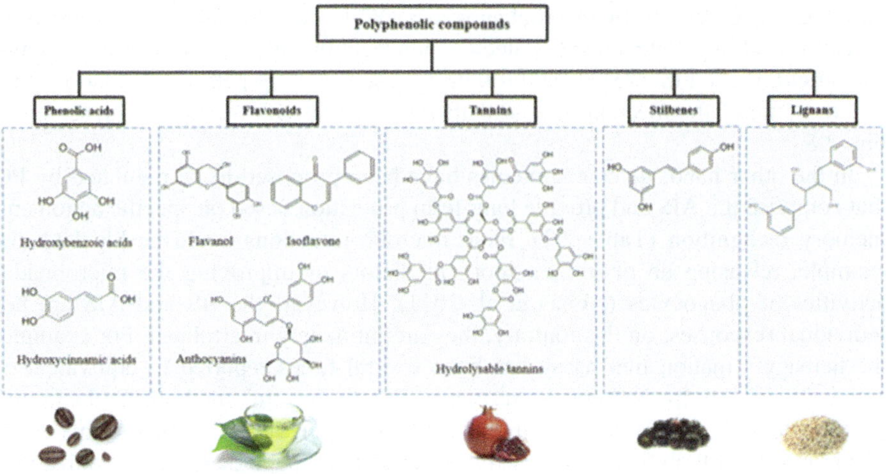

Fig. 2.4 Example of polyphenol classes and its representative molecules

by which these molecules exert beneficial effects; however, derived from the variable structures of its subfamilies (anthocyanins, tannins, phenolic acids, etc.), other biological actions have been recognized; they include metabolic regulation and modulation of several cellular pathways related with cardiometabolic health, improving cognition, anticancer effects, and immunomodulatory properties, among others (Fraga et al. 2019; Keservani et al. 2016). The immunomodulatory effects of PC have been considered an alternative as adjuvant for many disorders in which antioxidant stress and inflammatory events are present and interrelated with altered functions of the immune system. In this sense, several mechanisms have been described, and they include the regulation of both IIS and AIS.

2.3.1 Innate and Adaptive Immunity Regulated by Polyphenolic Compounds

Advances in science have shown that PC can cause epigenetic changes on different immune conditions; initially, PC can regulate IIS (Table 2.2). These activities include the regulation of phagocytic action of myeloid cells (neutrophils, macrophages) after the activation of the signaling pathways of mitogen-activated protein kinases (MAPK) and nuclear factor κB (NF-κB) (Boothapandi and Ramanibai 2019; Tanaka et al. 2018), the modulation of the antigen-presenting cells such as DC (Yoshimura et al. 2013), the activation of NK cells, as well as the release of diverse substances with inflammatory properties such as cytokines and chemokines (Kumar and Baradaran et al. 2020; Vugic et al. 2020). Noticeably, immunomodulation by PC could be divided into two main actions, immunostimulatory when the cellular response is boosted or immunosuppressant when the responses is diminished or suppressed. As example, pyrogallol-type green tea PC modulates the IIS stimulating the phagocytic activity of macrophages and DC in cell line studies. Surprisingly, epigallocatechin gallate, epigallocatechin, epicatechin gallate, epicatechin, catechin, and strictinin elicit this response through caspase signaling pathway (Monobe et al. 2010). Some of the mechanism and pathways related with the IIS are presented in Fig. 2.5.

In the other hand, diverse strategies have been proposed to be regulated by PC that comprise the AIS and provide long-term protection based on specific action and memory recognition (Table 2.2). First, microbe infections could be blocked, for example, releasing ab or inflammatory mediators or improving the microbicidal activities of phagocytes (Abbas et al. 2011). However, the IIS and AIS are not individual responses; on the contrary, they are intimately interrelated. For example, the hemagglutination titer assay has been several times reported as equivalent of active B-cell activity or humoral response. Employing this method, Guduchi and Brahmi PC-rich extracts have demonstrated to trigger the Ig synthesis; this event highly related with its subsequent activity over the pathogen marking (opsonization) and its final consumption by phagocytic cells (Husain et al. 2017). Thus, the *humoral*

Table 2.2 Immunomodulatory activities regulated by phenolic compounds in the IIS and AIS

Response	Mechanism	Action	Phenolic compound	Ref.
Innate	Inflammation	Recruitment of phagocytic cells and plasma proteins to destroy the microbes	Gallic acid Ellagic acid Morin CurcuminChlorogenic acidCaffeic acidResveratrol	de Moraes Alves et al. (2017), Jakhar et al. (2014), Keservani et al. (2016)
		Regulating antigen-presenting cells	Oenothein BGallic acidQuercetin	Jakhar et al. (2014), Yoshimura et al. (2013), Chan et al. (2015)
	Antiviral	Eliminating virus by NK and secreting cytokines such as IFN-γ and TNF-α	Naringenin Resveratrol Caffeic acid Ferulic acid P-coumaric acid Anthocyanins	Kim and Lee (2015), Kilani-Jaziri et al. (2017), Filardo et al. (2020) McAnulty et al. (2011)
Adaptive	Humoral	Production of ab that binds to extracellular microbesBlock the infect capacity over host cells	Aqueous extracts of Guduchi and Brahmi Hesperetin Kaempferol	Husain et al. (2017) Shih et al. (2020)
	Cell-mediated	Microbicidal abilities of phagocytes regulated by Th	Cy3G Apigenin Malvidin Caftaric acid Procyanidin dimers	Zhao et al. (2019) Zhao et al. (2020) Chen et al. (2020a, b)
		Boost and enhance the CD4+ or CD8+ T cells	Apigenin Catechin Flavonoids	Li et al. (2010) Gandhi et al. (2018) Kim et al. (2016)

Cy3G Cyanidin 3-O-glucoside; *DC* dendritic cells; *IFN-γ* interferon gamma; *NK* natural killer; *TNF-α* tumor necrosis factor alpha; *Th* helper T cells

response is regulated by molecules in the blood and mucosal secretions, which can bind to extracellular pathogens (and its toxins) and promote their elimination.

Complementary to the adaptive immunity, the regulation of T-cell lymphocytes and its "messenger molecules" is a well-documented line of action to elicit several cell responses. Interestingly, phenolic-rich extracts modulate lymphocyte subset as part of the *cell-mediated immunity.* First CD4+/CD8+ T cell ratio has been proposed as indicator of optimum IIS. Recent reports evidence that flavonoids could regulate the number and activities of T cells maintaining adequate immune responses. As example, several reports evidence the effectiveness of kaempferol, rutin, quercetin, or icariin as modulating agents of the IL-2, IL-4, and IFN-γ secretions mediated by

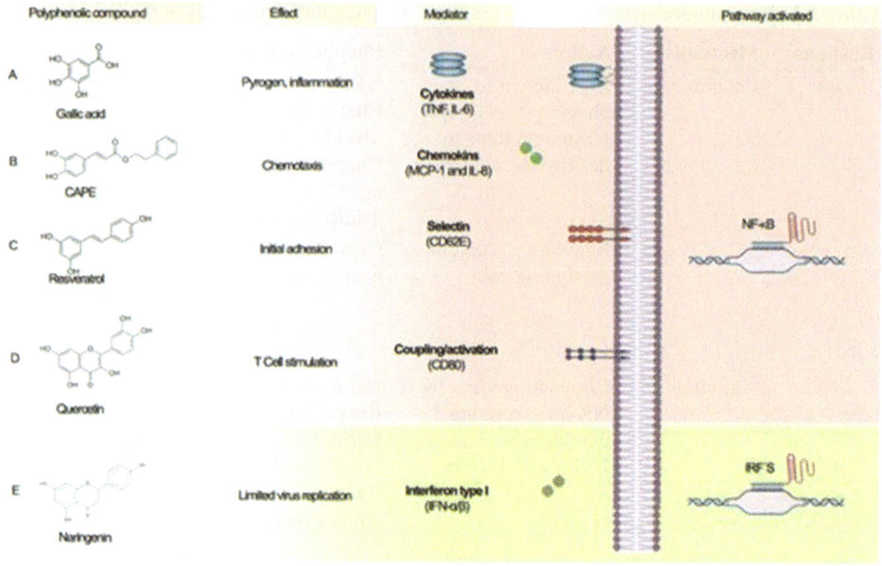

Fig. 2.5 PC as modulating agents over IIS. CAPE, caffeic acid phenethyl ester; CD, cluster of differentiation; IFN, interferons; IL, interleukin; IRF, interferon regulatory factors; MCP-1, monocyte chemoattractant protein-1; NF-kB, nuclear factor kappa-light-chain-enhancer of activated B cells. (**a**) Mirshekar et al. 2018; (**b**) Lawrence et al. 2006; (**c**) Xiao et al. 2013; (**d**) Li et al. 2019; (**e**) Wu et al. 2016. Abbas et al. 2011

Th (Gandhi et al. 2018). Although oral consumption of cyanidin 3-O glucoside (an anthocyanin present in berry fruits) ameliorates the symptoms of aggressive inflammatory responses against the exposition of antigens, the mechanisms proposed involve the reduced macrophage recruitment and a reduction of pro-inflammatory cytokines via signal transducer and activator of transcription (STAT1) and/or STAT3 phosphorylation (Zhao et al. 2020). MAPK and NF-kB signaling has been also related with alleviating Th responses by cyanidin-3-glucoside and chlorogenic acid (Zhao et al. 2019). On the other hand of the cell-mediated response, CD8+ lymphocytes eliminate cells infected with cytoplasmic antigens, thanks to their differentiation to cytotoxic T cells (CTLs). Thus, Lis et al. (2020) have showed that epicatechin stimulates the maturation of thymocytes toward CD8+ cells in a dose-dependent manner. In this sense, PC can abolish the reservoirs of infection by eliminating the infected cells.

2.3.2 *Overview of Immune Responses in Human Studies*

Several human studies have investigated the beneficial health effect of PC on immunomodulation (Table 2.3). Frequent consumption of PC-rich foods could

Table 2.3 Human evaluations of PC consumption over general health and autoimmune alterations

Food or PC	Treatment	N	Duration (weeks)	Main findings	Ref.
Grape powder	23 g	24	9	No differences in the proliferative responses of CD4+ or CD8+ T cells were observed between the grape and placebo powder groups	Zunino et al. (2014)
Cranberry juice	450 mL	23	10	Increase of γδ-T cells and NK cells proliferation	Nantz et al. (2013)
Quercetin	500 mg 1000 mg	38 40	12	No influence over the innate immune function	Heinz et al. (2010)
Bilberries	400 g	15	8	Decrease pro-inflammatory markers such as in IL-12	Kolehmainen et al. (2012)
Polyphenols in wine	1758 mg	34	4	Significant reduction of TNF-α, IL-6, and IFN-γ	Muñoz-González et al. (2014)
Ginkgoflavonglycosides	60 mg	12	12	The progression of vitiligo stopped	Szczurko et al. (2011)
Curcumin	3 g	26	4	Remission of ulcerative colitis	Lang et al. (2015)
Epigallocatechin-3-gallate	600 mg	18	12	Improves muscle metabolism in multiple sclerosis patients	Mähler et al. (2015)

γδ-T Gamma-delta cells; *NK* natural killer cells; *IL-12* interleukin 12; *IL-6* interleukin 6; *TNF-α* tumor necrosis factor alpha; *IFN-γ* interferon gamma

regulate changes in general health and several disorders such as autoimmune conditions. As example, regular bilberry consumption (400 g/d) decreased low-grade inflammation regulating the serum levels of IL-12, molecule responsible for the activation of the cell-mediated response of the AIS (Kolehmainen et al. 2012). This finding has in turn been evidenced by other sources such as green tea or wine (Muñoz-González et al. 2014; Cuevas et al. 2013). Additionally, both IIS and AIS have been regulated by PC present in cranberry juice. Nantz et al. (2013) reported an increase of gamma-delta cells (γδ-T) and NK cells after the consumption of 450 mL of cranberry juice. In counterpart, no effects have been reported by Zunino et al. (2014) over the differentiation of T cells regulated by grape powder consumption.

By its immunomodulatory activities, PC have been recently recognized as therapeutic agents by its positive effects as adjuvants in several autoimmune conditions.

As example, curcumin was informed to be effective inducing remission in patients with ulcerative colitis treated with mesalamine (Lang et al. 2015), flavonglycosides from Ginkgo biloba were effective to ameliorate the vitiligo lesions and favoring re-pigmentation (Szczurko et al. 2011), while epigallocatechin-3-gallate improves the symptoms of muscle weakness and excessive fatigue in multiple sclerosis patients (Mähler et al. 2015). Nowadays, other autoimmune conditions, such as sarcoidosis, rheumatoid arthritis, and Addison disease, are being evaluated (Khan et al. 2019).

2.4 Immunomodulatory Effect of Vitamins

VTs were not recognized until about the start of the twentieth century. Unlike other types of nutrients, VTs do not have structural functions and are not significant sources of energy. Their functionality lies in their action as cofactors, biological antioxidants, and hormones, all the above related to the host's immune response. VTs (*vital amines*) and their metabolites (vitamers) are organic compounds that are required in trace amounts in the diet because they cannot be synthesized in enough quantities by an organism. In this sense, the concept of vitamin is that "organic compound other than carbohydrates, lipids and proteins, which is a natural component in food and essential for normal physiological function, which, due to its absence, causes a specific deficiency syndrome, and which furthermore, it is not synthesized by the host in amounts adequate for physiological needs." Currently, 33 substances have been recognized as VTs. In addition, each one of them has its vitamers, which are compounds with a related chemical structure and qualitatively comparable metabolic activities (but not quantitatively). Also, some of them have their provitamins, compounds that when metabolized are precursors of a specific vitamin (Combs 2012).

VTs are important cofactors for the proper function of the immune system and have implications for both IIS and AIS. Fat-soluble vitamins A (VitA), D (VitD), and E (VitE) are the compounds with the greatest evidence proven on immune response, mainly VitD (Table 2.4). Also, some immunostimulating properties of the water-soluble VTs such as the B complex (VitB) and vitamin (VitC) C have been documented. In general, VitC, VitE, and members of the VitB complex can act in a relatively nonspecific manner in the immune system (e.g., as antioxidants), but VitA and VitD can influence the immune response in highly specific ways (Mora et al. 2008).

Table 2.4 Immunomodulatory response of fat-soluble and water-soluble vitamins

Vitamin	Study type	Immunomodulatory response	Ref.
Vitamin A	In vivo (mice)	↑ Regulation of the intestinal immune response through immunomodulatory actions on intestinal DC and lymphocytes	Zeng et al. (2013)
	In vivo (mice)	↑ Improve diarrhea and enteritis in IBS and could possibly inhibit inflammation of the colon and cancer development	Okayasu et al. (2016)
	In vitro	↑ Expression of the integrin α4β7 and the production of IL-5 and IL-13 and ILC2 cells	Ruiter et al. (2015)
	Clinical trial	↓ Reduced TNF-α and IL-10 production after vaccination against diphtheria, pertussis, and tetanus in newborns	Jørgensen et al. (2013)
Vitamin D	In vivo (rats)	↑ Protection against memory and learning dysfunction caused by inflammation, inhibiting oxidative stress and inflammation in the hippocampus	Mokhtari-Zaer et al. (2020)
	Clinical trial	↑ Ability of CD4+ CD25+ T cells to upregulate the programmed death receptor and reduces CD69 expression in Crohn's disease patients	Bendix et al. (2017)
	Clinical trial	↑ IL-8, IL-10, and IFN-γ levels in brain-dead subjects	Custódio et al. (2018)
	Clinical trial	↑ PBMC broad-spectrum pro-inflammatory cytokine response in pregnant women, with an average 1.7- to 2.1-fold increase in pro-inflammatory cytokines (GM-CSF, INF- γ, IL-1β, IL- 6, and IL-8)	Hornsby et al. (2018)
Vitamin C	In vitro	↑ Active signaling pathways (NOD and TLRs) to promote the production of INF, activate and balance T cells, and regulate inflammatory response by inhibiting the MAPK signaling pathways	Chen et al. (2020a, b)
Vitamin E	Clinical trial	↓ Allergy and IL-28 production	Elenius et al. (2017)

DC dendritic cells; *IBS* irritable bowel syndrome; *IL* interleukin; *ILC2* type 2 innate lymphoid cells; *TNF-α* tumor necrosis factor alpha; *IFN-γ* interferon γ; *PBMC* blood mononuclear cell; *GM-CSF* granulocyte-macrophage colony-stimulating factor; *NOD* nucleotide-binding oligomerization domain-like receptors; *TLRs* toll-like receptors; *INF* interferon; *MAPK* mitogen-activated protein kinases

2.4.1 Immunomodulatory Properties of Liposoluble Vitamins

2.4.1.1 Vitamin A

VitA is an essential organic compound that influences growth, epithelial cell composition, and the immune system. In nature, VitA occurs in the alcohol form known as retinol, in the aldehyde form known as retinal, and in its acidic form known as retinoic acid (RA) (Combs 2012). VitA can be obtained through the diet either as provitamin A (carotenoids [pigments in green, orange, and yellow vegetables]) or as preformed VitA (retinol and retinyl esters [animal tissues such as milk, egg, and fish]). Once ingested, VitA is converted into RA in the liver, and subsequently, RA is connected to the RA receptors (RXRs) in immune cells and here initiates the complex molecular mechanisms for the immunological activation, mainly through the expression of genes and regulatory cytokines (Harirchian et al. 2019). However, the metabolism of VitA within DC and macrophages will depend on the immunological context and will be modulated through an interaction with available cytokines, inflammatory signals, and signals with the TLRs of immune cells (Erkelens and Mebius 2017). According to in vivo and in vitro studies, VitA has an immunostimulating response by regulating both the IIS and the AIS (Huang et al. 2018). RA influences AIS as it increases cytotoxicity and T-cell proliferation, regulates the maturation and differentiation of B cells, inhibits apoptosis of B cells, and modulates antigen presentation through direct mechanisms in the DC. Furthermore, in the presence of inflammatory stimuli such as TNF, RA promotes DC maturation through receptor RXRs. In other immune responses, RA promotes Th2-cell differentiation and blocks Th1 expression (Mora et al. 2008). Also, RA acts on immune cells increasing lymphocyte mitogenesis and improving monocyte/macrophage phagocytosis, altering the level of expression of Bcl-2 genes (Huang et al. 2018), modifies the immune system through the cytosine IL-1 (Elitsur et al. 1997), and modulates inflammatory responses by contributing to the synthesis of ab such as IgA through the differentiation of B cells (Mora et al. 2008), in addition to promoting DC to express CD103 and in turn Treg cells (Coombes et al. 2007). RA can function as a pro- and anti-inflammatory molecule, this depending on the microenvironment and the phenotype of the affected cells (Erkelens and Mebius 2017). In this sense, according to a meta-analysis of clinical trials, the gene expression of inflammatory cytokines (IL-17, IFN-g, and T-bet) and anti-inflammatory cytokines (TGF-b and FOXP3) decreases and increases significantly due to supplementation with VitA in patients with autoimmune diseases (multiple sclerosis and atherosclerosis) (Harirchian et al. 2019).

2.4.1.2 Vitamin D

VitD is a fat-soluble compound classified as a steroid hormone. It can be obtained through the diet (salmon, tuna, cod liver oil, milk, etc.), but it is mainly biosynthesized from 7-dehydrocholesterol in the skin exposed to ultraviolet light (Adorini and Penna 2008). VitD is enzymatically hydroxylated in the liver to produce 25-hydroxyvitamin (calcidiol), which is subsequently hydroxylated in the kidneys to produce 1,25-dihydroxyvitamin D also known as calcitriol, which is the biologically active circulating form of VitD (Chandran et al. 2020). The functionality of VitD has been commonly related to the regulation of mineral homeostasis and bone tissue; however, it also has functionalities related to the immune system, mainly with the prevention of infectious and autoimmune diseases (Bivona III et al. 2017).

VitD has an immunomodulatory response by enhancing IIS and an inhibitory effect on AIS, the latter positively related to the prevention/treatment of autoimmune

Table 2.5 Vitamin D immunostimulating action

Immune cell type	Action
DC	↓ Maturation ↓ IL-12 ↑ IL-10
Macrophages and monocytes	↑ IL-1 ↑ Proliferation ↑Cathelicidin (an antimicrobial peptide) ↑ VDR and the cytochrome P450 protein CYP27B1
T cells	↓ Cytotoxicity ↓ IL-2, IFN-γ, IL-17 ↓ Proliferation ↓ CD4+/CD8+ ratio ↑IL-4 and IL-10 ↑ Treg cell generation
B cells	↓ Proliferation ↓ IgG and IgM production ↑ VDR and CYP24A1
DC	↓ Maturation ↓ IL-12 ↑ IL-10

DC dendritic cells; *IL* interleukin; *VDR* vitamin D receptor. Source: Mora et al. (2008)

diseases. The immunomodulatory effect of VitD on IIS was first recognized in 1986 when VitD treatment was found to inhibit the pathogen *Mycobacterium tuberculosis* in human monocytes (Rook et al. 1986). Subsequently, receptors for vitamin D (VDR) were recognized in immune cells such as DC, macrophages, and lymphocytes. It was established that calcitriol binds to VDR and translocates to the nucleus regulating the differentiation of macrophages, DC, and T helper (Triantos et al. 2020). Calcitriol also binds to VDR of PBMC to modulate the expression of genes that encode molecules related to the antiviral and anti-inflammatory response, such as defensins, antimicrobial peptides (cathelicidin), IL-10, and the inhibitor of NF-kB. On the other hand, macrophages, DC, and B and T cells, through VDR, can produce calcitriol, which has autocrine and paracrine activity (Adorini and Penna 2008). Table 2.5 summarizes the main immunomodulatory responses to VitD in immune cells.

The potentiation of IIS by VitD is also related to the activation of TLRs and the antibacterial response (Adorini and Penna 2008) since VitD promotes production of cathelicidin and β-defensin and enhances the capacity for autophagy via TLR activation in monocytes/macrophages. About its inhibitory effect on AIS, it inhibits the differentiation and maturation of DC and B lymphocytes and weakens antigen presentation. VitD also increases Th2 cytokine production and the efficiency of Treg lymphocytes but suppresses the secretion of Th1 and Th17 cytokines (Skrobot et al. 2018). In this sense, VitD deficiency has been associated with an increase in pro-inflammatory cytokines (IL-1, IL-2, IL-6, IL-8, TNF, and INF-γ) and a decrease in anti-inflammatory cytokines (IL-4, IL-10, and IL-13) (Weir et al. 2020). In addition, VitD deficiency is related to several common inflammatory disorders such as the progression of liver fibrosis and cirrhosis and has an immunomodulatory action on cancer and autoimmune diseases such as inflammatory arthritis, type 1 diabetes, and multiple sclerosis (Mele et al. 2020).

VitD has been shown to have a potent immune response against various viruses, including HIV, and possibly for coronavirus disease 2019 (COVID-19). Through retrospective observational studies, it has been preliminarily suggested that the development of COVID-19 by the SARS-CoV-2 coronavirus could be favored by VitD deficiency in the geographic regions where the disease broke out (Chandran et al. 2020). COVID-19 affects the immune system by producing a systemic inflammatory response or cytokine release. Based on this, it has been suggested that the immunostimulating effect of VitD to decrease the pro-inflammatory cytokine IL-6 could counteract the development of COVID-19, since this pro-inflammatory cytokine is related to the clinical feature of this viral disease (Orrù et al. 2020). Similarly, it has been suggested that VitD could reduce the thrombotic complications of COVID-19 by effecting an anti-inflammatory immune response and reducing severe inflammatory episodes (Weir et al. 2020). However, in vitro and in vivo studies in animals and humans are necessary to reach a robust conclusion. On the other hand, in vivo studies show that calcitriol increases the expression of the CAMP antiviral gene and decreases the percentage of CD4 + cells infected by HIV, in addition to having a direct effect on the response of T cells, since they decrease the production of IL-2 and IFN-γ in a dose-dependent manner (Gonzalez et al. 2019). Also, within the immunostimulating responses of VitD, the antimicrobial properties stand out. VitD has been shown to have a strong synergistic effect with drugs on the induction of antimicrobial peptides in the lung epithelium of cell lines and macrophages and in healthy human immune cells and acts as an antimicrobial adjuvant to effectively treat tuberculosis (Rekha et al. 2018).

2.4.1.3 Vitamin E

The molecule commonly known as VitE is a fat-soluble compound first discovered in 1922 by Evans and Bishop. Tocopherols and tocotrienols, which include alpha (α), beta (β), gamma (γ), and delta (δ) structures, are the common types of VitE. Dietary sources of VitE include vegetable oils and, to a lesser extent, seeds and cereal grains. Once taken and absorbed, it is found in higher concentration in immune cell and is one of the most effective nutrients known to modulate immune function (Lewis et al. 2019). VitE is a powerful lipophilic antioxidant that has both innate and adaptive immunomodulatory properties. VitE supplementation has been shown to enhance the function of the immune system particularly in older individuals by improving age-associated deteriorations (Bivona III et al. 2017).

The immunomodulatory mechanism of VitE has been explored in animal and human studies. It has been reported that VitE has an indirect and direct immunoenhancing effect on immune cells, mainly based on its antioxidant capacity. For example, VitE directly protects cell membrane integrity against oxidative damage, which allows the correct transduction of cellular signals and the synthesis of molecules of immunological interest, mainly in T cells (Wang and Quinn 1999). On the other hand, indirectly, VitE modulates pro-inflammatory cytokines and prostaglandin E2 (PGE2), the latter related to the suppression of T cells and to affect both the IIS and the AIS (Meydani et al. 2005). Similarly, VitE supplementation reduces the production of other inflammatory markers such as TNF-α and IL-6 (Lewis et al. 2019). VitE and its derivatives improve the immune response mainly by modulating the production of cytokines. For instance, α-tocopherol reduces inflammation, inhibiting the production of INF-γ (cytokines Th 1) (Xue et al. 2016) and IL-6. Furthermore, VitE supplementation increases the level of IL-10 (El-Feki et al. 2016), the cytokine with the greatest anti-inflammatory response. Various studies in humans have shown that VitE supplementation causes a higher production of IL-2, a higher activity of NK, and a decrease in the production of IL-6, in addition to changes in DC function, such as a lower production by IL-12 (Lee and Han 2018). Also, VitE has been shown to modulate the immune system in response to *Streptococcus* pneumoniae infection, specifically by limiting the migration of neutrophils and reducing the production of pro-inflammatory cytokines (Ghanem et al. 2015). Administration of VitE has also been shown to enhance cytokine production when influenza and retrovirus infection occur. Furthermore, it has been shown as an excellent adjuvant in immunization with tetanus toxoid, by increasing the production of IFN-γ and IL-4 (Bivona III et al. 2017).

2.4.2 Immunomodulatory Properties of Water-Soluble Vitamins

VitB complex is essential for the healthy balance of the immune systems. VitB is mainly found in cereals and cereal-based foods. These VTs have been extensively related to biological functionalities related to the regulation of nutrient metabolism (glucose, proteins, and lipids), cognitive development, temperature regulation, the nervous system, and the immune system. VTs B1 (VitB1), B3 (VitB3), B6 (VitB6), and B9 (VitB9) have various immune-stimulating responses. For instance, VitB1 or thiamin prevents the expression of pro-apoptotic proteins (Bcl-2) and the activation of caspase-3, intervenes in the regulation of the mitochondrial membrane and apoptosis, and suppresses oxidative stress by the activation of anti-inflammatory processes related to NF-kB. It is also related to the activation and migration of monocytes, T cells, and NK cells. In this sense, VitB1 deficiency is related to autoimmune diseases such as encephalomyelitis and with the overexpression of inflammatory mediators (IL-6 and IL-1) and TNF (Spinas et al. 2015). On the other hand, VitB3 (niacin, nicotinic acid, nicotinamide) can module the IIS. VitB3 protects against *Staphylococcus aureus* infections and can reduce the production of pro-inflammatory cytokines (IL-1, IL-6, and TNF-α) and downregulates NF-B activation by macrophages. For its part, VitB6 has been implicated in the regulation of immune responses that are associated with a wide range of diseases, including inflammation and various cancers. The serum ab (IgG, IgM) production in VitB6-deficient mice decreased and could be recovered to normal level after a short term of VitB6 supplementation. In this sense, the deficiency of VitB6 can enhance the inflammatory response in DC, retard growth, inhibit lymphocyte proliferation, and lead to decreased IL-2 and increased IL-4 (Mikkelsen et al. 2017). On the other hand, VitB9 is also involved in some immunomodulatory responses, for example, it has been shown that VitB9 is necessary for the differentiation and maintenance of Treg cells in the intestine (Kunisawa et al. 2012).

VitC (ascorbic acid) is a water-soluble compound that cannot be synthesized by humans and has a fundamental impact on IIS and AIS. VitC accumulates in leukocytes, in concentrations of 50–100-fold higher than in the plasma. Therefore, VitC that is present in leukocytes is rapidly utilized during infection. VitC stores in patients with viral infections increases and downregulates the production of pro-inflammatory cytokines (Bivona III et al. 2017). In addition, VitC is necessary for the proliferation of T cells and NK cells (van Gorkom et al. 2019). On the other hand, increasing levels of oxidants lead to activation of NF-κB, triggering a signaling cascade, with the result of further production of oxidative species and inflammatory mediators. In this sense, VitC acts as an antioxidant that can scavenge reactive oxygen species (ROS) (Shakoor et al. 2020). Also, VitC has direct antioxidant capacity and contributes to the cell protection from the damaging effects of ROS, e.g., during immune activation. For instance, VitC was shown to protect neutrophils from ROS generated during phagocytosis and prevent oxidative damage on lymphocytes and DNA (Wintergerst et al. 2006).

VitC supplementation confers a defense mechanism against viral respiratory diseases such as pneumonia and tuberculosis. VitC may have a defense mechanism against COVID-19 disease. The clinical characteristic of this viral disease is characterized by a rapid elevation in the blood of pro-inflammatory cytokines such as IL-1β and TNF-α and subsequently the secretion of IL-6 and IL-8, which results in a highly inflammatory state. In this sense, VitC is effective in reducing the levels of pro-inflammatory cytokines such as TNF-α. According to clinical trials 1 g/day of VitC increases PMBC secretion of IL-10 to control inflammatory (Shakoor et al. 2020). VitC also has a mechanism of action against influenza caused by the H1N1 virus; administration of 125–250 mg/kg decreased the lethality of the virus in animals and upregulates the mitochondrial antiviral response (Cai et al. 2015). The intravenous administration of VitC alone or in combination with other drugs has also been promoted for the therapeutic management of different types of cancer (Boretti and Banik 2020). Furthermore, VitC, being a cofactor, has an important role in the modulation of hydroxylase enzymes that regulate hypoxia-inducible factors and the hypoxic cellular response that results from inflammatory clinical pictures due to the oxidative metabolism of tumor tissues (Ang et al. 2018).

2.5 Immunomodulatory Response of Probiotics

The term *probiotic* comes from the Greek word *biotikos* meaning "for life." The concept of probiotics was initially based on the observations of Elie Metchnikoff in 1907 who observed that the regular consumption of dairy fermented with lactic acid bacteria (LAB) was associated with enhanced health and longevity in Bulgarian elderly; due to this, the term probiotic has been directly related to gut beneficial bacteria. Nowadays, probiotics are defined by the *World Health Organization* as "living microorganisms that, when administered in sufficient quantities, confer a benefit to the health of the host" (Corona-Hernandez et al. 2013), expanding their benefits beyond gastrointestinal (GI) health. Nevertheless, the term probiotic is in constant transition, and scientists have been tasked with introducing new subdivisions and concepts. Within these new "biotic" subdivisions, the set of probiotics that are aimed at improving the immune system are considered as "immunobiotics".

The most common probiotics used by the food industry are LAB, namely, *Lactobacillus* spp. and *Bifidobacterium* spp. The genus *Lactobacillus* was firstly described in 1901 and comprised 261 species by 2015, but in 2020, it was reclassified into 25 genera based on 16S rRNA-based phylogenic analysis to facilitate the understanding of common mechanisms that could mediate strain-specific probiotic benefits (Zheng et al. 2020a, b). For example, while *Lactobacillus acidophilus* remains in the emended *Lactobacillaceae* genus, *L. rhamnosus* is now a *Lacticaseibacillus* specie (Table 2.6). As if this were not enough, other bacilli (e.g., *Bacillus* spp.), nonpathogenic *Escherichia coli*, and yeasts (e.g., *Saccharomyces boulardii*) are also considered probiotic organisms. Most probiotic bacteria have been isolated from fermented milk and other fermented foods (Fig. 2.6), and it has been established that a food product must contain a minimum of 10^7 colony-forming units per gram of food (CFU/g) to ensure sufficient bioavailable bacteria and exert a functional effect within the body (Peng et al. 2020; Corona-Hernandez et al. 2013); however, common intestinal inhabitants such as *Akkermansia muciniphila* (Table 2.6) are now considered "next-generation" beneficial microbes (Azad et al. 2018).

Lactobacillus spp. are Gram-positive, homofermentative, thermophilic, and nonspore-forming rods belonging to the *Firmicutes* phyla that ferment a relatively broad spectrum of carbohydrates and have the strain-specific ability to ferment extracellular fructans, starch, or glycogen (Zheng et al. 2020a, b); conversely, *Bifidobacterium* spp. are heterofermentative, Gram-positive, nonspore-forming, nonmotile, and catalase-negative anaerobic bacteria belonging to the phylum of *Actinobacteria*. These and other probiotics used for human consumption must be of human origin and nonpathogenic and survive GI transit to confer their health benefits. Particularly, they must have the ability to withstand the low pH of the stomach, bile salts, and pancreatin in the small intestine. In addition, certain probiotic characteristics are required to colonize the intestines, such as hydrophobicity, self-aggregation, and co-aggregation (Khalil et al. 2018).

Table 2.6 Probiotics with immunomodulatory activity and selected characteristics

Genus	Strains	Characteristics
Akkermansia	*A. muciniphila*	Mucin-degrading
Bacillus	*B. clausii, B. coagulans, B. subtilis*	Spore-forming, natural GI resistance
Bacteroides	*P. jensenii, P. freudenreichii*	Uses lactate, propionate producer
Bifidobacterium	*B. animalis, B. breve, B. bifidum, B. lactis*	Bacteriocin producer, gastric acid, and bile-tolerant
Enterococcus	*E. faecium*	Wide pH resistance
Escherichia	*E. coli*	Inducer of human b-defensin
Faecalibacterium	*F. prausnitzii*	Produces butyrate and anti-inflammatory factors
Lacticaseibacillus[a]	*L. rhamnosus, L. casei, L. paracasei*	Exopolysaccharide production
Lactiplantibacillus[a]	*L. plantarum*	Ubiquitous in plant foods
Lactobacillus	*L. acidophilus, L. delbrueckii, L. helveticus*	High intestinal adherence
Lactococcus	*L. lactis, L. reuteri, L. casei*	Strict homolactic producer
Levilactobacillus[a]	*L. brevis*	Produces anti-inflammatory factors
Ligilactobacillus[a]	*L. salivarius*	Bacteriocin producer
Limosilactobacillus[a]	*L. fermentum, L. reuteri*	Bacteriocin producer
Micrococcus	*M. luteus*	Biofilm producer
Pediococcus	*P. acidilactici, P. pentosaceus*	Bacteriocin producer
Peptostreptococcus	*P. productus*	Mucin-degrading
Saccharomyces	*S. cerevisiae, S. boulardii*	Polyamine production, enhances GI enzyme activity
Streptococcus	*S. thermophilus*	High lactase activity

Source: Azad et al. (2018), Zheng et al. (2020ab), [a]previously *Lactobacillus*

Probiotics play a key role in maintaining immunity by modulating both IIS and AIS (Peng et al. 2020; Schaafsma and Slavin 2015). Probiotics significantly influence human health through the detoxification of xenobiotics, vitamin K biosynthesis, metabolic effects of fermentation (postbiotic) of prebiotics (symbiotic effect), and competition with pathogenic microbes for nutrients and binding sites in cell epithelial mucosa (Hardy et al. 2013). Thereby, the probiotics are effective in the prevention or treatment of several GI disorders such as infectious diarrhea, antibiotic-related diarrhea, irritable bowel syndrome (IBS), or Crohn's disease (Linares et al. 2017).

Probiotics exert their immunomodulatory activity in many and very different ways; however, these bioactivities can be framed into gut integrity-promoting activities, direct stimulation of IIS and AIS immune cells, and production of microbial metabolites and substrates (pharmabiotics: postbiotics and paraprobiotics) with immunostimulatory properties (*immunoprobiotics*; Fig. 2.6).

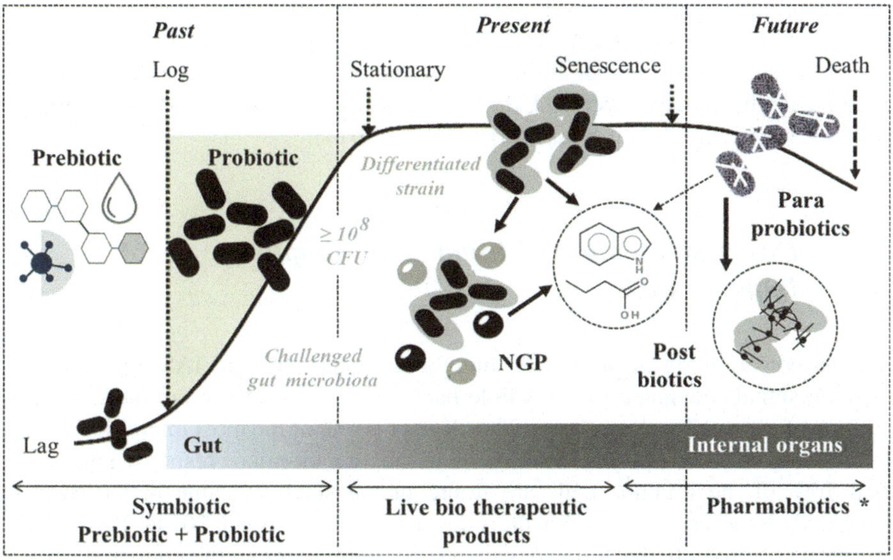

Fig. 2.6 "Biotic" spectrum of gut beneficial microbes. Probiotics (black ellipses) are live microorganisms that, when administered in adequate amounts (~10^7 to 10^8 colony-forming units (CFU) per gram of food, probiotic gray zone), confer a health benefit on the host, and when co-administered with probiotic growth-promoting substrates (prebiotics) such as nondigestible carbohydrates, fatty acids, and phenolic compounds, a symbiotic (a.k.a. symbiotic) combination is produced. When reaching maturity (differentiation at stationary phase), metabolically active probiotics are efficient producers of bioactive bacterial products (e.g., exopolysaccharides and bacteriocins) and "healthy" metabolites (e.g., indole and short-chain fatty acids), collectively known as postbiotics that challenge the metabolic capacity of host's microbiota to produce unhealthy microbial end products and increasing the odd for probiotic colonization. Next-generation probiotics (NGP) refer to live microorganisms fulfilling the probiotic definition but that were identified based on comparative microbiota analyses. Lastly, senescent-death (ghost) probiotic cells (postbiotics, crossed-gray ellipses, right) may also be a source of bioactive molecules. From a nutraceutical perspective, improved probiotic strains and NGP are major drivers in formulating live biotherapeutic products (LBP), and postbiotics and paraprobiotics are the present future of pharmabiotics

2.5.1 Gut Integrity-Promoting Activities

Besides improving physical barriers (e.g., mucin layer) that protect intestinal cells from foreign toxic substances, probiotics also compete with pathogens for nutrients. For example, *L. rhamnosus* strain GG and *L. plantarum* can inhibit the adherence of enteropathogenic *E. coli* in the GI tract (Galdeano et al. 2019). Probiotics also modulate toxin receptors avoiding their entry to epithelial cells by mean of enzymatic mechanisms and production of specific microbial metabolites. For example, *S. boulardii* produces polyamines which counteract and degrade *Clostridium difficile* toxins by blocking their corresponding cell receptor. This barrier effect in the intestine is also due to the production of broad-spectrum inhibitory bacteriocins and microbial metabolites resulting from the fermentation of prebiotics and the production of SCFAs. SCFAs lower the intestinal pH and discourage bacterial growth. Also, SCFAs can act at the level of the signaling pathways that lead to an increase in the mucus layer and the production of defensins and tight junction proteins, improving the intestinal barrier (Butel 2014).

2.5.2 Direct Stimulation of Innate and Adaptive Immune Cells

Regular consumption of probiotics directly stimulates IIS and IAS, inducing networking signals mediated by the whole bacteria, their metabolic products, or components of their cell wall structure. The GI tract is one of the most microbiologically active ecosystems and plays a crucial role in the direct immune response since more than 70% of the immune cells are found in the small intestine in the form of lymphoid tissue associated with the intestine (GALT) (Butel 2014). Particularly, fragments of probiotic bacteria are capable of internalizing in the epithelial cells of the intestine, and therefore, the epithelial cells initiate a complex network of signals that stimulate immune cells associated with the lamina propria and activate the IIS and the cytokines released by T cells. Therefore, once administered, oral probiotic bacteria interact with the TLRs present in immune cells, inducing the production of different cytokines or chemokines (Galdeano et al. 2019).

Many probiotics also interact with macrophages and DC to induce cytokine production that triggers specific immunostimulatory responses. For example, *L. casei* is capable of stimulating macrophages to produce IL-12 and IL-10, both involved in the immune response against infections and cancer and/or differentiation of Th17/Treg cells to reduce intestinal inflammation (Kaji et al. 2018; Wang et al. 2017). *Lactobacillus* spp. also reduce the production of other pro-inflammatory cytokines such as TNF-α and IL-8 (Tuo et al. 2018), while other probiotic strains including *S. thermophilus*, *L. rhamnosus*, *L. acidophilus*, *L. casei*, *B. bifidum*, and *B. longum* stimulate human PBMC to produce anti-inflammatory cytokines such as TNF-α, IFN-γ, IL-1β, IL-6, IL-10, and IL-1ra (Djaldetti and Bessler 2017). This AIS-stimulatory effect of probiotics is restricted not only to small intestinal cells but also to peripheral ones; in this sense, oral probiotic supplementation has been found to have an effect on salivary anti-inflammatory cytokines, primarily IL-1β and IL-8 (Ebrahimpour-Koujan et al. 2020). *L. gasseri* TMC0356 acts against H1N1 flu influenza virus by stimulating both local and systemic immune responses and the expression of IL-12, IL-15, and IL-21 in Peyer's patch and IFN-c, TNF-α, IL-12a, IL-12rbl, IL-2rb, and perforin 1 in lungs. In this sense, excellent review on the direct immunostimulatory effect of probiotics has been published elsewhere (Nwobodo and Ugwu 2020; Azad et al. 2018; Llewellyn and Foey 2017).

2.5.3 Production of Microbial Metabolites and Substrates

The term probiotics fits into a bigger category called *pharmabiotics*. Pharmabiotics are defined as "bacterial cells of human origin, or their products, with a proven pharmacological role in health or disease." In this category, it can include postbiotics that are "non-viable bacterial products or metabolic products from microorganisms that have biologic activity in the host" and paraprobiotics or inactivated probiotics that are "non-viable microbial cells or crude cell extracts which when administered confer a benefit on the human or animal consumer", all of them with an immunostimulatory effect.

Among postbiotics, heteropolysaccharides (HePSs) and homopolysaccharides (HoPSs) stand as the most studied. While HePSs are synthesized intracellularly, HoPSs come from the cell wall enzymatic degradation of mesophilic (e.g., *L. lactis* subsp. *lactis*, *L. rhamnosus*, *L. sakei*, *L. casei*) and thermophilic (*L. delbrueckii* subsp. *bulgaricus*, *L. acidophilus*, *L. helveticus*, *S. thermophilus*) LAB (Ripari 2019). Their main functions include protection against toxic compounds and osmotic stress and to allow adhesion. Also, exopolysaccharides promote the growth of the microbiota as the microbiota ferments these substrates to convert them into SCFAs, resulting in a modulation of the host's immune response (Chaisuwan et al. 2020). In addition, exopolysaccharides reduce inflammatory processes (Chen et al. 2019). The exopolysaccharides use TLR4 and TLR2 receptor on the macrophage cell surface as a receptor and stimulate the cells via the NF-κB signaling pathway. Genes, cytokines, chemokines, enzymes, and adhesion molecules are synergistically involved in IIS and AIS, inflammation, proliferation, and anti-apoptosis. Exopolysaccharides also bind to macrophage TLR4 or TLR2 and macrophages via MAPK pathways, inducing their proliferation and phagocytic activity and enhancing the immune response against pathogenic microorganisms. Other postbiotics exert antimicrobial activity such as bacteriocins (Khalil et al. 2018); these bacterial metabolites of proteinaceous nature have bactericidal or bacteriostatic activity against closely related species (narrow spectrum) or across genera (broad spectrum). LAB (e.g., *Lactobacillus*, *Lactococcus*, and *Pediococcus*) produce bacteriocins in the ribosomes. Lactacin B (*Lactobacillus*), plantaricin 423 (*L. plantarum*), pediocin ST18 (*P. pentosaceus*), and nisin Q (*L. lactis*) are just some examples (Gaspar et al. 2018).

Paraprobiotics are nonliving microorganisms, regularly obtained by inactivating microbes using ultraviolet rays, high pressure, ultrasonication, ionization, and heat treatments (de Almada et al. 2016). Heated *L. gasseri* TMC0356 has a better anti-inflammatory effect when compared to the living organism; such activity is mediated by IL-12 secreted by macrophages (Miyazawa et al. 2011). Cell components of *L. gasseri* appear to be fundamental in the protection against viral infections, increasing the immune responses of the intestine and respiratory system (Kawase et al. 2012). Another paraprobiotic, *L. casei*, seems to have a remarkable immunomodulatory effect, since it modulates the response of macrophages by improving the expression of pro-inflammatory cytokines and the transcription of TLRs (TLR2,

TLR3, TLR4, and TLR9) (Wang et al. 2013). In vitro evaluations of *L. plantarum* b240 (another relevant paraprobiotic), has shown the hability to downregulate the systematic infectious activity of the pathogen *Salmonella enterica* serovar *Typhimurium* in various organs such as the Peyer's patches, mesenteric lymph nodes, spleen, liver, and blood of mice, based on in vitro studies (Ishikawa et al. 2010).

It is noteworthy that paraprobiotic-formulated foods also possess immunomodulatory activity. Skim milk supplemented with heat-inactivated *L. rhamnosus* HN001 enhances IIS function characterized by an increased phagocytic activity in the blood and peritoneum of mice associated with cell debris (Gill and Rutherfurd 2001). Also, yogurt with inactivated *L. bulgaricus*, *S. thermophilus*, and *L. acidophilus* effectively improves the epithelial barrier function in human intestinal Caco-2 cells mainly due to the production of nitric oxide induced by pro-inflammatory cytokines (Zeng et al. 2016).

2.5.4 In Situ Immunoenhancing Response of Probiotics by Fermented Foods

Scientific evidence has increasingly shown that during the fermentation of dairy products by probiotic bacteria, it results in the in situ production of bioactive with immunostimulant activity. Bioactive compounds with immunostimulant response are formed in situ during manufacture of fermented dairy foods. For instance, some strains have produced VTs during fermentation procedure. *Lactobacillus* and *Bifidobacterium* can provide B VTs during dairy fermentation. *L. casei KNE-1* produce thiamine and riboflavin in fermented milk drinks. *B. infantis* CCRC14633 and *B. longum* B6 strains produce riboflavin and thiamine during soymilk fermentation. Also, during milk fermentation, lactic acid bacteria can transform milk proteins into biologically active peptides, that it can exert a wide range of effects, such as antimicrobial, antihypertensive, antithrombotic, immunomodulatory, and antioxidative. On the other hand, bacteriocins as antimicrobial molecules are among the beneficial peptides intrinsically synthesized by certain LAB during milk fermentation. They may function in the GI tract as biotherapeutic agents facilitating the competition of probiotic strains and/or inhibition of pathogens (Linares et al. 2017).

2.5.5 Live Biotherapeutic Products as Immune Probiotics

Designer functional foods with live or inactivated beneficial microbes and by-products have increased in recent years. The purpose is to satisfy the demands of the current healthy consumer, offering immunoprobiotics with preventive or biotherapeutic function. Fermentation is a cheap biotechnological process used worldwide to produce several functional foods and beverages, and LAB have been widely used as starter cultures to produce a wide range of in-market products, and dairy-based products such as fermented milks and yogurt are by far the most produced (Linares et al. 2017). Strains of *Bifidobacterium* and *Lactobacillus* are widely used in fermented foods (Butel 2014). Probiotic microorganisms mostly used are *L. acidophilus*, *L. gasseri*, *L. helveticus*, *L. johnsonii*, *L. (para) casei*, *L. reuteri*, *L. plantarum*, *L. rhamnosus*, and *L. fermentum*. Genera of *Bifidobacterium* are also used, such as *B. bifidum*, *B. longum*, *B. animalis*, and *B. breve*. The use of these beneficial microbes has been optimized in several cheese varieties. Kefir is another milk-fermented product that has health-promoting bacteria. Other non-fermented dairy foods such as low-fat ice cream, chocolate mousse, coconut flan, or infant milk formula have also been supplemented with probiotic strains (Linares et al. 2017).

Fermented dairy. Fermented milks are probably the most studied immunoprobiotics with a wide range of metabolic benefits. The main immunostimulant effects of these functional foods are related to the modulation of the expression of cytokines, ab, antitumor, antibacterial, and antiviral activity. Both in vitro and in vivo studies in animals and humans support these immunostimulant responses. For instance, a recent study showed that 15 *Lactobacillus* strains isolated from artisanal cheese and added to a fermented milk beverage modulate the immune system due to the fermented enhanced IL-10 levels and upregulated IL-6 and IgA levels in serum samples of rats (Santiago-López et al. 2018). Similarly, *L. plantarum* (B7) from Brazilian artisanal cheese was added to a fermented milk and evaluated for its action against *Salmonella enterica serovar Typhimurium* in mice; in vivo administration of fermented milk showed a protective effect demonstrated by lower bacterial translocation, histological lesions, and cumulative mortality, while reduction in IFN-c and IL-6 expression and lower *Salmonella* spp. levels were observed in mice ileum after treatment (Acurcio et al. 2017). Also, milk fermented with *L. helveticus* decreases the growth rate of mammary tumors; the effect was mediated by increased apoptosis and decreased production of pro-inflammatory cytokines, IL-6, implicated in estrogen synthesis (Rachid et al. 2006). On the other hand, drinking fermented milk may be an approach to raise immune response against influenza virus type H1N1. A study showed that when healthy adults consumed a fermented milk containing *L. casei* 431 and acquires the vaccination, higher seroconversion rates for H1N1 were observed (Trachootham et al. 2017).

Several in vivo and in vitro studies have evidenced immunostimulatory effects of probiotic yogurts. Such responses are mainly due to the modulation of inflammatory processes associated with IBS, the maintenance of the intestinal barrier, and the regulation of tumorigenesis. For instance, low-fat yogurt inoculated with *S. thermophilus* and *L. delbrueckii* ssp. *bulgaricus* can maintain intestinal barrier integrity and decrease pro-inflammatory cytokine IL-8 in vitro (Zhai et al. 2019). A study in humans showed that the consumption of probiotic yogurt inoculated with *Bifidobacterium* spp. and *Lactobacillus* spp. had effects on the inflammatory processes related to the development of IBS. Consumption of yogurt with probiotics caused a significant decrease in the expression of pro-inflammatory cytokines IL-1β and TNF-α. In addition, the anti-inflammatory cytokines IL-6 and IL-10 were upregulated (Shadnoush et al. 2013). Similarly, the administration of probiotic yogurt has been related to the prevention of tumorigenesis due to a modulation of pro-inflammatory cytokines (Urbanska et al. 2010).

Nondairy fermented foods. LAB have been used to produce nondairy fermented foods from vegetables, cereals, fruits, cocoa, and starchy root crops. Artisanal cereal-based beverages such as "Boza" (barley, oats, rye, millet, maize, wheat, or rice), "Bushera" (sorghum, millet), "kvas" (rye bread, rye, and barley malt/flour), or "pozol" (maize) are fermented with LAB, and some of them have shown immunoenhancing properties (e.g., cytokine production) besides exhibiting a better biological/nutritional value (Ripari 2019; Panghal et al. 2018).

2.6 Immunomodulatory Response of Prebiotics

The health benefits of prebiotics with a focus on the immune system have been increasingly documented. Prebiotics are a category of functional ingredients based on nondigestible carbohydrates and fermentable substrates that selectively promote the growth of colonic bacteria (probiotics) and confer several benefits for human health. The term "prebiotic," based on the term "probiotic," was defined for the first time in 1995. After several definitions, Gibson in 2004 redefined the term prebiotic "to that selectively fermented ingredient that allows specific changes, both in composition and/or activity in the GI microbiota that confers benefits on the well-being and health of the host." Indispensable criteria have been established to classify a prebiotic ingredient. Scientific demonstration is required that the ingredient (1) resists gastric acidity, hydrolysis by mammalian enzymes, and GI absorption, (2) is fermented by the intestinal microflora (probiotics), and (3) selectively stimulates gut growth and/or activity (Gibson et al. 2004).

Almost all oligosaccharides and polysaccharides in food have prebiotic activity. Oligosaccharide-type prebiotics include lactose, lactulose, raffinose, stachyose, fructooligosaccharides (FOS), galactooligosaccharides (GOS), xylooligosaccharides (XOS), and mannooligosaccharides (MOS) (Shokryazdan et al. 2017; Lim et al. 2005). Prebiotic fibers include nondigestible carbohydrates that have three or more monomeric units. These can be starchy-type (resistant starch) and non-starchy-type (inulin, pectin, cellulose, and xylan), which are not digested in the upper GI system, nor are they absorbed in the human small intestine (Meijerink et al. 2018). Oligosaccharides, particularly FOS and GOS, stand out over other types of prebiotics as they selectively stimulate the growth of *Bifidobacterium* (Gibson et al. 2017).

2.6.1 Immunostimulating Response of Prebiotics

Prebiotics exert an immunostimulating response indirectly or directly (Fig. 2.7). The indirect way is dependent on microbial fermentation. VTs, bacteriocins, and SCFAs (acetic, propionic, and butyric) are the products of this fermentation (Meijerink et al. 2018). SCFAs influence the immune system by binding to G-protein-coupled receptors on immune cells within GALT (Shokryazdan et al. 2017; Hartog et al. 2015). Butyrate inhibits the production of the pro-inflammatory cytokines IL-2 and IFN-γ, while acetate and propionate increase the production of the anti-inflammatory cytokine IL-10 (Cavaglieri et al. 2003). In addition, prebiotics promote the growth and development of probiotic microorganisms such as *Lactobacillus* and

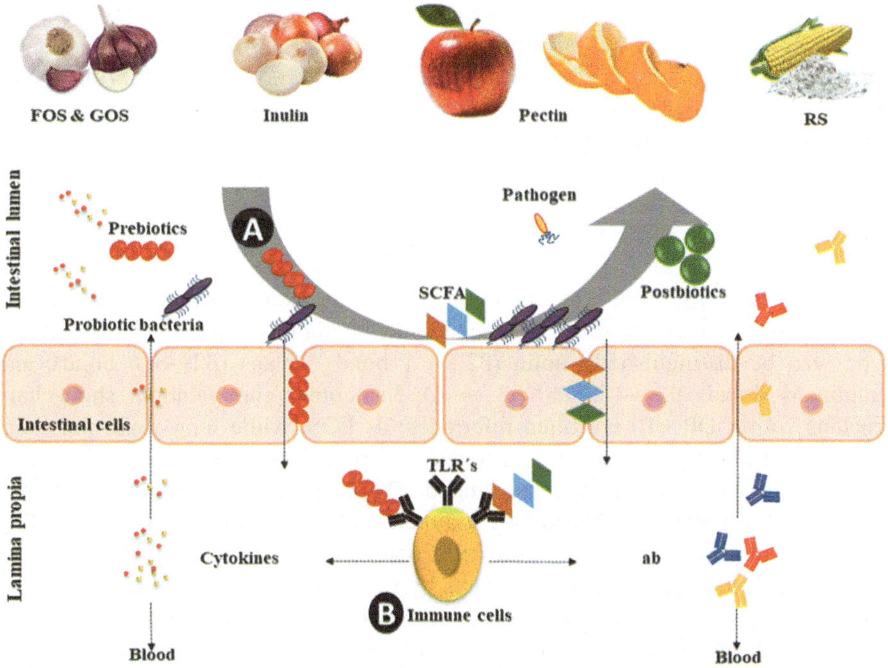

Fig. 2.7 Immunostimulating response of prebiotics. The main prebiotics reported are fructooligosaccharides (FOS) and galactooligosaccharides (GOS), inulin, pectin, and resistant starch (RS). Prebiotics have a direct and indirect immunostimulating response. (**a**) The indirect immunostimulatory response is dependent on microbial fermentation. Prebiotic compounds are fermented by probiotic microbes in the gut, resulting in the production of short-chain fatty acids (SCFAs), which can interact with immune cells in the lamina propria to trigger the secretion of cytokines and antibodies (ab). In addition, the fermentation of prebiotics causes the reproduction of probiotic bacteria and the production of metabolites (postbiotics) with immunostimulating response. (**b**) The direct immunostimulatory response of prebiotics is characterized by the molecular interaction of these with immune cell receptors (TLRs) and the production of different cytokines and antibodies that have effects both on the epithelial cells of the intestine, on the GALT, and on the systemic level

Bifidobacterium, which in turn decrease the level of invading pathogens such as *Helicobacter*, *Vibrio cholerae*, and *Enterococcus* (Shokryazdan et al. 2017).

Regarding to the direct immunostimulatory response, prebiotics interact directly with immune cells and can stimulate the secretion of anti-inflammatory cytokines IFN-γ and IL-10 by CD4 + cells. Shokryazdan et al. (2017) reported a comprehensive review on the production of various cytokines as part of the immunostimulatory response of various prebiotics (Shokryazdan et al. 2017). Prebiotics bind and interact directly with the TLRs of the immune cells to trigger the immune response and cause an increase in the production of mucin, the main barrier proteins in the intestine. Some indigestible oligosaccharides, mainly FOS and GOS, act as TLR4 ligands in both epithelial cells and monocytes, thus regulating cytokine secretion, related with the IIS (Valcheva and Dieleman 2016). On the other hand, some prebiotics (starchy, cellulosic, hemicellulosic, gums) have been associated with immunomodulatory cellular responses since they activate PBMC (Meijerink et al. 2018), critical components for the immune system and to fight infections. Also, the administration of prebiotics such as inulin in combination with probiotics (*symbiotic effect*) increases the serum level of IgA, an antibody widely found in the mucosa and with action against pathogens in the intestine (Frece et al. 2009).

2.6.1.1 Fructans

Fructans are polymers of fructose (1–70 units) that can be found attached to a terminal molecule of sucrose, and depending on their chemical structure, three types can be distinguished: inulin (β2 → 1 bond), levans (β2 → 6 bond), and graminans (bonds β2 → 1 and β2 → 6). In another classification, short-chain fructans with a DP <10 are often referred to as FOS, while long-chain polymers with DP > 10 belong to the name of inulin-type fructans (Dobrange et al. 2019).

Inulin-type fructans are the most studied. The response of the immune system is modulated by inulin-type fructans by a direct mechanism through their interaction with membrane receptors of immune cells, mainly of IIS, or by an indirect mechanism through their fermentation products by intestinal bacteria. Studies in healthy humans show that the administration of fructans increases at least three times the content of fecal *Bifidobacterium*, the total concentration of SCFAs (especially propionate and butyrate) and the cytokine IL-4 in the blood, in addition to the cytokine GM-SCF and CD282 +/TLR2 + (Clarke et al. 2016).

DP has an important effect on the immunomodulatory activity of fructans. Short-chain fructans like FOS induce greater anti-inflammatory activity and prevent epithelial cell disruption due to their interaction with TLRs (Vogt et al. 2014), particularly with TLR2 and, to a lesser extent, TLR4, TLR5, TLR7, and TLR8 (Dobrange et al. 2019). On the other hand, short-chain fructans are better at inducing and regulating cytokine production in human PBMC (lymphocytes or monocytes), compared to long-chain fructans (Vogt et al. 2013). However, long-chain inulins promote splenocyte production in a dose-dependent manner (Kumar et al. 2015) and can delay the development of autoimmune diabetes (type 1) by modulating pancreatic and intestinal immunity, barrier function, and microbiota homeostasis. Physiological mechanisms exerted include a decrease in Th17 (inflammatory) cells, stimulation of the secretion of the regulatory cytokine IL-10, and a decrease in the secretion of the inflammatory cytokine IL-1β (Chen et al. 2017). Fructans also act as signaling molecules. After binding to TLRs, fructans trigger signaling pathways dependent on the NF-kB, resembling a vaccination effect. On the other hand, the antiviral properties of fructans are attributed to the ability to improve the production of nitric oxide (NO, an inhibitor of viral replication), immunostimulating factors (IL-1β, IL-6, IL-10, IFN-γ), and TNF-α (Dobrange et al. 2019).

Several foods and agro-industrial subproducts are a potential source of fructans with immunomodulatory activity, mainly garlic, onion, and agave. The fructans of *Agave* (*A. tequilana*) can modify the evolution of autoimmune systemic lupus in murine models by suppressing the cytokines IL-1β, IL-6, TNF-α, IFN-γ, and IL-10 (Gutiérrez Nava et al. 2017). They have also shown immunomodulatory activity through the activation and differentiation of cells of the immune system and the proliferation of probiotics. They modulate the expression of the CD69 ab, cell cycle progression, and NO production and the expression of transcription factors for lymphocyte differentiation (Moreno-Vilet et al. 2014). Studies on the effects of agave fructans on immune cells suggest that they exert direct effects through receptor-ligand interactions and subsequent cytokine production (Vogt et al. 2014). Onion (*Allium cepa*) is a food rich in fructans, and it has been documented that onion FOS (50 μg/ml) increases the proliferation of splenocytes/thymocytes up to ~three times and improves (~ 2.5 times) the production of NO in peritoneal cells of Wistar rats, the latter related to the activation of macrophages and important in the inflammatory response to eliminate antigens (Kumar et al. 2015). For their part, fructans from garlic (*A. sativum*) have been shown to have antigenicity against ovalbumin through the production of Ig (Chandrashekar and Venkatesh 2012), exhibit mitogenic activity in splenocytes, and promotes macrophage activation in vitro (Chandrashekara and Venkatesh 2016).

2.6.1.2 Pectin

Another group of prebiotics with remarkable documentation are pectins. Pectins are heteropolysaccharides found in cell walls and belong to the group of acidic polysaccharides. Pectins are made up of polymers rich in D-galacturonic acid and contain significant amounts of L-rhamnose, D-arabinose, and D-galactose (Naqash et al. 2017). Structurally, the linear region consists of 1,4-α-D-galacturonan units that are linked together with one or two α-L-rhamnopyranose residues via 1,2-linkages. The branched region consists of a substituted rhamnogalacturonan backbone with side chains rich in neutral sugars such as arabinogalactan, arabinan, galactooligosaccharides, and arabinooligosaccharides (Popov and Ovodov 2013).

The immunostimulatory response of pectins lies primarily in selectively stimulating the growth and activity of *Bifidobacterium* and *Lactobacillus* probiotics. Furthermore, the colonic fermentation of pectins results in the generation of

SCFAs and in a series of health effects such as the inhibition of pathogenic bacteria (Babbar et al. 2016). Another immunostimulatory response to pectins lies in the activation and regulation of phagocytic leukocytes, B and T lymphocytes, and thrombocytes and the suppression of tumor cells (Popov and Ovodov 2013).

The structural characteristics of pectins significantly influence the immunomodulating properties. For instance, pectins with less than 75% galacturonic acid increase the immunostimulatory response by stimulating macrophage activity. However, others have reported that highly esterified pectic polysaccharides have shown antitumor activity due to their ability to suppress suppressor cells of the immune system (Dong et al. 2018). The branched region of pectins increased phagocytosis and ab production. And the linear structure determines the stimulating interaction between pectin and immune cells (Popov and Ovodov 2013). The immunomodulatory activity of pectin is also related to the presence of arabinogalactan structures linked to the rhamnogalacturonan domain (Georgiev et al. 2017).

The powerful effects of pectins and some pectic polysaccharides on the IIS and AIS are widely recognized for their antitumor response. Pectins have been shown to have an antitumor effect due to their binding to carcinogens in the intestine. Pectins inhibit carcinogenesis in the colon by increasing apoptosis, stimulating the proliferation of colon cells, reducing the activity of β-glucuronidase, inhibiting tumor development and the expression of oncogenes, decreasing the cell cycle, and inactivating caspase-3 (Naqash et al. 2017; Popov and Ovodov 2013). Highly esterified pectic polysaccharides (from *Curcuma kwangsiensis*) decreased the expression of myeloid-derived suppressor cells, the latter related to the suppression of the immune system, since they inhibit T cells (CD4 +, CD8 +) and promote tumor growth (Dong et al. 2018).

Pectins are found in fruits, mainly in citrus peels and apple pomace. The exhaustive review carried out by Naqash et al. (2017) mentions that pectins obtained from various natural sources showed immunomodulatory and anticancer activity due to the inhibition of metastasis, angiogenesis, and tumor growth (Naqash et al. 2017). For example, apple pectin showed anti-breast cancer effects in vitro and in vivo by inhibiting cell growth and attachment, fragmenting chromatin, and blocking the sub-G1 phase of the cell cycle (Delphi and Sepehri 2016). Citrus pectins also play an important role. Lemon pectins with different degrees of esterification have been reported to increase pancreatic cell viability through inhibiting the apoptotic effect of streptozotocin on β cells, the production of free radicals, and the effect of the inflammatory cytokines IL-1β, IFN-γ, and TNF-α, all of the above, factors that are highly related to pro-inflammatory processes and apoptosis of the pancreatic islets and the development of type 1 and 2 diabetes mellitus (Hu et al. 2020).

2.6.1.3 Resistant Starch

Resistant starch (RS) is another prebiotic with a remarkable immunostimulating property. Starch is a polysaccharide produced by many plants as the main storage of energy and is the second most abundant biomass material in nature after cellulose. It is found in semicrystalline granules (1–100 μm) made up of amylose and amylopectin molecules, in linear and branched structures, respectively (Escobar-Puentes et al. 2019). Previously, it was believed that the starch provided by the diet was fully digested during various digestive phases. However, over the years, it has been recognized that some starch fractions may be slowly digestible or not digestible.

RS refers to starch that is not digested within the first 120 minutes in the upper GI tract and therefore passes into the large intestine, where it can act as a fermentative substrate. There are five types of RS: type 1 RS, which is physically inaccessible to digestive enzymes because it is trapped in the cell wall of the food matrix, for example, whole grains; type 2 RS, which refers to native high amylose starch granules or starch found in potatoes, unripe plantains, and some legumes; type 3 RS, which results after cooking and retrogradation of starchy foods such as bakery products and pasta; type 4 RS, which results from subjecting starch to chemical modifications such as etherification and esterification; and the recently added to the classification, type 5 RS, which results from the inclusion of various molecules, commonly lipidic, in the helical structure of amylose (Jiang et al. 2020).

The beneficial health properties of RS lie mainly in its therapeutics for metabolic diseases. The immunostimulating response of RS is related to the regulation of the GI microbiota and its anticancer properties. The consumption of RS increases the level of beneficial bacteria such as *Lactobacillus* and *Bifidobacterium* (Zheng et al. 2020a, b). Furthermore, the fermentation of RS by colonic bacteria increases the production of SCFAs, the SCFA receptor GPR43, and causes a reduction in inflammatory and carcinogenic processes in the colon (Yin and Zhao 2017). The intake of RS downregulates the expression of genes associated with carcinogenesis in the colon, while the expression of proteins (caspase-3 and caspase-9) and pro-apoptotic genes is downregulated (Wang et al. 2018). The RS improves biochemical markers related to inflammatory processes (IFN-IL-6, TNF-, IL-12). In addition, RS induces regulatory mechanisms of the expression of pro-apoptotic proteins and genes (Bax, caspasse-9, and caspase-3) and anti-apoptotic proteins (Bcl-2, p53) and the modulation of gene expression related to inflammation (NF -κB, COX-2, iNOS) (Wang et al. 2017). In a recent meta-analysis, it was found that high RS intake significantly reduced inflammatory serum biomarkers in healthy volunteers, including IL-6 and TNF-α levels, immunological effects highly related to SCFA production during microbial fermentation (Sheikhhossein et al. 2020).

RS also has immunostimulating effects by interacting directly with specialized receptors on immune cells. Direct interaction with the immune system is regulated by binding of RS to PRRs, specifically TLRs, which are expressed in intestinal cells and especially in immune cells in the intestine. Several structural characteristics of RS such as chain length influence its interaction with immune cells. For example, the amylose chain length in type 3 RS supports its immunological role by interacting with the TLRs of lymphocytes in a dose-dependent manner, inducing T-cell polarization, in addition to stimulating DC for the expression of various cytokines related to anti-inflammatory processes (Lépine et al. 2019). Similarly, others have shown that RS modulates human intestinal immune cells in a manner dependent on physical morphology (Bermudez-Brito et al. 2015).

2.7 Immunomodulatory Response of Bioactive Peptides

When dealing with acute or chronic immunocompromised illnesses, IIS and IAS become imbalanced (Dutra 2017; Fig. 2.2). For example, the chronic inflammation initiated by TNF-α and IFN-γ is followed by a dysregulated recruitment of macrophages and naïve DC and an accumulation of foreign microbial activating antigens during the development of IBS, while the failure to eliminate foreign antigens, bacterial translocation, and an inadequate humoral response may result in autoimmune diseases (Santiago-López et al. 2016). In these and other pathological conditions, exogenous immunomodulatory aids are needed to reactivate and boost the immune system by regulating many immune effector cells and molecules (Cai et al. 2020). Immunomodulatory dietary peptides (DIMPs), also known as immunopeptides, comprise different classes of peptides such as anticancer and anti-inflammatory peptide (Bechaux et al. 2019). Some food-derived immunomodulatory agents have been discussed in depth in preceding sections, and now is the turn of DIMPs. Before going any further on the specific health effects of DIMPs, the human body produces its own immunomodulatory peptides whose amino acid sequences have used the search for alternative naturally derived or synthetic DIMPs. For example, without diminishing the role of other HDPs, the human cationic antimicrobial peptide (hCAP-18) is abundant in neutrophils, macrophages, lymphocytes, and epithelial, cells and its proteolysis at c-termini generates a 37-residue α-helical peptide (LL-37) which along with PRRs builds a bridge between IIS and AIS by enhancing cell recruitment (e.g., neutrophils, eosinophils) and activation and endocytic capability of DC (naïve T-cell activators) and upregulates cytokine and chemokine [monocyte chemoattractant protein-1 (MCP-1/CCL-2), C-X-C motif chemokine ligand 8 (CXCL-8), and transforming growth factor-β (TGF-β)] production that mediates Th1 polarization and TLR-mediated inflammatory response from immune cells, epithelial cells, keratinocytes, and other cell types (Cai et al. 2020). Interestingly, up to 0.6 mg. kg^{-1} of LL-37 can be produced from transgenic barley seeds by using a low-cost molecular farming technology (Holásková et al. 2018).

DIMPs typically consist of sequences of 2–43 amino acids of both hydrophobic and hydrophilic nature, naturally encrypted and inactive in the parent protein sequence (Maestri et al. 2016). From a structural function stand point, immunopeptides interact on the opioid receptors that regulate the peripheral immune system, and those with arginine (R), tryptophan (W), and phosphoserine in the N- or C-terminal residues exert high reactivity in immune cells, while glutamine (Q) residues recognize opioid receptors located on the surface of immune cells (Orona-Tamayo et al. 2019). DIMPs can be released from a wide range of protein sources by enzymatic hydrolysis or microbial fermentation or by the action of proteases and peptidases during their GI journey (Sun et al. 2020; Reyes-Díaz et al. 2018; Santiago-López et al. 2016). These exogenous peptides can be administered to humans, and they can perform their biological activity in the same way as endogenous peptides (Bechaux et al. 2019). LAB-mediated fermentation is an efficient and cost-effective method for generating food grade hydrolyzed products including DIMP and probiotics (Chatterjee et al. 2018). Immunomodulatory peptides produced by enzymatic hydrolysis or microbial fermentation from animal (Reyes-Díaz et al. 2018) and vegetable (Orona-Tamayo et al. 2019; Rizzello et al. 2016) matrices are resistant to GI peptidases, are readily absorbed actively by gut peptide transporters (e.g., Pept-1) [paracellular, transcytosis, and passive transcellular transport mechanisms; Xu et al. 2019], and can exert preliminary immunomodulatory effects by directly affecting TLRs (Cai et al. 2020).

The physicochemical nature of the original protein matrix and further processing (e.g., liquid, purees, gels) modifies the chemical interaction of DIMP with other components of the food matrix, affecting their GI digestibility and their first-pass bioavailability (Sun et al. 2020) in such a way that most products have been mostly introduced to the functional market as ready-to-drink or semisolid goods (Shimizu and Hettiarachchy 2012). For example, the in vivo bioaccessibility of DIMPs from dairy products depends on the original matrix, grinding level, and mixing along the GI tract, more than enzymatic proteolysis itself (Barbé et al. 2014). As for plant proteins, resistance to gastro-duodenal digestion is even higher due to the presence of nondigestible macromolecules (e.g., dietary fiber) and their compartmentalization, which impacts not only their bioaccesibility but also its residual immunogenicity (De Angelis et al. 2017).

Once absorbed, DIMP interact with the gut-associated lymphoid tissue (GALT) which is the largest lymphoid tissue in the body and includes Peyer's patches, the appendix, and isolated lymphoid follicles. GALT includes several immune cells such as B and T lymphocytes, macrophages, antigen-presenting cells (e.g., dendritic cells and intraepithelial lymphocytes), and HDPs involved in cell signaling. Dendritic cells located in the subepithelial region of the Peyer's patches transport immunoreactive peptides to the mesenteric lymph nodes (MLNs) via the afferent lymphatics, and MLNs are the key site for oral tolerance induction. DIMPs are further translocated to the bloodstream (portal bioavailability) at nano-to-picomolar amounts and transported until reaching the target organ to exert their functions (Sun et al. 2020; Keservani et al. 2017). During this journey, particularly under pathological conditions, DIMP exert many protective effects, going from initial stimulation of immune cells (proliferation and activation) and cytokine production to the induction of immunological tolerance. It should be noted that omics sciences and bioinformatics (data mining) have allowed us to discover alternative sources of DIMP and their potential mechanisms (Nielsen et al. 2017; Maestri et al. 2016); the most common sources, immunomodulatory actions, and amino acid sequence of selected DIMP are reported in Table 2.7 and described in the following paragraphs.

Table 2.7 Selected immunomodulatory dietary peptides (DIMP)

Source	Amino acid sequence	Action
Maize	YA	Antioxidant and ACE inhibition
Cow's milk	YG	Lymphocyte proliferation
Cow's milk	LLY	Stimulate phagocytosis, antigen-dependent T-cell proliferation, antibody production
Cow's milk	YGG	Lymphocyte proliferation
Cow's milk	GLF	Increases phagocytic activity
Soybean	MITL	TNF-α expression
Maize	LMCH	Antioxidant and ACE inhibition
Common bean	LKBM	Reduced cytokine production
Common bean	LKBY	Reduced cytokine production
Blood	YPWT	Antioxidant
Thymus*	RKDVY	Bactericidal
Soybean	IYVFVR	Antioxidant and IL-8 inhibition
Cow's milk	PGPIPN	Stimulate phagocytosis, antibody production
Cow's milk	VEPIPY	Stimulate phagocytosis
Cow's milk	TTMPLW	Murine macrophage stimulation
Amaranth	SSDEIKE	Anti-inflammatory and IgE activation
Cow's milk	YPFPGPI	Stimulates/suppresses lymphocyte proliferation
Soybean	IYVVDLR	Antioxidant and IL-8 inhibition
Cow's milk	VKEAMAPK	Antioxidant and antimicrobial
Cow's milk	LAYFYPEL	Increases antigen-specific IFN-γ production from CD8+ T cells
Soybean	MITLAIPVN	Stimulate phagocytosis
Cow's milk	YQQPVLGPVR	Stimulates/suppresses lymphocyte proliferation
Soybean	MITLAIPVNKPGR	Stimulate phagocytosis
Cow's milk	YQEPVLGPVRGPFPIIV	Bone marrow macrophage stimulation
Cow's milk	LLDAQSAPLRVYVEELKP	Enhances CD11b+/CD103+ dendritic cells and CD25+ Foxp3+ T cells in mesenteric lymph nodes
Cow's milk	RPKHPIKHQGLPQEVLNENLLRF	Stimulation of phagocytosis/immune response

(continued)

Table 2.7 (continued)

Source	Amino acid sequence	Action
Cow's milk	RELEELNVPGEIVESLSSSEESITR	Mitogenic activity
Cow's milk	RELEELNVPGEIVESLSSSEESITRINK	Mitogenic activity

Source: Amigo et al. (2020), Cao et al. (2019), Orona-Tamayo et al. (2019), Chatterjee et al. (2018), Nielsen et al. (2017), Lucarini (2017), Maestri et al. (2016), Rizzello et al. (2016), Sun and Jenssen (2012), Nagpal et al. (2011). Synthetic (*)

2.7.1 Immunomodulatory Peptides from Animal Foods

Milk. Human milk is superior to other kinds of milk in terms of immune-modulating components, including antigens, cytokines, immunoglobulins, polyunsaturated fatty acids, oligosaccharides, exosomes, chemokines, and a wide range of immune cells, whose role in the IIS and IAS of the newborn has been extensively examined (Simon et al. 2015). Particularly, the protein fraction of breast milk contains lactoferrin and other proteins with immunomodulatory peptide sequences that are bioaccessible during GI transit, and together, they modulate mucosal immunity by nurturing and improving the functionality of lymphoid tissues (GALT and MLNs) and by modulating the newborn's GI microbiome. However, DIMP from cow's milk proteins produced by enzymatic hydrolysis and fermentation with LAB are novel ingredients to design food products for the functional adult market (Reyes-Díaz et al. 2018; Santiago-Lopez et al. 2016). Nielsen et al. (2017) compiled a comprehensive database of milk protein-derived bioactive peptides from which cow's milk emerges as the most studied source, a fact that has prompt the identification of the same bioactive peptides among species due to the high-sequence homology within milk proteins among species. Also, according to *Coherent Market Insights*, the global bioactive protein and peptide market will reach US$ 88.3 billion in revenues by 2027 and a compound annual growth rate (CAGR) of 8.2% between 2020 and 2027, being DIMP-based products one of the most important growing segments.

Cow's milk proteins comprise 80% caseins and 20% whey proteins, respectively. The casein fraction consists of s1-, β-, s2-, and k-CN families (38:38:11:13 by weight), while whey proteins are basically α-lactalbumin (~25%) and β-lactalbumin (~65%), bovine serum albumin (~8%), lactoferrin, and immunoglobulins. The biological value of cow's milk proteins as a source of indispensable amino acids has been extensively recognized, but such biological value goes beyond this matter. Proteins and peptides from cow's milk could potentially be used in the production of immunostimulating and immunosuppressant agents for both prophylaxis and treatment of infectious diseases and immune-related illnesses (Nielsen et al. (2017), Lucarini (2017), Sun and Jenssen (2012), Nagpal et al. (2011)). Colostrinin™ is a complex consisting of low-molecular-weight (0.5–3 kDa) proline-rich (~25%) polypeptides (PRP) firstly identified in sheep colostrum that are potent inducers of T-cell maturation and differentiation [Th and natural killer cells (NK)], Th1-/Th2-modulating activity, radical-scavenging (vs. ROS, NOS), and activation of certain cytokines [IFN-γ, TNF-α, and interleukins 6 and 10 (IL-6, IL-10)] and interrupts several pathophysiological processes involved in the progression of Alzheimer's disease (Ashraf et al. 2019) and other age-related disorders (Boldogh and Kruzel 2008). Two peptides derived from α-lactalbumin [fractions 18–19 (YG) and 18–20 (YGG)] stimulate lymphocyte proliferation, and two from β-casein named β-casomorphin-7 (fraction 60–66; YPFPGPI) and β-casokinin-10 (fractions 193–202; YQQPVLGPVR) suppress (low concentration) or promote (high concentration) lymphocyte proliferation (Table 2.7; Nielsen et al. (2017); Lucarini 2017; Nagpal et al. 2011).

It can be ruled out that the smaller the DIMP, the greater the chance of finding the same sequence in different parental proteins. For instance, the dipeptide YG previously mentioned can be also found in residues 38–39 of k-casein. Also, certain peptides have multiple functions, such as the hexapeptide TTMPLW derived from α-S-casein which exhibits antimicrobial, angiotensin-converting enzyme (ACE) inhibitory, and macrophage-stimulating activity (Nielsen et al. 2017); other milk-derived immunopeptides enhance phagocytosis, antioxidant/antimicrobial, antibody production, enhanced cytokine production, dendritic cell/MLN stimulation, and mitogenic activity (Nielsen et al. 2017; Lucarini 2017; Sun and Jenssen 2012; Nagpal et al. 2011). Such diversity in immunomodulatory actions exerted by milk-derived peptides makes them attractive candidates for the development of antiviral drugs. Certain peptides derived from β-casein [e.g., HLPLP and WSVPQPK] have demonstrated potent antioxidant and ACE inhibitory activity (Shahidi and Zhong 2008), a fact that could be useful in the treatment of hypertension and immunosuppression-associated severe acute respiratory syndrome coronavirus 2 (SARS-CoV-2; Mancia et al. 2020).

Dairy hydrolysates. Production of DIMPs from fresh whole milk during milk secretion and storage is attributed to its own (e.g., plasmin and cathepsin D) and microbial enzymes. Conversely, enzymatic proteolysis with exogenous GI (e.g., Alcalase, chymotrypsin) or microbial enzymes (from starter and nonstarter cultures) results in many kinds of artisanal/industrialized dairy products such as fermented milk, yogurt, kefir, and buttermilk (Albenzio et al. 2017; Santiago-López et al. 2015). The immunomodulatory action of hydrolyzed milk has been tested in vitro using immortalized immune cell lines and in vivo (rodent models) but scarcely in human clinical trials. Abelson murine leukemia virus-transformed macrophages (RAW 264.7), human monocytes U937 and THP-1, and human T lymphoblasts (Jurkat) have been used to test the immunomodulatory properties (mainly cytokine release ability) of milk-derived peptides (Chalamaiah et al. 2018). O'Sullivan et al. (2019) studied the immunomodulatory activity of a casein hydrolysate (Flavourzyme®, FlavorPro Whey, and trypsin) fraction <5 kDa in concanavalin A-stimulated Jurkat T lymphoblasts and lipopolysaccharide-stimulated RAW264.7 macrophages observing a low production/expression of IL-6, IL-1β, and NF-kB subunit p65. Although the synthetic peptides usually are shorter than natural proteins, the antiviral immune-regulating properties of many of these synthetic derivatives appear to be similar for the entire proteins (Sun and Jenssen 2012). Lastly, fermented dairy products are commonly used as the most efficient vehicle to deliver probiotics and bioactive peptides (Santiago-López et al. 2016) with synergistic effects on cytokine production, phagocytic activity, IgA production, T-cell function, and NK cell activity, and different studies have investigated the therapeutic effects of probiotic-fermented milk on cancer, allergies, and infectious and GI disorders such as IBS (Santiago-López et al. 2015). It is noteworthy that multifunctional activity (e.g., antioxidant, immunomodulatory, or antimicrobial) of LAB-derived immunopeptides is strain-specific and molecular size-specific (Amigo et al. 2020; Aguilar-Toalá et al. 2017), and so, this will allow the formulation of a wide range of fermented milks whose effect on health could be different from other similar products on the market.

Meat and other animal-derived peptides. Enzymatic hydrolysis and fermentation (curation) of parent proteins present in beef, pork, mutton, chicken, duck, marine organisms, and their by-products (e.g., skin collagen, blood protein) have been recently proposed as valuable sources of DIMP (Xing et al. 2019; Albenzio et al. 2017). Convection drying, curing, and ripening of these animal protein sources also promote the synthesis of bioactive peptides with antioxidant, opiate-like, and antimicrobial activity all of them involved in immunocompromised illnesses such as obesity and cancer. Thymopentin (TP5), a synthetic water-soluble pentapeptide derived from thymopoietin (residues 32–36), exhibits bactericidal and protective effect on dextran sulfate (DSS)-induced colitis by triggering the production of IL-22 in both IIS and AIS lymphocytes (Cao et al. 2019). Alternative sources of DIMP from animal origin that include whey, egg, fish, chicken, shellfish, sheep, frog, bivalves, and their by-products (e.g., blood) have been recently reviewed by Bechaux et al. (2019) and Chalamaiah et al. (2018). For instance, Amigo et al. (2020) performed a comprehensive in vitro and in silico analysis of the multifunctionality of milk- and blood-derived peptides whose amino acid sequence matched previously described opioid peptides, reporting that four peptides had the ability to protect Caco-2 and RAW264.7 cells from the oxidative damage (RYLGYLE, YLGYLE, YFYPEL, YPWT) caused by chemicals, stimulate mucin expression and secretion (YLGYLE, YFYPEL), and had anticancer, antioxidant, and opioid activity ($IC_{50} = 1.2–45\ \mu M$).

2.7.2 Immunomodulatory Peptides from Vegetable Foods

Legumes. Approximately ~35–40% of raw soybeans (*Glycine max* L.) seeds are good quality protein, and it is also considered as a major source of bioactive peptides with relevant bioactivities such as ACE inhibitory, appetite, hypolipidemic, and immunomodulatory activity (Chatterjee et al. 2018). Soymetides 13 (MITLAIPVNKPGR), 9 (MITLAIPVN), and 4 (MITL) from beta-conglycinin enhance phagocytosis activity of human macrophages, but soymetide 4 promotes TNF-α expression to a better extent than soymetide 13 (Kumar et al. 2017). Zheng et al. (2020a, b) reported that certain two low-molecular-weight peptides

[IYVVDLR (877 Da), IYVFVR (795 Da)] from Alcalase-hydrolyzed soybeans have cytoprotective effects in H_2O_2-stressed Caco-2 cells, characterized by a pronounced downregulation of intracellular ROS generation and lipid peroxidation and upregulation of glutathione synthesis, catalase and glutathione reductase activity, and IL-8 inhibition. Lunasin is a 43-amino acid chromatin-binding peptide derived from the 2S albumin fraction of soybean, barley, wheat, and *Amaranthus* that exhibits anticancer (by inhibiting acetylation of histones) and immunomodulatory activities (Fernández-Tomé and Hernández-Ledesma 2019). Thermolysin hydrolysates from yellow pea protein isolate (0% protein, dry weight basis) stimulate phagocyte activity in LPS/IFN-γ-activated RAW 264.7 NO (−) macrophages and IgA+/cytokine-positive murine intestinal cells (Ndiaye et al. 2012). Lastly, Chen et al. (2019) reported the anti-inflammatory effect of small peptides (LKBM, LKBY) obtained from common bean (*Phaseolus vulgaris* L.) in Caco-2 and Caco-2/EA. hy926 monolayers characterized by a strong inhibition of TNF-α pro-inflammatory mediators of the NF-kB and mitogen-activated protein kinase signal cascades in vascular endothelial cells; the authors suggested that these peptides can be absorbed and possibly have systemic inhibition on inflammatory responses in vascular endothelial cells, indicating potential preventive effects on vascular diseases. Other relevant bioactivities and molecular mechanisms of legume peptides have been recently reviewed by Reyes-Díaz et al. (2019).

Cereals and pseudocereals. Despite being low in protein content (~7–14 g.100 g^{-1}) and sometimes in quality, cereal and pseudocereals are also important sources of DIMP. For example, protein fractions present in maize kernels are mostly prolamins or zeins (40%) and glutelins (30%), and two peptides with antioxidant (ABTS> DPPH> $O_2^{\cdot-}$) and ACE inhibitory activity have been identified in Alcalase-/neutrase-hydrolzyed zein fractions (YA, LMCH), but the authors observed that such bioactivities were strongly related to the molecular weight and hydrophobicity of the identified peptides (Tang et al. 2010). Gluten represents 80% of all wheat proteins and is rich in glutamine (Q), important in the process of cell division in the GI and immune systems, while Q-containing peptides may also act as immunostimulants and NK cell activators (Santiago-López et al. 2016). Other cereal grains from which immunopeptides have been isolated include barley (EVSLNSGYY), buckwheat (DVWY, FQ, VVG, GPP), kamut (KDVTDM, GVSNAAVVAGGH, DAQEFKR), oat (VPP, IPP, LQP, LLP), rice (FRDEHKK, KHDRGDEF, GYPMYPLPR), rye (VSP, LCPVHRAADL), and wheat (VY, IY, TF, LQP, IQP, LRP) whose bioactivity has been reviewed by Maestri et al. (2016). As for pseudocereals, Moronta et al. (2016) reported that the heptapeptide SSDEIKE from amaranth contains sequences with anti-inflammatory effect.

2.8 Perspectives

Immunocompromised illnesses are probably the highest public health burden worldwide, causing high morbidity and mortality rates. While recognizing the strong interplay between immune system derangements with infectious diseases, chronic noncommunicable diseases such as cardiovascular diseases, type 2 diabetes mellitus, and cancer are themselves immunocompromised diseases. Foods can also play a significant role in fortifying and balancing immune responses in primary and secondary prevention, but certain food components exhibit specific roles in modulating key actors of both IIS and AIS cascades. Polyphenolic antioxidants, vitamins, complex polysaccharides, immunobiotics, and immunopeptides were briefly reviewed in this chapter, but others are yet to come with the advance in metabolomic studies.

Acknowledgments The authors wish to thank the National Council for Science and Technology for the financial support through the postdoctoral scholarship (177611).

References

Abbas AK, Lichtman AH, Pober JS (2011) Innate immunity. In: Cellular and molecular immunology, 7th edn. Elsevier, Philadelphia

Acurcio LB, Bastos RW, de Cicco Sandes SH, de Carvalho Guimarães AC, Alves CG, dos Reis DC, Wuyts S, Nunes ÁC, Cassali GD, Lebeer S, de Souza MR, Nicoli JR (2017) Protective effects of milk fermented by Lactobacillus plantarum B7 from Brazilian artisanal cheese on a Salmonella enterica serovar Typhimurium infection in BALB/c mice. Journal of Functional Foods 33:436–445

Adorini L, Penna G (2008) Control of autoimmune diseases by the vitamin D endocrine system. Nat Clin Pract Rheumatol 4(8):404–412

Aguilar-Toalá JE, Santiago-López L, Peres CM, Peres C, Garcia HS, Vallejo-Cordoba B et al (2017) Assessment of multifunctional activity of bioactive peptides derived from fermented milk by specific Lactobacillus plantarum strains. J Dairy Sci 100(1):65–75

Ahola AJ, Lassenius MI, Forsblom C, Harjutsalo V, Lehto M, Groop PH (2017) Dietary patterns reflecting healthy food choices are associated with lower serum LPS activity. Sci Rep 7(1):1–9

Albenzio M, Santillo A, Caroprese M, Della Malva A, Marino R (2017) Bioactive peptides in animal food products. Foods 6(5):35

Alwarawrah Y, Kiernan K, MacIver NJ (2018) Changes in nutritional status impact immune cell metabolism and function. Front Immunol 9:1055

Amigo L, Martínez-Maqueda D, Hernández-Ledesma B (2020) In Silico and in vitro analysis of multifunctionality of animal food-derived peptides. Foods 9(8):991

Ang A, Pullar JM, Currie MJ, Vissers MC (2018) Vitamin C and immune cell function in inflammation and cancer. Biochem Soc Trans 46(5):1147–1159

Ashraf GM, Tarasov VV, Makhmutova A, Chubarev VN, Avila-Rodriguez M, Bachurin SO, Aliev G (2019) The possibility of an infectious etiology of Alzheimer disease. Mol Neurobiol 56(6): 4479–4491

Aslam MF, Majeed S, Aslam S, Irfan JA (2017) Vitamins: key role players in boosting up immune response–a mini review. Vitamins and Minerals 6:153

Azad M, Kalam A, Sarker M, Wan D (2018) Immunomodulatory effects of probiotics on cytokine profiles. Biomed Res Int:10

Babbar N, Dejonghe W, Gatti M, Sforza S, Elst K (2016) Pectic oligosaccharides from agricultural by-products: production, characterization and health benefits. Crit Rev Biotechnol 36(4): 594–606

Barbé F, Le Feunteun S, Rémond D, Ménard O, Jardin J, Henry G et al (2014) Tracking the in vivo release of bioactive peptides in the gut during digestion: mass spectrometry peptidomic characterization of effluents collected in the gut of dairy matrix fed mini-pigs. Food Res Int 63:147–156

Bechaux J, Gatellier P, Le Page JF, Drillet Y, Sante-Lhoutellier V (2019) A comprehensive review of bioactive peptides obtained from animal byproducts and their applications. Food Funct 10(10):6244–6266

Bendix M, Greisen S, Dige A, Hvas CL, Bak N, Jørgensen SP, Dahlerup JF, Deleuran B, Agnholt J (2017) Vitamin D increases programmed death receptor-1 expression in Crohn's disease. Oncotarget 8(15):24177

Bermudez-Brito M, Rösch C, Schols HA, Faas MM, de Vos P (2015) Resistant starches differentially stimulate toll-like receptors and attenuate proinflammatory cytokines in dendritic cells by modulation of intestinal epithelial cells. Mol Nutr Food Res 59(9):1814–1826

Bivona JJ III, Patel S, Vajdy M (2017) Induction of cellular and molecular immunomodulatory pathways by vitamin E and vitamin C. Expert Opin Biol Ther 17(12):1539–1551

Boehm T (2011) Design principles of adaptive immune systems. Nat Rev Immunol 11(5):307–317

Boldogh I, Kruzel ML (2008) Colostrinin™: an oxidative stress modulator for prevention and treatment of age-related disorders. J Alzheimers Dis 13(3):303–321

Boothapandi M, Ramanibai R (2019) Immunomodulatory effect of natural flavonoid chrysin (5, 7-dihydroxyflavone) on LPS stimulated RAW 264.7 macrophages via inhibition of NF-κB activation. Process Biochem 84:186–195

Boretti A, Banik BK (2020) Intravenous vitamin C for reduction of cytokines storm in acute respiratory distress syndrome. Pharma Nutrition 12:100190

Butel MJ (2014) Probiotics, gut microbiota and health. Med Mal Infect 44(1):1–8

Cai J, Li X, Du H, Jiang C, Xu S, Cao Y (2020) Immunomodulatory significance of natural peptides in mammalians: promising agents for medical application. Immunobiology 225(3):151936

Cai Y, Li YF, Tang LP, Tsoi B, Chen M, Chen H, Chen X-M, Tan R-R, Kurihara H, He RR (2015) A new mechanism of vitamin C effects on a/FM/1/47 (H1N1) virus-induced pneumonia in restraint-stressed mice. Biomed Res Int 2015:675149

Cao Q, Gao X, Lin Y, Yue C, Wang Y, Quan F, Quan F, Zhang Z, Liu X, Lu Y, Zhan Y, Yang H, Li X, Di Qin LB, Hao K, Yang H (2019) Thymopentin ameliorates dextran sulfate sodium-induced colitis by triggering the production of IL-22 in both innate and adaptive lymphocytes. Theranostics 9(25):7490

Cavaglieri CR, Nishiyama A, Fernandes LC, Curi R, Miles EA, Calder PC (2003) Differential effects of short-chain fatty acids on proliferation and production of pro-and anti-inflammatory cytokines by cultured lymphocytes. Life Sci 73(13):1683–1690

Chaisuwan W, Jantanasakulwong K, Wangtueai S, Phimolsiripol Y, Chaiyaso T, Techapun C, Phongthai S, You SG, Regenstein J, Seesuriyachan P (2020) Microbial exopolysaccharides for immune enhancement: fermentation, modifications and bioactivities. Food Biosci 35:100564

Chalamaiah M, Yu W, Wu J (2018) Immunomodulatory and anticancer protein hydrolysates (peptides) from food proteins: a review. Food Chem 245:205–222

Chan BCL, Li LF, Hu SQ, Wat E, Wong ECW, Zhang VX, Lau CBS, Wong CK, Hon KLE, Hui PCL, Leung PC (2015) Gallic acid is the major active component of cortex moutan in inhibiting immune maturation of human monocyte-derived dendritic cells. Molecules 20(9):16388–16403

Chandra RK (1974) Rosette-forming T lymphocytes and cell-mediated immunity in malnutrition. Br Med J 3(5931):608–609

Chandran M, Maung AC, Mithal A, Parameswaran R (2020) Vitamin D in COVID-19: dousing the fire or averting the storm?–a perspective from the Asia-Pacific. Osteoporosis and Sarcopenia 6(3):97–105
Chandrashekar PM, Venkatesh YP (2012) Fructans from aged garlic extract produce a delayed immunoadjuvant response to ovalbumin antigen in BALB/c mice. Immunopharmacol Immunotoxicol 34(1):174–180
Chandrashekara PM, Venkatesh YP (2016) Immunostimulatory properties of fructans derived from raw garlic (Allium sativum L.). Bioact Carbohydr Diet Fibre 8(2):65–70
Chatterjee C, Gleddie S, Xiao C-W (2018) Soybean bioactive peptides and their functional properties. Nutrients 10:1211
Chen K, Chen H, Faas MM, de Haan BJ, Li J, Xiao P, Zhang H, Diana J, de Vos P, Sun J (2017) Specific inulin-type fructan fibers protect against autoimmune diabetes by modulating gut immunity, barrier function, and microbiota homeostasis. Mol Nutr Food Res 61(8):1601006
Chen F, He D, Yan B (2020a) Apigenin attenuates allergic responses of ovalbumin-induced allergic rhinitis through modulation of Th1/Th2 responses in experimental mice. Dose-Response 18(1): 1559325820904799
Chen L, Hu C, Hood M, Zhang X, Zhang L, Kan J, Du J (2020b) A novel combination of vitamin C, curcumin and Glycyrrhizic acid potentially regulates immune and inflammatory response associated with coronavirus infections: a perspective from system biology analysis. Nutrients 12(4):1193
Chen YC, Wu YJ, Hu CY (2019) Monosaccharide composition influence and immunomodulatory effects of probiotic exopolysaccharides. Int J Biol Macromol 133:575–582
Childs CE, Calder PC, Miles EA (2019) Diet and immune function. Nutrients 11(8):1933
Chinen J, Shearer WT (2010) Secondary immunodeficiencies, including HIV infection. J Allergy Clin Immunol 125(2):S195–S203
Clarke ST, Green-Johnson JM, Brooks SP, Ramdath DD, Bercik P, Avila C, Inglis GD, Green J, Yanke LJ, Selinger LB, Kalmokoff M (2016) β2-1 Fructan supplementation alters host immune responses in a manner consistent with increased exposure to microbial components: results from a double-blinded, randomised, cross-over study in healthy adults. Br J Nutr 115(10):1748–1759
Combs GF (2012) Vitamin C. In: The vitamins. Academic Press, pp 233–259
Coombes JL, Siddiqui KR, Arancibia-Cárcamo CV, Hall J, Sun CM, Belkaid Y, Powrie F (2007) A functionally specialized population of mucosal CD103+ DCs induces Foxp3+ regulatory T cells via a TGF-β–and retinoic acid–dependent mechanism. J Exp Med 204(8):1757–1764
Corona-Hernandez RI, Álvarez-Parrilla E, Lizardi-Mendoza J, Islas-Rubio AR, de la Rosa LA, Wall-Medrano A (2013) Structural stability and viability of microencapsulated probiotic bacteria: a review. Compr Rev Food Sci Food Saf 12(6):614–628
Cuevas A, Saavedra N, Salazar LA, Abdalla DS (2013) Modulation of immune function by polyphenols: possible contribution of epigenetic factors. Nutrients 5(7):2314–2332
Cunningham-Rundles S, McNeeley DF, Moon A (2005) Mechanisms of nutrient modulation of the immune response. J Allergy Clin Immunol 115(6):1119–1128
Custódio G, Schwarz P, Crispim D, Moraes RB, Czepielewski M, Leitão CB, Rech TH (2018) Association between vitamin D levels and inflammatory activity in brain death: a prospective study. Transpl Immunol 48:65–69
Daoud AK, Tayyar MA, Fouda IM, Harfeil NA (2009) Effects of diabetes mellitus vs. in vitro hyperglycemia on select immune cell functions. J Immunotoxicol 6(1):36–41
de Almada CN, Almada CN, Martinez RC, Sant'Ana AS (2016) Paraprobiotics: evidences on their ability to modify biological responses, inactivation methods and perspectives on their application in foods. Trends Food Sci Technol 58:96–114
De Angelis E, Pilolli R, Bavaro SL, Monaci L (2017) Insight into the gastro-duodenal digestion resistance of soybean proteins and potential implications for residual immunogenicity. Food Funct 8:1599–1610
de Moraes Alves MM, Brito LM, Souza AC, Queiroz BCSH, de Carvalho TP, Batista JF, Oliveira JSDSM, de Mendonça IL, de Sousa Lira SR, Chaves MH, Gonçalves JCR (2017) Gallic and

ellagic acids: two natural immunomodulator compounds solve infection of macrophages by Leishmania major. Naunyn Schmiedeberg's Arch Pharmacol 390(9):893–903
Delphi L, Sepehri H (2016) Apple pectin: a natural source for cancer suppression in 4T1 breast cancer cells in vitro and express p53 in mouse bearing 4T1 cancer tumors, in vivo. Biomed Pharmacother 84:637–644
Dinarello CA, Nold-Petry C, Nold M, Fujita M, Li S, Kim S, Bufler P (2016) Suppression of innate inflammation and immunity by interleukin-37. Eur J Immunol 46(5):1067–1081
Djaldetti M, Bessler H (2017) Probiotic strains modulate cytokine production and the immune interplay between human peripheral blood mononuclear cells and colon cancer cells. FEMS Microbiol Lett 364(3):fmx014
Dobrange E, Peshev D, Loedolff B, Van den Ende W (2019) Fructans as immunomodulatory and antiviral agents: the case of Echinacea. Biomol Ther 9(10):615
Dong CX, Zhang WS, Sun QL, Jiang YJ, Fu Z, Zhang Y, Du J, Sun Y-X, Tao N, Yao Z (2018) Structural characterization of a pectin-type polysaccharide from Curcuma kwangsiensis and its effects on reversing MDSC-mediated T cell suppression. Int J Biol Macromol 115:1233–1240
Dutra RC (2017) Kinin receptors: key regulators of autoimmunity. Autoimmun Rev 16(2):192–207
Ebrahimpour-Koujan S, Milajerdi A, Larijani B, Esmaillzadeh A (2020) Effects of probiotics on salivary cytokines and immunoglobulins: a systematic review and meta-analysis on clinical trials. Sci Rep 10(1):1–11
Elenius V, Palomares O, Waris M, Turunen R, Puhakka T, Rückert B, Vuorinen T, Allander T, Vahlberg T, Akdis M, Camargo CA Jr, Akdis CA, Jartti T (2017) The relationship of serum vitamins a, D, E and LL-37 levels with allergic status, tonsillar virus detection and immune response. PLoS One 12(2):e0172350
El-Feki MA, Amin HM, Abdalla AA, Fesal M (2016) Immunomodulatory and anti-oxidant effects of alpha-Lipoic acid and vitamin E on lipopolysaccharide-induced liver injury in rats. Middle East Journal of Applied Sciences 6(03):460–467
Elitsur Y, Neace C, Liu X, Dosescu J, Moshier JA (1997) Vitamin a and retinoic acids immunomodulation on human gut lymphocytes. Immunopharmacology 35(3):247–253
Enciso J, Mendoza L, Pelayo R (2015) Normal vs. malignant hematopoiesis: the complexity of acute leukemia through systems biology. Front Genet 6:290
Erkelens MN, Mebius RE (2017) Retinoic acid and immune homeostasis: a balancing act. Trends Immunol 38(3):168–180
Escobar-Puentes AA, Rincón S, García-Gurrola A, Zepeda A, Martínez-Bustos F (2019) Preparation and characterization of succinylated nanoparticles from high-amylose starch via the extrusion process followed by ultrasonic energy. Food Bioprocess Technol 12(10):1672–1682
Fernández RG, Leehan JA, Pastrana RF, Muñiz RO (2005) Effect of malnutrition on K+ current in T lymphocytes. Clin Diagn Lab Immunol 12(7):808–813
Fernández-Riejos P, Najib S, Santos-Alvarez J, Martín-Romero C, Pérez-Pérez A, González-Yanes C, Sánchez-Margalet V (2010) Role of leptin in the activation of immune cells. Mediat Inflamm 2010:1–10
Fernández-Tomé S, Hernández-Ledesma B (2019) Current state of art after twenty years of the discovery of bioactive peptide lunasin. Food Res Int 116:71–78
Filardo S, Di Pietro M, Mastromarino P, Sessa R (2020) Therapeutic potential of resveratrol against emerging respiratory viral infections. Pharmacol Ther 214:107613
Fraga CG, Croft KD, Kennedy DO, Tomás-Barberán FA (2019) The effects of polyphenols and other bioactives on human health. Food Funct 10(2):514–528
Frece J, Kos B, Svetec IK, Zgaga Z, Beganovic J, Lebos A, Suskovic J (2009) Synbiotic effect of Lactobacillus helveticus M92 and prebiotics on the intestinal microflora and immune system of mice. J Dairy Res 76(1):98
Galdeano CM, Cazorla SI, Dumit JML, Vélez E, Perdigón G (2019) Beneficial effects of probiotic consumption on the immune system. Ann Nutr Metab 74(2):115–124

Gandhi GR, Neta MTSL, Sathiyabama RG, Quintans JDSS, de Silva AMDO, de Souza Araújo AA, Narain N, Júnior LJQ, Gurgel RQ (2018) Flavonoids as Th1/Th2 cytokines immunomodulators: a systematic review of studies on animal models. Phytomedicine 44:74–84

Gaspar C, Donders GG, Palmeira-de-Oliveira R, Queiroz JA, Tomaz C, Martinez-de-Oliveira J, Palmeira-de-Oliveira A (2018) Bacteriocin production of the probiotic Lactobacillus acidophilus KS400. AMB Express 8(1):153

Georgiev YN, Paulsen BS, Kiyohara H, Ciz M, Ognyanov MH, Vasicek O, Rise F, Denev PN, Yamada H, Lojek A, Kussovski V, Barsett H, Krastanov AI, Yanakieva IZ, Kratchanova MG (2017) The common lavender (Lavandula angustifolia mill.) pectic polysaccharides modulate phagocytic leukocytes and intestinal Peyer's patch cells. Carbohydr Polym 174:948–959

Ghanem ENB, Clark S, Du X, Wu D, Camilli A, Leong JM, Meydani SN (2015) The α-tocopherol form of vitamin E reverses age-associated susceptibility to streptococcus pneumoniae lung infection by modulating pulmonary neutrophil recruitment. J Immunol 194(3):1090–1099

Gibson GR, Hutkins R, Sanders ME, Prescott SL, Reimer RA, Salminen SJ, Scott K, Stanton C, Swanson KS, Cani PD, Verbeke K, Reid G (2017) Expert consensus document: the international scientific Association for Probiotics and Prebiotics (ISAPP) consensus statement on the definition and scope of prebiotics. Nat Rev Gastroenterol Hepatol 14(8):491

Gibson GR, Probert HM, Van Loo J, Rastall RA, Roberfroid MB (2004) Dietary modulation of the human colonic microbiota: updating the concept of prebiotics. Nutr Res Rev 17(2):259–275

Gill HS, Rutherfurd KJ (2001) Viability and dose–response studies on the effects of the immunoenhancing lactic acid bacterium Lactobacillus rhamnosus in mice. Br J Nutr 86(2): 285–289

Gonzalez SM, Aguilar-Jimenez W, Trujillo-Gil E, Zapata W, Su RC, Ball TB, Rugeles MT (2019) Vitamin D treatment of peripheral blood mononuclear cells modulated immune activation and reduced susceptibility to HIV-1 infection of CD4+ T lymphocytes. PLoS One 14(9):e0222878

Gonzalez H, Hagerling C, Werb Z (2018) Roles of the immune system in cancer: from tumor initiation to metastatic progression. Genes Dev 32(19–20):1267–1284

Gutiérrez Nava ZJ, Jiménez-Aparicio AR, Herrera-Ruiz ML, Jiménez-Ferrer E (2017) Immunomodulatory effect of Agave tequilana evaluated on an autoimmunity like-SLE model induced in Balb/c mice with pristane. Molecules 22(6):848

Hardy H, Harris J, Lyon E, Beal J, Foey AD (2013) Probiotics, prebiotics and immunomodulation of gut mucosal defences: homeostasis and immunopathology. Nutrients 5(6):1869–1912

Harirchian MH, Mohammadpour Z, Fatehi F, Firoozeh N, Bitarafan S (2019) A systematic review and meta-analysis of randomized controlled trials to evaluating the trend of cytokines to vitamin a supplementation in autoimmune diseases. Clin Nutr 38(5):2038–2044

Hartog A, Belle FN, Bastiaans J, de Graaff P, Garssen J, Harthoorn LF, Vos AP (2015) A potential role for regulatory T-cells in the amelioration of DSS induced colitis by dietary non-digestible polysaccharides. J Nutr Biochem 26(3):227–233

Hastreiter AA, Dos Santos GG, Santos EWC, Makiyama EN, Borelli P, Fock RA (2020) Protein malnutrition impairs bone marrow endothelial cells affecting hematopoiesis. Clin Nutr 39(5): 1551–1559

Heinz SA, Henson DA, Nieman DC, Austin MD, Jin F (2010) A 12-week supplementation with quercetin does not affect natural killer cell activity, granulocyte oxidative burst activity or granulocyte phagocytosis in female human subjects. Br J Nutr 104(6):849–857

Holásková E, Galuszka P, Mičúchová A, Šebela M, Öz MT, Frébort I (2018) Molecular farming in barley: development of a novel production platform to produce human antimicrobial peptide LL-37. Biotechnol J 13(6):1700628

Hornsby E, Pfeffer PE, Laranjo N, Cruikshank W, Tuzova M, Litonjua AA et al (2018) Vitamin D supplementation during pregnancy: effect on the neonatal immune system in a randomized controlled trial. J Allergy Clin Immunol 141(1):269–278

Hu S, Kuwabara R, Beukema M, Ferrari M, de Haan BJ, Walvoort MT et al (2020) Low methyl-esterified pectin protects pancreatic β-cells against diabetes-induced oxidative and inflammatory stress via galectin-3. Carbohydr Polym 249:116863

Huang Z, Liu Y, Qi G, Brand D, Zheng SG (2018) Role of vitamin a in the immune system. J Clin Med 7(9):258

Husain A, Kaushik A, Awasthi H, Singh DP, Khan R, Mani D (2017) Immunomodulatory and antioxidant activities of fresh juice extracts of Brahmi and Guduchi. Indian J Tradit Knowl 16(3):498–505

Ishikawa H, Kutsukake E, Fukui T, Sato I, Shirai T, Kurihara T, Okada N, Danbara H, Toba M, Kohda N, Maeda Y, Matsumoto T (2010) Oral administration of heat-killed Lactobacillus plantarum strain b240 protected mice against Salmonella enterica Serovar Typhimurium. Biosci Biotechnol Biochem 74(7):1338–1342

Jakhar R, Paul S, Chauhan AK, Kang SC (2014) Morin hydrate augments phagocytosis mechanism and inhibits LPS induced autophagic signaling in murine macrophage. Int Immunopharmacol 22(2):356–365

Jiang F, Du C, Jiang W, Wang L, Du SK (2020) The preparation, formation, fermentability, and applications of resistant starch. Int J Biol Macromol 150:1155–1161

Jørgensen MJ, Fisker AB, Sartono E, Andersen A, Erikstrup C, Lisse IM, Yazdanbakhsh M, Aaby P, Benn CS (2013) The effect of at-birth vitamin a supplementation on differential leucocyte counts and in vitro cytokine production: an immunological study nested within a randomised trial in Guinea-Bissau. Br J Nutr 109(3):467–477

Kaji R, Kiyoshima-Shibata J, Tsujibe S, Nanno M, Shida K (2018) Probiotic induction of interleukin-10 and interleukin-12 production by macrophages is modulated by co-stimulation with microbial components. J Dairy Sci 101(4):2838–2841

Kawase M, He F, Kubota A, Yoda K, Miyazawa K, Hiramatsu M (2012) Heat-killed Lactobacillus gasseri TMC0356 protects mice against influenza virus infection by stimulating gut and respiratory immune responses. FEMS Immunology & Medical Microbiology 64(2):280–288

Keservani RK, Sharma AK, Kesharwani RK (2016) Nutracetical and functional foods for cardiovascular health, food process engineering: emerging technologies and innovations in food engineering, Edited by Murlidhar Meghwal, Megh R. Goyal, Apple Academic Press, CRC Press, Taylor and Francis Vol 5

Keservani RK, Sharma AK, Kesharwani RK (2017) An overview and therapeutic applications of nutraceutical and functional foods. In: Recent advances in drug delivery technology, pp 160–201

Keservani RK, Sharma AK, Kesharwani RK (2020a) Nutraceuticals and dietary supplements: applications in health improvement and disease management. CRC Press

Keservani RK, Sharma AK, Kesharwani RK (2020b) Delivery approaches and Nanosystems, volume 2 (drug targeting aspects of nanotechnology). Apple Academic Press, p 2017

Khalil ES, Manap A, Yazid M, Mustafa S, Alhelli AM, Shokryazdan P (2018) Probiotic properties of exopolysaccharide-producing Lactobacillus strains isolated from Tempoyak. Molecules 23(2):398

Khan H, Sureda A, Belwal T, Çetinkaya S, Süntar İ, Tejada S, Devkota HP, Ullah H, Aschner M (2019) Polyphenols in the treatment of autoimmune diseases. Autoimmun Rev 18(7):647–657

Kilani-Jaziri S, Mokdad-Bzeouich I, Krifa M, Nasr N, Ghedira K, Chekir-Ghedira L (2017) Immunomodulatory and cellular anti-oxidant activities of caffeic, ferulic, and p-coumaric phenolic acids: a structure–activity relationship study. Drug Chem Toxicol 40(4):416–424

Kim JH, Lee JK (2015) Naringenin enhances NK cell lysis activity by increasing the expression of NKG2D ligands on Burkitt's lymphoma cells. Arch Pharm Res 38(11):2042–2048

Kim YH, Won YS, Yang X, Kumazoe M, Yamashita S, Hara A, Takagaki A, Goto K, Nanjo F, Tachibana H (2016) Green tea catechin metabolites exert immunoregulatory effects on CD4+ T cell and natural killer cell activities. J Agric Food Chem 64(18):3591–3597

Kolehmainen M, Mykkanen O, Kirjavainen PV, Leppanen T, Moilanen E, Adriaens M, Laaksonen DE, Hallikainen M, Puupponen-Pimia R, Pulkkinen L et al (2012) Bilberries reduce low-grade inflammation in individuals with features of metabolic syndrome. Molecular Nutrition and Food Research 56:1501–1510

Kumar VP, Prashanth KH, Venkatesh YP (2015) Structural analyses and immunomodulatory properties of fructo-oligosaccharides from onion (Allium cepa). Carbohydr Polym 117:115–122
Kumar SJ, Rajendra S, Singh ANS (2017) Health prospects of bioactive peptides derived from seed storage proteins. In: State-of-the-art Technologies in food science. Apple Academic Press, Florida, pp 24–35
Kunisawa J, Hashimoto E, Ishikawa I, Kiyono H (2012) A pivotal role of vitamin B9 in the maintenance of regulatory T cells in vitro and in vivo. PLoS One 7(2):e32094
Lang A, Salomon N, Wu JC, Kopylov U, Lahat A, Har-Noy O, Ching JY, Cheong PK, Avidan B, Gamus D, Kaimakliotis I (2015) Curcumin in combination with mesalamine induces remission in patients with mild-to-moderate ulcerative colitis in a randomized controlled trial. Clin Gastroenterol Hepatol 13(8):1444–1449
Larbi A, Cexus O, Bosco N (2018) Nutrition as a tool to reverse immunosenescence? In: Immunity and Inflammation in Health and Disease. Academic Press, pp 319–337
Lawrence DM, Seth P, Durham L, Diaz F, Boursiquot R, Ransohoff RM, Major EO (2006) Astrocyte differentiation selectively upregulates CCL2/monocyte chemoattractant protein-1 in cultured human brain-derived progenitor cells. Glia 53(1):81–91
Lee GY, Han SN (2018) The role of vitamin E in immunity. Nutrients 10(11):1614
Lépine AF, de Hilster RH, Leemhuis H, Oudhuis L, Buwalda PL, de Vos P (2019) Higher chain length distribution in Debranched Type-3 resistant starches (RS3) increases TLR signaling and supports dendritic cell cytokine production. Mol Nutr Food Res 63(2):1801007
Lewis ED, Meydani SN, Wu D (2019) Regulatory role of vitamin E in the immune system and inflammation. IUBMB Life 71(4):487–494
Li W, Kim TI, Kim JH, Chung HS (2019) Immune checkpoint PD-1/PD-L1 CTLA-4/CD80 are blocked by Rhus verniciflua Stokes and its active compounds. Molecules 24(22):4062
Li RR, Pang LL, Du Q, Shi Y, Dai WJ, Yin KS (2010) Apigenin inhibits allergen-induced airway inflammation and switches immune response in a murine model of asthma. Immunopharmacol Immunotoxicol 32(3):364–370
Li P, Yin YL, Li D, Kim SW, Wu G (2007) Amino acids and immune function. Br J Nutr 98(2): 237–252
Lim CC, Ferguson LR, Tannock GW (2005) Dietary fibres as "prebiotics": implications for colorectal cancer. Mol Nutr Food Res 49(6):609–619
Lim WH, Kireta S, Leedham E, Russ GR, Coates PT (2007) Uremia impairs monocyte and monocyte-derived dendritic cell function in hemodialysis patients. Kidney Int 72(9):1138–1148
Linares DM, Gomez C, Renes E, Fresno JM, Tornadijo ME, Ross RP, Stanton C (2017) Lactic acid bacteria and bifidobacteria with potential to design natural biofunctional health-promoting dairy foods. Front Microbiol 8:846
Lis M, Szczypka M, Suszko-Pawłowska A, Sokół-Łętowska A, Kucharska A, Obmińska-Mrukowicz B (2020) Hawthorn (crataegus monogyna) phenolic extract modulates lymphocyte subsets and humoral immune response in mice. Planta Med 86(02):160–168
Llewellyn A, Foey A (2017) Probiotic modulation of innate cell pathogen sensing and signaling events. Nutrients 9(10):1156
Lucarini M (2017) Bioactive peptides in milk: from encrypted sequences to nutraceutical aspects. Beverages 3(3):41
Maestri E, Marmiroli M, Marmiroli N (2016) Bioactive peptides in plant-derived foodstuffs. J Proteome 147:140–155
Mähler A, Steiniger J, Bock M, Klug L, Parreidt N, Lorenz M, Zimmermann BF, Krannich A, Paul F, Boschmann M (2015) Metabolic response to epigallocatechin-3-gallate in relapsing-remitting multiple sclerosis: a randomized clinical trial. Am J Clin Nutr 101(3):487–495
Mancia G, Rea F, Ludergnani M, Apolone G, Corrao G (2020) Renin–angiotensin–aldosterone system blockers and the risk of Covid-19. N Engl J Med 382(25):2431–2440
McAnulty LS, Nieman DC, Dumke CL, Shooter LA, Henson DA, Utter AC, Milne G, McAnulty SR (2011) Effect of blueberry ingestion on natural killer cell counts, oxidative stress, and inflammation prior to and after 2.5 h of running. Appl Physiol Nutr Metab 36(6):976–984

Meijerink M, Rösch C, Taverne N, Venema K, Gruppen H, Schols HA, Wells JM (2018) Structure dependent-immunomodulation by sugar beet arabinans via a SYK tyrosine kinase-dependent signaling pathway. Front Immunol 9:1972

Mele C, Caputo M, Bisceglia A, Samà MT, Zavattaro M, Aimaretti G, Pagano L, Prodam F, Marzullo P (2020) Immunomodulatory effects of vitamin D in thyroid diseases. Nutrients 12(5): 1444

Meydani SN, Han SN, Wu D (2005) Vitamin E and immune response in the aged: molecular mechanisms and clinical implications. Immunol Rev 205(1):269–284

Mikkelsen K, Stojanovska L, Prakash M, Apostolopoulos V (2017) The effects of vitamin B on the immune/cytokine network and their involvement in depression. Maturitas 96:58–71

Mirshekar MA, Sarkaki A, Farbood Y, Naseri MKG, Badavi M, Mansouri MT, Haghparast A (2018) Neuroprotective effects of gallic acid in a rat model of traumatic brain injury: behavioral, electrophysiological, and molecular studies. Iran J Basic Med Sci 21(10):1056

Miyazawa K, He F, Kawase M, Kubota A, Yoda K, Hiramatsu M (2011) Enhancement of immunoregulatory effects of Lactobacillus gasseri TMC0356 by heat treatment and culture medium. Lett Appl Microbiol 53(2):210–216

Mokhtari-Zaer A, Hosseini M, Salmani H, Arab Z, Zareian P (2020) Vitamin D3 attenuates lipopolysaccharide-induced cognitive impairment in rats by inhibiting inflammation and oxidative stress. Life Sci 253:117703

Monobe M, Ema K, Tokuda Y, Maeda-Yamamoto M (2010) Enhancement of phagocytic activity of macrophage-like cells by pyrogallol-type green tea polyphenols through caspase signaling pathways. Cytotechnology 62(3):201–203

Mora JR, Iwata M, Von Andrian UH (2008) Vitamin effects on the immune system: vitamins a and D take Centre stage. Nat Rev Immunol 8(9):685–698

Moreno-Vilet L, Garcia-Hernandez MH, Delgado-Portales RE, Corral-Fernandez NE, Cortez-Espinosa N, Ruiz-Cabrera MA, Portales-Perez DP (2014) In vitro assessment of agave fructans (Agave salmiana) as prebiotics and immune system activators. Int J Biol Macromol 63:181–187

Moronta J, Smaldini PL, Fossati CA, Añon MC, Docena GH (2016) The anti-inflammatory SSEDIKE peptide from Amaranth seeds modulates IgE-mediated food allergy. J Funct Foods 25:579–587

Muñoz-González I, Espinosa-Martos I, Rodríguez JM, Jiménez-Girón A, Martin-Alvarez PJ, Bartolomé B, Moreno-Arribas MV (2014) Moderate consumption of red wine can modulate human intestinal inflammatory response. J Agric Food Chem 62(43):10567–10575

Nagpal R, Behare P, Rana R, Kumar A, Kumar M, Arora S et al (2011) Bioactive peptides derived from milk proteins and their health beneficial potentials: an update. Food Funct 2(1):18–27

Nantz MP, Rowe CA, Muller C, Creasy R, Colee J, Khoo C, Percival SS (2013) Consumption of cranberry polyphenols enhances human γδ-T cell proliferation and reduces the number of symptoms associated with colds and influenza: a randomized, placebo-controlled intervention study. Nutr J 12(1):161

Naqash F, Masoodi FA, Rather SA, Wani SM, Gani A (2017) Emerging concepts in the nutraceutical and functional properties of pectin–a review. Carbohydr Polym 168:227–239

Ndiaye F, Vuong T, Duarte J, Aluko RE, Matar C (2012) Anti-oxidant, anti-inflammatory and immunomodulating properties of an enzymatic protein hydrolysate from yellow field pea seeds. Eur J Nutr 51(1):29–37

Nielsen SD, Beverly RL, Qu Y, Dallas DC (2017) Milk bioactive peptide database: a comprehensive database of milk protein-derived bioactive peptides and novel visualization. Food Chem 232:673–682

Nwobodo DC, Ugwu MC (2020) Immunomodulatory potentials of probiotics: a review. Asian Journal of Immunology:1–15

Okayasu I, Hana K, Nemoto N, Yoshida T, Saegusa M, Yokota-Nakatsuma A et al (2016) Vitamin a inhibits development of dextran sulfate sodium-induced colitis and colon cancer in a mouse model. Biomed Res Int 2016:4874809

Orona-Tamayo D, Valverde ME, Paredes-López O (2019) Bioactive peptides from selected latin american food crops–a nutraceutical and molecular approach. Crit Rev Food Sci Nutr 59(12): 1949–1975

Orrù B, Szekeres-Bartho J, Bizzarri M, Spiga AM, Unfer V (2020) Inhibitory effects of vitamin D on inflammation and IL-6 release. A further support for COVID-19 management? Eur Rev Med Pharmacol Sci 24(15):8187–8193

O'Sullivan SM, O'Callaghan YC, O'Keeffe MB, FitzGerald RJ, O'Brien NM (2019) Immunomodulatory activity of 5 kDa permeate fractions of casein hydrolysates generated using a range of enzymes in Jurkat T cells and RAW264. 7 macrophages. Int Dairy J 91:9–17

Palm NW, Medzhitov R (2009) Pattern recognition receptors and control of adaptive immunity. Immunol Rev 227(1):221–233

Panghal A, Janghu S, Virkar K, Gat Y, Kumar V, Chhikara N (2018) Potential non-dairy probiotic products–a healthy approach. Food Biosci 21:80–89

Pawelec G (2018) Age and immunity: what is "immunosenescence"? Exp Gerontol 105:4–9

Peng M, Tabashsum Z, Anderson M, Truong A, Houser AK, Padilla J et al (2020) Effectiveness of probiotics, prebiotics, and prebiotic-like components in common functional foods. Compr Rev Food Sci Food Saf 19(4):1908–1933

Popov SV, Ovodov YS (2013) Polypotency of the immunomodulatory effect of pectins. Biochem Mosc 78(7):823–835

Rachid M, Matar C, Duarte J, Perdigon G (2006) Effect of milk fermented with a Lactobacillus helveticus R389 (+) proteolytic strain on the immune system and on the growth of 4T1 breast cancer cells in mice. FEMS Immunology & Medical Microbiology 47(2):242–253

Rekha RS, Mily A, Sultana T, Haq A, Ahmed S, Kamal SM et al (2018) Immune responses in the treatment of drug-sensitive pulmonary tuberculosis with phenylbutyrate and vitamin D 3 as host directed therapy. BMC Infect Dis 18(1):303

Reyes-Díaz A, Del-Toro-Sánchez CL, Rodríguez-Figueroa JC, Valdéz-Hurtado S, Wong-Corral FJ, Borboa-Flores J, González-Osuna MF, Perez-Perez LM, González-Vega RI (2019) Legume proteins as a promising source of anti-inflammatory peptides. Current Protein and Peptide Science 20(12):1204–1217

Reyes-Díaz A, González-Córdova AF, Hernández-Mendoza A, Reyes-Díaz R, Vallejo-Cordoba B (2018) Immunomodulation by hydrolysates and peptides derived from milk proteins. Int J Dairy Technol 71(1):1–9

Ripari V (2019) Techno-functional role of exopolysaccharides in cereal-based, yogurt-like beverages. Beverages 5(1):16

Rizzello CG, Tagliazucchi D, Babini E, Rutella GS, Saa DLT, Gianotti A (2016) Bioactive peptides from vegetable food matrices: research trends and novel biotechnologies for synthesis and recovery. J Funct Foods 27:549–569

Rook GA, Steele J, Fraher L, Barker S, Karmali R, O'riordan J, Stanford J (1986) Vitamin D3, gamma interferon, and control of proliferation of Mycobacterium tuberculosis by human monocytes. Immunology 57(1):159

Rothwell JA, Pérez-Jiménez J, Neveu V, Medina-Ramon A, M'Hiri N, Garcia Lobato P, Manach C, Knox K, Eisner R, Wishart D, Scalbert A (2013) Phenol-explorer 3.0: a major update of the phenol-explorer database to incorporate data on the effects of food processing on polyphenol content. Database 2013:bat070. https://doi.org/10.1093/database/bat070

Ruiter B, Patil SU, Shreffler WG (2015) Vitamins a and D have antagonistic effects on expression of effector cytokines and gut-homing integrin in human innate lymphoid cells. Clin Exp Allergy 45(7):1214–1225

Rytter MJH, Kolte L, Briend A, Friis H, Christensen VB (2014) The immune system in children with malnutrition–a systematic review. PLoS One 9(8):e105017

Santiago-López L, Hernández-Mendoza A, Garcia HS, Mata-Haro V, Vallejo-Cordoba B, González-Córdova AF (2015) The effects of consuming probiotic-fermented milk on the immune system: a review of scientific evidence. Int J Dairy Technol 68(2):153–165

Santiago-López L, Hernández-Mendoza A, Mata-Haro V, Vallejo-Cordoba B, González-Córdova AF (2018) Immune response induced by fermented milk with potential probiotic strains isolated from artisanal Cocido cheese. Food Agric Immunol 29(1):911–929
Santiago-López L, Hernández-Mendoza A, Vallejo-Cordoba B, Mata-Haro V, González-Córdova AF (2016) Food-derived immunomodulatory peptides. J Sci Food Agric 96(11):3631–3641
Schaafsma G, Slavin JL (2015) Significance of inulin fructans in the human diet. Compr Rev Food Sci Food Saf 14(1):37–47
Shadnoush M, Hosseini RS, Mehrabi Y, Delpisheh A, Alipoor E, Faghfoori Z, Mohammadpour N, Moghadam JZ (2013) Probiotic yogurt affects pro-and anti-inflammatory factors in patients with inflammatory bowel disease. Iranian journal of pharmaceutical research: IJPR 12(4):929
Shahidi F, Zhong Y (2008) Bioactive peptides. Journal of AOAC international 91(4):914–931
Shakoor H, Feehan J, Al Dhaheri AS, Ali HI, Platat C, Ismail LC, Apostolopoulos V, Stojanovska L (2020) Immune-boosting role of vitamins D, C, E, zinc, selenium and omega-3 fatty acids: could they help against COVID-19? Maturitas 143:1–9
Sheikhhossein F, Amini MR, Shahinfar H, Djafari F, Safabakhsh M, Shab-Bidar S (2020) Effects of chromium supplementation on inflammatory biomarkers: a systematic review and dose-response meta-analysis of randomized controlled trials. European Journal of Integrative Medicine 37:101147
Shih CH, Wang WH, Chen CM, Ko WC (2020) Hesperetin-5, 7, 3′-O-Trimethylether dually inhibits phosphodiesterase 3/4 and Methacholine-induced airway Hyperresponsiveness in sensitized and challenged mice. Drug Des Devel Ther 14:519
Shimizu T, Hettiarachchy N (2012) Food-derived bioactive peptides in the market. In: Food proteins and peptides-chemistry, functionality, interactions, and commercialization. CRC Press Boca Raton, FL, pp 375–392
Shokryazdan P, Jahromi MF, Navidshad B, Liang JB (2017) Effects of prebiotics on immune system and cytokine expression. Med Microbiol Immunol 206(1):1–9
Simon AK, Hollander GA, McMichael A (2015) Evolution of the immune system in humans from infancy to old age. Proc R Soc B 282:20143085
Singh MV, Chapleau MW, Harwani SC, Abboud FM (2014) The immune system and hypertension. Immunol Res 59(1–3):243–253
Skrobot A, Demkow U, Wachowska M (2018) Immunomodulatory role of vitamin D: a review. Advances in Experimental Medicine and Biology–Clinical and Experimental Biomedicine 2: 13–23
Spinas E, Saggini A, Kritas SK, Cerulli G, Caraffa A, Antinolfi P, Pantalone A, Frydas A, Tei M, Speziali A, Saggini R, Pandolfi F, Conti P (2015) Crosstalk between vitamin B and immunity. J Biol Regul Homeost Agents 29(2):283–288
Sun X, Acquah C, Aluko RE, Udenigwe CC (2020) Considering food matrix and gastrointestinal effects in enhancing bioactive peptide absorption and bioavailability. J Funct Foods 64:103680
Sun H, Jenssen H (2012) Milk derived peptides with immune stimulating antiviral properties. Milk Protein. Intech. Rijeka, Croatia, pp 45–82
Szczurko O, Shear N, Taddio A, Boon H (2011) Ginkgo biloba for the treatment of vitilgo vulgaris: an open label pilot clinical trial. BMC Complement Altern Med 11(1):21
Takele Y, Adem E, Getahun M, Tajebe F, Kiflie A, Hailu A, Raynes J, Mengesha B, Ayele TA, Shkedy Z, Lemma M, Diro E, Toulza F, Modolell M, Munder M, Müller I, Kropf P (2016) Malnutrition in healthy individuals results in increased mixed cytokine profiles, altered neutrophil subsets and function. PLoS One 11(8):e0157919
Tanaka M, Kishimoto Y, Sasaki M, Sato A, Kamiya T, Kondo K, Iida K (2018) Terminalia bellirica (Gaertn.) Roxb. Extract and gallic acid attenuate LPS-induced inflammation and oxidative stress via MAPK/NF-κB and Akt/AMPK/Nrf2 pathways. Oxidative Med Cell Longev 2018
Tang X, He Z, Dai Y, Xiong YL, Xie M, Chen J (2010) Peptide fractionation and free radical scavenging activity of zein hydrolysate. J Agric Food Chem 58(1):587–593
Trachootham D, Chupeerach C, Tuntipopipat S, Pathomyok L, Boonnak K, Praengam K et al (2017) Drinking fermented milk containing Lactobacillus paracasei 431 (IMULUS™) improves

immune response against H1N1 and cross-reactive H3N2 viruses after influenza vaccination: a pilot randomized triple-blinded placebo controlled trial. J Funct Foods 33:1–10

Triantos C, Kalafateli M, Aggeletopoulou I, Diamantopoulou G, Spantidea PI, Michalaki M, Vourli G, Konstantakis C, Assimakopoulos SF, Manolakopoulos S, Gogos C, Kyriazopoulou V, Mouzaki A, Thomopoulos K (2020) Vitamin D-related immunomodulation in patients with liver cirrhosis. Eur J Gastroenterol Hepatol 32(7):867–876

Tuo Y, Song X, Song Y, Liu W, Tang Y, Gao Y, Jiang S, Qian F, Mu G (2018) Screening probiotics from Lactobacillus strains according to their abilities to inhibit pathogen adhesion and induction of pro-inflammatory cytokine IL-8. J Dairy Sci 101(6):4822–4829

Urbanska AM, Paul A, Bhahena J, Prakash S (2010) Suppression of tumorigenesis: modulation of inflammatory cytokines by oral administration of microencapsulated probiotic yogurt formulation. Int J Inflamm 2010

Valcheva R, Dieleman LA (2016) Prebiotics: definition and protective mechanisms. Best Pract Res Clin Gastroenterol 30(1):27–37

van Gorkom GN, Lookermans EL, Van Elssen CH, Bos GM (2019) The effect of vitamin C (ascorbic acid) in the treatment of patients with cancer: a systematic review. Nutrients 11(5):977

Vogt LM, Meyer D, Pullens G, Faas MM, Venema K, Ramasamy U et al (2014) Toll-like receptor 2 activation by β2→ 1-fructans protects barrier function of T84 human intestinal epithelial cells in a chain length–dependent manner. J Nutr 144(7):1002–1008

Vogt L, Ramasamy U, Meyer D, Pullens G, Venema K, Faas MM et al (2013) Immune modulation by different types of β2→ 1-fructans is toll-like receptor dependent. PLoS One 8(7):e68367

Vugic L, Colson N, Nikbakht E, Gaiz A, Holland OJ, Kundur AR, Singh I (2020) Anthocyanin supplementation inhibits secretion of pro-inflammatory cytokines in overweight and obese individuals. J Funct Foods 64:103596

Wang X, Quinn PJ (1999) Vitamin E and its function in membranes. Prog Lipid Res 38(4):309–336

Wang Q, Wang P, Xiao Z (2018) Resistant starch prevents tumorigenesis of dimethylhydrazine-induced colon tumors via regulation of an ER stress-mediated mitochondrial apoptosis pathway. Int J Mol Med 41(4):1887–1898

Wang Y, Xie J, Wang N, Li Y, Sun X, Zhang Y, Zhang H (2013) Lactobacillus casei Zhang modulate cytokine and toll-like receptor expression and beneficially regulate poly I: C-induced immune responses in RAW264. 7 macrophages. Microbiol Immunol 57(1):54–62

Wang Y, Zhou Z, Diao Y, Strappe P, Blanchard C (2017) The potential role of p53 and MAPK pathways in the hepatotoxicity of deep-fried oil and in resistant starch-induced protection. Eur J Lipid Sci Technol 119(8):1600296

Weir EK, Thenappan T, Bhargava M, Chen Y (2020) Does vitamin D deficiency increase the severity of COVID-19? Clin Med 20(4):e107

Wintergerst ES, Maggini S, Hornig DH (2006) Immune-enhancing role of vitamin C and zinc and effect on clinical conditions. Ann Nutr Metab 50(2):85–94

Wu LH, Lin C, Lin HY, Liu YS, Wu CYJ, Tsai CF et al (2016) Naringenin suppresses neuroinflammatory responses through inducing suppressor of cytokine signaling 3 expression. Mol Neurobiol 53(2):1080–1091

Xia S, Li XP, Cheng L, Han MT, Zhang MM, Shao QX et al (2015) Fish oil–rich diet promotes hematopoiesis and alters hematopoietic niche. Endocrinology 156(8):2821–2830

Xiao J, Song J, Hodara V, Ford A, Wang XL, Shi Q, Chen L, VandeBerg JL (2013) Research article protective effects of resveratrol on TNF- -induced endothelial cytotoxicity in baboon femoral arterial endothelial cells. J Diabetes Res 2013:185172

Xing L, Liu R, Cao S, Zhang W, Guanghong Z (2019) Meat protein based bioactive peptides and their potential functional activity: a review. Int J Food Sci Technol 54(6):1956–1966

Xu Q, Hong H, Wu J, Yan X (2019) Bioavailability of bioactive peptides derived from food proteins across the intestinal epithelial membrane: a review. Trends Food Sci Technol 86:399–411

Xue H, Ren H, Zhang L, Sun X, Wang W, Zhang S et al (2016) Alpha-tocopherol ameliorates experimental autoimmune encephalomyelitis through the regulation of Th1 cells. Iran J Basic Med Sci 19(5):561

Yang H, Sun Y, Cai R, Chen Y, Gu B (2020) The impact of dietary fiber and probiotics in infectious diseases. Microb Pathog 140:103931

Yin DT, Zhao XH (2017) Impact of exogenous strains on in vitro fermentation and anti-colon cancer activities of maize resistant starch and xylo-oligosaccharides. Starch-Stärke 69(11–12): 1700064

Yoshimura M, Akiyama H, Kondo K, Sakata K, Matsuoka H, Amakura Y, Teshima R, Yoshida T (2013) Immunological effects of oenothein B, an ellagitannin dimer, on dendritic cells. Int J Mol Sci 14(1):46–56

Zeng J, Jiang J, Zhu W, Chu Y (2016) Heat-killed yogurt-containing lactic acid bacteria prevent cytokine-induced barrier disruption in human intestinal Caco-2 cells. Ann Microbiol 66(1): 171–178

Zeng R, Oderup C, Yuan R, Lee M, Habtezion A, Hadeiba H, Butcher EC (2013) Retinoic acid regulates the development of a gut-homing precursor for intestinal dendritic cells. Mucosal Immunol 6(4):847–856

Zhai Z, Wang J, Huang B, Yin S (2019) Low-fat yogurt alleviates the pro-inflammatory cytokine IL-1β-induced intestinal epithelial barrier dysfunction. J Dairy Sci 102(2):976–984

Zhao J, Ma J, Zhang Q, Tian J, Wang Y, Meng X (2020) Cyanidin-3-glucoside attenuates silica-induced pulmonary inflammatory responses by modulating T cell immune responses and STAT1/STAT3 signaling. J Funct Foods 68:103911

Zhao J, Zang J, Lin Y, Wang Y, Li D, Meng X (2019) Polyphenol-rich blue honeysuckle extract alleviates silica-induced lung fibrosis by modulating Th immune response and NRF2/HO-1 MAPK signaling. J Funct Foods 53:176–186

Zheng B, Wang T, Wang H, Chen L, Zhou Z (2020a) Studies on nutritional intervention of rice starch-oleic acid complex (resistant starch type V) in rats fed by high-fat diet. Carbohydr Polym 243:116637

Zheng J, Wittouck S, Salvetti E, Franz CM, Harris HM, Mattarelli P et al (2020b) A taxonomic note on the genus Lactobacillus: description of 23 novel genera, emended description of the genus Lactobacillus Beijerinck 1901, and union of Lactobacillaceae and Leuconostocaceae. Int J Syst Evol Microbiol 70(4):2782–2858

Zunino SJ, Peerson JM, Freytag TL, Breksa AP, Bonnel EL, Woodhouse LR, Storms DH (2014) Dietary grape powder increases IL-1β and IL-6 production by lipopolysaccharide-activated monocytes and reduces plasma concentrations of large LDL and large LDL-cholesterol particles in obese humans. Br J Nutr 112(3):369–380

Chapter 3
Immunoregulatory Bioactive Phytoconstituents: Recent Trends and Future Challenges

Sreeharsha Nagaraja, Vandana Gawande, Amita Joshi, and Swati Pund

Abstract Realizing the indispensable role of the immune system in maintaining the disease-free state, concept of immunomodulation is gaining increased public perception. Rising incidences of life-threatening infections need improvement in disease resistance ability of individuals by immune stimulation. In healthy state, immunostimulants act as prophylactic on both innate and adaptive immune responses and protect body against infections, allergy, as well as life-threatening diseases like cancer. On the contrary, immunosuppressants are useful in autoimmune inflammatory diseases and transplant recipients for prevention of graft rejection. Wide range of synthetic or recombinant immunomodulators currently in use, like levamisole, tacrolimus, and thalidomide, are associated with numerous side effects. Microbiological or recombinant synthetic cytokines too exhibit severe adverse reactions. Plant-derived nutraceuticals belonging to alkaloids, flavonoids, glycosides, sterols, terpenoids, etc. not only protect the plant but also target oxidative stress in the human body and show beneficial outcome in the management of chronic inflammation, diabetes, cancer, and cardiovascular diseases and in developing immunity. The primary focus of this chapter is to provide a deep insight into scientific data on immunomodulatory properties of plant-derived nutraceuticals, mechanism of action, and challenges and clinical restraints.

S. Nagaraja
Department of Pharmaceutical Sciences, College of Clinical Pharmacy, King Faisal University, Al-Ahsa, Saudi Arabia

Department of Pharmaceutics, Vidya Siri College of Pharmacy, Bengaluru, Karnataka, India

V. Gawande
Department of Pharmaceutical Chemistry, Sinhgad Institute of Pharmacy, Pune, Maharashtra, India

A. Joshi
Biobay, Ahmedabad, Gujarat, India

S. Pund (✉)
Biobay, Ahmedabad, Gujarat, India

Nanomedicine Laboratory, Department of Bioscience and Bioengineering, Indian Institute of Technology-Bombay, Mumbai, Maharashtra, India

R. K. Kesharwani et al. (eds.), *Nutraceuticals and Functional Foods as Immunomodulators*, https://doi.org/10.1007/978-981-19-2507-8_3

Keywords Immunomodulation · Phytochemicals · Resveratrol · Quercetin · Curcumin · Gingerol

3.1 Immunomodulators

The primary function of the immune system is to defend the body against infected pathogens and harmful antigens and carry out these functions through its two main branches, viz., nonspecific immunity and antigen-specific immunity. The immediate defense system of the body that protects the body from invading pathogens is known as the nonspecific immunity, while a more complex and antigen-specific or cell-mediated immunity that also results in an immunological memory due to a prior exposure to the antigen is known as antigen-specific immunity (Mohamed et al. 2017). The reaction of the body against any foreign body/pathogen is known as immune response. There are certain molecules that can alter the immune response, positively or negatively, and such molecules are known as immunomodulators (Nair et al. 2019). Depending on this effect, the molecules can be classified as stimulant, suppressor, or adjuvant. Accordingly, in the condition like an autoimmune disorder, wherein the body immune cells do not differentiate between self-cells and foreign cells and start eliminating their own body cells, an immune-suppressor molecule is required. In such a condition, an immune-suppressor molecule will help to restore or normalize the immune system. Even in conditions like organ transplant, immune suppressors are administered so as to avoid transplant rejection. While in conditions like AIDS wherein the immune system of the body is disengaged, an immune stimulator activates/boosts the immune system toward normality. Immuno-adjuvants are essentially used to augment the effect of a vaccine. An overview of immunomodulators and their functions is represented in Fig. 3.1.

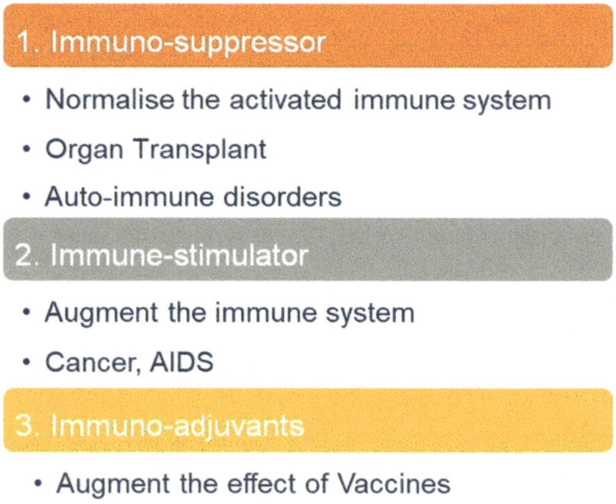

Fig. 3.1 Types of immunomodulators and their applications

All the three categories of the immunomodulators employ one or more of the activities, viz., apoptosis, protein synthesis, antigen presentation, targeting various transcription factors and immune mediators, etc., to exert their actions (Nair et al. 2019). Immunomodulator molecules either inhibit or proliferate the immune-responsive cells, viz., lymphocytes, macrophages, neutrophils, natural killer (NK) cells, and cytotoxic T lymphocytes. The immunomodulators can act by specific or nonspecific action (Bascones-Martinez et al. 2014). Immunomodulators that act on a specific immunity primarily act on the immune system of the cells depending on the attacking antigen or immunogen, with discriminatory specificity for immune response, while nonspecific immunomodulators stimulate or suppress the immune response, without activating any specific immune cell for a specific antigen. Nonspecific immunomodulators are further subdivided into three types (Bascones-Martinez et al. 2014):

- Type I, which acts on normal immune system
- Type II, which acts on immunosuppressed immune system
- Type III, which acts on functionally normal and immunosuppressed immune system

Cancer is one of the diseases wherein immunomodulators have been most commonly used with an aim to prompt the immune system to respond against tumor cells. The chief immune cells involved in body's defense against tumor cells are NK, dendritic cells, macrophages, polymorph nuclear cells, mast cells, and cytotoxic T cells. Dendritic cells are the main antigen-presenting cells (APC) and form a link between innate and adaptive immune systems. Dendritic cells capture and present the tumor-associated antigen to naïve T cells through major histocompatibility complex, classes I and II. This results in activation, cloning, and proliferation of T lymphocytes into CD4+ helper T cell and CD8+ cytotoxic T cells. Simultaneously, pro-inflammatory cytokines such as IFN-γ and IL-12 are released, thereby improving the tumor uptake by APCs and activation of Th1 response. Th1 response eventually activates the cytotoxic T lymphocytes (Wu et al. 2009a). Various immunomodulators currently used in clinical practice are tabulated in Table 3.1.

It can be observed from Table 3.1 that currently used immunomodulators have generalized effect on the immune system and several adverse effects. Therefore, safer, effective alternative having specific mechanism of immunomodulatory effect is need of the hour.

3.2 Immunomodulatory Phytochemicals

Currently, the pharmaceutical research and new drug development involve finding new molecular targets for specific diseases and focus mainly on single-molecule studies. However, the new drug development process is a time-consuming, slow, and expensive business. Chances of discovering a potent drug with high selectivity and

Table 3.1 Few commonly used synthetic immunomodulators, their mechanism of action, and associated side effects

Drug	Immunological effect	Adverse effects	References
Cyclosporine A	Inhibits gene transcription in T-cell lymphocytes for IL-2	Tremors, kidney damage, hypertension, hypertrichosis, hyperkalemia, and hypomagnesemia	Rosmarin et al. (2010) and Kim et al. (2018)
Tacrolimus	Inhibits T-cell proliferation by binding to FK506-binding protein	Angina pectoris, cardiac arrhythmias, hypertension, alopecia, pruritus, rash, new-onset diabetes mellitus after transplant, electrolyte imbalance, metabolic acidosis, weight gain, abdominal pain, nausea, vomiting, diarrhea, urinary tract infection, hepatotoxicity, arthralgia, muscle cramps, blurred vision, visual disturbance, otalgia, otitis media, tinnitus, nephrotoxicity	Randomised trial (1994) and Pham et al. (2011)
Sirolimus	Interferes with signals from growth factor receptors, such as the IL-2 receptor and blocks T-cell proliferation	Increased susceptibility to infection, graft loss, lymphoma, malignancy, hepatic artery thrombosis in liver transplant patients, bronchial anastomotic dehiscence in lung transplant patients, hypersensitivity reactions, exfoliative dermatitis, angioedema, hypertriglyceridemia, hypercholesterolemia, decline in renal function, proteinuria, interstitial lung disease	Jewett and Tseng (2017)
Azathioprine	Inhibits DNA and RNA synthesis by interfering purine synthesis and suppressing de novo purine synthesis Blocks lymphocyte proliferation and PL-2 production	Nausea, leukopenia, pancreatitis, risk of lymphoma, fever, fatigue arthralgias/myalgia, hepatotoxicity, nephrotoxicity, hypersensitivity	Morris (2014) and Jewett and Tseng (2017)
Cyclophosphamide	Alkylating agent and cross-links cellular macromolecules like DNA, RNA, and proteins	Hemorrhagic cystitis, amenorrhea, alopecia, myelosuppression, nausea, vomiting, cardiotoxicity, lung toxicity, hepatotoxicity, and secondary malignancies	Ogino and Tadi (2020) and Jewett and Tseng (2017)

(continued)

Table 3.1 (continued)

Drug	Immunological effect	Adverse effects	References
Mofetil mycophenolate (prodrug) Hydrolyses to mycophenolic acid	Mycophenolate inhibits inosine monophosphate dehydrogenase which eventually inhibits lymphocyte proliferation	Abdominal or stomach cramps or pain; black; tarry stools; bladder pain; bleeding gums; bloating or swelling of the face, arms, hands, lower legs, or feet; blood in the urine or stools; bloody or cloudy urine; burning; crawling; itching; numbness; prickling; "pins and needles"; or tingling feelings	Jewett and Tseng (2017) and Ruiz and Kirk (2015)
Leflunomide	Inhibits pyrimidine synthesis, block T-cell and B-cell proliferation	Diarrhea, nausea, headache, rash, stomach upset, abnormal liver tests	Haraoui (2015)
Muromonab	Binds and blocks the CD3 complex of the T-cell receptor	Increased risk of infection, reduction in the number of blood cells needed for clotting, chest pain, dizziness, fever and chills, shortness of breath, stomach upset, and trembling or shaking of the hands within a few hours after taking the first dose	Sgro (1995)

low toxicity are extremely low. Therefore, scientists are looking for new drug candidates from the well-established and proven alternative medicines like plants, algae, and mushrooms, having better patient tolerance and acceptance. Salicin, cocaine, codeine, digitoxin, quinine, vincristine, vinblastine, pilocarpine, paclitaxel, and artemisinin are few representative isolated phytoconstituents being successful today as drugs for the treatment of various diseases. The search for natural products of plant origin as new leads for development of potent and safe immunosuppressant and immunostimulant agents is gaining much major research interest. Wide varieties of plants either whole or parts of plants have shown immunomodulatory mechanism of action, thus having application as successful antioxidant, anti-inflammatory, hepatoprotective, cardiotonic, and several other medicinal benefits. Very popular examples of such plants are *Curcuma longa*, *Withania somnifera*, *Ocimum sanctum*, *Aloe vera*, *Andrographis paniculata*, *Asparagus racemosus*, *Boerhavia diffusa*, and *Echinacea angustifolia* (Kumar et al. 2012). Isolated phytoconstituents like glycosides, coumarins, polyphenolics, flavonoids, polysaccharides, and alkaloids have been reported to be responsible for the plant-immunomodulating properties (Kesharwani and Misra 2010; Mishra et al. 2019) (Fig. 3.2).

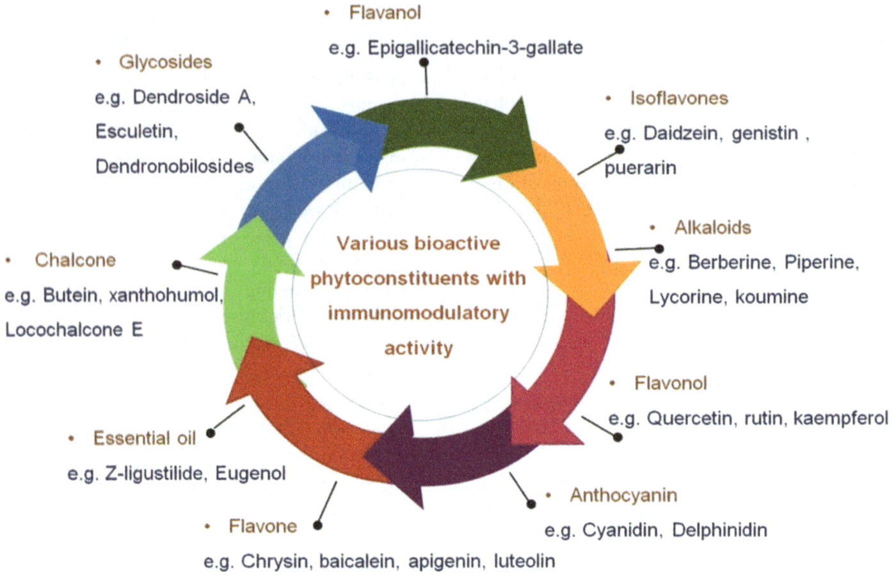

Fig. 3.2 Various classes of immunomodulatory phytoconstituents and examples thereof

The literature of the last decade suggests that marine algae metabolites, namely, lectins, laminarans, polyphenols, and sulfated polysaccharides like fucoidans and carrageenans, have shown biological activities and applications for the treatment of immunodeficiency diseases (Besednova et al. 2019). Lectins, carbohydrate-binding proteins without any enzymatic activity, have therapeutic potential. Cyanobacteria such as cyanovirin-N, microvirin, microcrystalline viridis-lectin scytovirin, *Oscillatoria agardhii* lectin, and griffithsin inhibit the infection of cells with HIV and also prevent the transmission of the pathogen from infected cells to uninfected CD4+ T lymphocytes (Singh and Walia 2018). Eckol, a phlorotannin from marine brown algae, exhibited in vivo antitumor effect in a sarcoma 180 (S180) xenograft-bearing animal model (Zhang et al. 2019a, b). Eckol showed antitumor activity by upregulation of Caspase-3 and Caspase-9 and downregulation of Bcl-2, Bax, EGFR, and p-EGFR. Microalgae biosynthesize a large number of diverse bioactive metabolites, and therefore fractions of microalgae or pure compounds isolated from microalgae have shown anticancer, anti-inflammatory, immunomodulatory properties (Martínez Andrade et al. 2018; Riccio and Lauritano 2019) and beneficial effects in neurodegenerative diseases as well (Bule et al. 2018). Compounds like sulfated polysaccharides, sulfolipids, polyunsaturated fatty acids, and astaxanthins from microalgae are potential immunomodulators. Extracts from *Alexandrium tamarense*, *Chaetoceros calcitrans*, *Chaetoceros socialis*, and *Thalassiosira weissflogii* have shown activation of IL-6 and human peripheral blood mononuclear cells (Cutignano et al. 2015). Food supplement of condensed water-soluble extract of commercially available spray-dried *Spirulina* sp. activates human innate immune system augmenting interferon production and NK cytotoxicity (Hirahashi et al. 2002). Sulfated polysaccharides from *Tribonema* sp. showed immunostimulation and anticancer activities on RAW264.7 macrophage cells and HepG2 cells and significant immunomodulatory activity by stimulating macrophage cells, upregulating IL-6, IL-10, and TNF-α (Chen et al. 2019). Manzo et al. (2017) synthesized a sulfoglycolipid using microalgae-derived sulfolipids as lead compounds and used it as vaccine adjuvant to trigger dendritic cell activation and improve immune response against cancer cell. The synthetic sulfoglycolipid stimulated the production of the pro-inflammatory cytokines IL-12 and INF-γ and increased the expression levels of IL-1α, IL-1β, IL-18, and IL-27.

Increasing body of scientific literature now suggests that, apart from plants, many phytochemicals from edible and nonedible mushrooms too have diverse biological actions (Xu et al. 2012). Mushrooms are considered as a healthy food as they contain high protein, low fat, and essential amino acids like lysine and leucine and essential micronutrient vitamins such as thiamine, riboflavin, niacin, biotin, ascorbic acid, and vitamin D (Cardwell et al. 2018; Liu et al. 2019; Sliva 2004). In addition, mushrooms also contain polysaccharides, polysaccharopeptides, phenolic compounds, and terpenoids. Polysaccharides, particularly β-D-glucans, are responsible for the immunomodulatory activity. In many cultures, mushrooms are used as a food to stimulate the immune system to cure diabetes and cancer. Mallard et al. (2019) studied immune modulation by a medicinal mixture of three different mushrooms, namely, reishi, shiitake, and maitake. The extracts of *Ganoderma lucidum* (reishi), *Lentinula edodes* (shiitake), and *Grifola frondosa* (maitake), alone and in combination, were evaluated for their effect on the expression of cytokines IL-1α, IL-6, IL-10, and TNF-α in human macrophages with and without lipopolysaccharide (LPS) stimulation. All the extracts were found to be highly potent immunostimulators as observed from very low effective concentrations (<100 μg/mL). Combined extract formula exhibited lowest effective concentration for TNF-α expression in LPS-stimulated macrophages compared to the individual extracts, suggesting a potential synergism. Therefore, a combination of these mushrooms is widely used in nutraceutical supplements (Keservani et al. 2016a, b, c, 2020).

Aqueous extracts from the vegetative submerged mycelia of these three mushrooms showed antitumor activity in vivo in rats with implanted kidney cancer (Vetchinkina et al. 2016; Keservani et al. 2016a, b, c, 2018). Polysaccharide fraction of *Ganoderma lucidum* has been extensively studied for identification of cellular immunostimulatory mechanism. It exerts antitumor effects from human neutrophils by inhibiting spontaneous and Fas-mediated apoptosis through activation of the phosphatidylinositol 3-kinase (PI3K)/Akt pathway, independent of the extracellular signal-regulated kinase pathway. It also stimulates phagocytosis and chemotaxis of neutrophils through the mitogen-activated protein kinase and protein kinase C pathways (Sliva 2004).

Considering the potential of immunoregulatory effect of plants and mushrooms, intellectual property rights are being issued to protect the new and unique key characteristics. The majority of patents and patent applications are of plant extract or extracts of combinations of plants. Being mixtures of discrete components, extracts usually display synergistic effects and multiple therapeutic actions than their isolated constituents. However, issues pertaining to quality, safety, and efficacy, viz., authentication, variability due to geographic location, seasonal changes, soil, weather, and microbial and other creature attacks, must be addressed. Advanced chromatographic fingerprinting by HPLC, HPTLC hyphenated with spectrometric detections like NMR, and MS in view of phytochemical profiling is of utmost importance to account for batch-to-batch variations. Lack of toxicity data is also of significant concern; assessing microbial content or mycotoxins would be of great help. Thus, rigorous quality control and extensive physicochemical and biological profiling can cater the need of developing potential herbal compositions into effective therapeutic agents (Gibbons 2003). Among Ayurvedic medicines, ashwagandha (Indian medicinal herb) holds the first US patent (US 2010/0285064 A1) for acting as a vaccine adjuvant. Its capability of invoking T-cell-dependent immune responses against weaker antigens such as bacterial polysaccharide has been studied. Process of preparing withanolide-rich fraction (withanolide A, withaferin A, withanolide B, withanoside IV, withanoside V, and 12-deoxy-withastramonolide of ashwagandha roots in high yields) has been patented by Jadhav et al. (2010). Adjuvant effect was studied on meningococcal A polysaccharide vaccine, and IgG estimation was carried out. Vaccine containing the adjuvant ashwagandha turned out to be more immunogenic and induced significant IgG response either if used immediately after combining with vaccine or when kept overnight with vaccine. However, higher adjuvant activity was seen when kept overnight. Additionally, significant coadjuvant effect was observed with alum, thereby suggesting its chance inclusion in adjuvant-containing formulations (Keservani et al. 2016a). Summary of few patents from the past two decades on phytoextracts as immunomodulator is presented in Table 3.2.

3.3 Isolated Phytochemicals as Immunoregulators

The possibility to enhance endogenous immune functions with the use of herbal bioactive constituents/herbal nutraceuticals is extensively investigated (Trung and An 2018). Several herbal nutraceuticals have been reported to modulate in immune functions; however, their potentials as immune boosters are not systematically investigated in well-controlled human clinical studies. Widely investigated isolated phytoconstituents having potent effects on innate and adaptive immunity in preclinical research, namely, resveratrol, epigallocatechol-3-gallate (EGCG), quercetin, curcumin, gingerol, and genistein, are discussed here.

3.3.1 Resveratrol

Resveratrol is trans-3,4′,5-trihydroxystilbene, a non-flavonoid polyphenolic phytoalexin derived from the skin of grapes and other fruits that demonstrate a broad range

Table 3.2 Phytoextracts as immunomodulator

References and patent number	Brief description of the invention	Patented herb/herbs and their parts or chemical constituents	Immunomodulatory effect claimed
Oosthuizen (2015) WO 2015/049643 Al	A crude or purified ethanolic extract and its pharmaceutical compositions having immunomodulatory and hepatoprotective activity for the treatment of upper respiratory bacterial infections and immunodeficiency disorders	*Euclea natalensis* preferably shoots	Switching of Th2 immune response to a Th1 immune response Increase in three out of four Th1 cytokines, decrease in one of the Th2 cytokines
Pylypchuk (2011) US 7.964.221 B2	Medicinal herb compositions for the treatment of infectious diseases Dzherelo (oral-immunomodulating agent produced from 26 plant materials)	Elecampane rhizome (*Inula* sp.), fennel fruit (*Foeniculum* sp.), juniper berry (*Juniperus* sp.), licorice root (*Glycyrrhiza* sp.), oregano herb (*Origanum* sp.), marigold flowers (*Calendula* sp.), rose hips (*Rosa* sp.), thyme (*Thymus* sp.), etc.	Increase in total and CD4 lymphocyte counts
Mitra et al. (2009) US 2009/0136602A1	Natural immunostimulant composition, methods for obtaining, preparing, and treating diseases related to immunodeficiency, e.g., cancer conditions, hepatitis B, HIV	*Symplocos racemosa* bark *Prosopis glandulosa* leaves	Cancer: surge in pro-inflammatory cytokine TNF-C-hastening apoptosis Increased macrophage activity, phagocytosis of the apoptosis cells HIV: increase in cell-mediated immunity by initiating the clonal expansion of the lymphocytes Hepatitis B: increase in cell-mediated response and humoral response and aid in scavenging of the affected cells
Banerjee et al. (2009) US 2009/0175964 A1	Edible composition for enhanced immunity	Theanine or a source of theanine Shankhpushpi, shatavari, or a mixture thereof	In vitro activation of macrophage in cell culture assays leading to higher phagocytosis index

(continued)

Table 3.2 (continued)

References and patent number	Brief description of the invention	Patented herb/herbs and their parts or chemical constituents	Immunomodulatory effect claimed
Rangel (2008) US 2008/ 0118582 A1	Phyto-nutraceutical composition for prevention and treatment of chronic degenerative diseases	*Energy-enhancing herbs* *Eleutherococcus senticosus*, *Panax ginseng*, *Panax quinquefolius*, *Pfaffia paniculata*, *Rhodiola rosea*, and *Schisandra chinensis* *Bio-intelligence modulators* *Andrographis paniculata*, *Astragalus membranaceus* root, *Coriolus versicolor*, *Echinacea* spp., *Ganoderma lucidum* mushroom, *Grifola frondosa* mushroom, *Hydrastis canadensis* root, *Lentinus edodes*, *Morinda citrifolia*, *Petiveria alliacea*, *Sutherlandia frutescens*, *Uncaria tomentosa*, *Vitex agnus* cactus *Organizational improver* Fulvic acid, *Hydrocotyle asiatica*, *Opuntia ficus indica*, shark cartilage	Bio-intelligence modulator plants regulate the neuroendocrine systems, immunological systems, and cellular processes *Eleutherococcus senticosus* stimulates production of T-lymphocyte and natural killer cells *Grifola frondosa* increases production of cytokines IL-1 and IL-2 *Hydrastis canadensis* increases production of immunoglobulins G and M and stimulates the phagocytic capacity of macrophages
Holt (2007) US 2007/ 0196381A1	Herbal compositions, methods of stimulating immunomodulation, and enhancement of immunomodulating agents using the herbal composition	Shiitake mushroom, *Aloe vera* leaf, brewer's yeast, *Coriolus* mushroom, active hexose correlate compound, *Andrographis Paniculata* *Eleutherococcus Senticosus*	Strong direct activation of NK cells by induction of CD69
Patwardhan and Kapadi (2004) US 2004/ 0033273A1	Medicinal fractions having immunostimulant and antitumor activity for reversing natural or drug-induced immunosuppression or myelosuppression and a process for their manufacturing	*Withania somnifera* root	Increase in white cell total counts and antibody titers

(continued)

Table 3.2 (continued)

References and patent number	Brief description of the invention	Patented herb/herbs and their parts or chemical constituents	Immunomodulatory effect claimed
De Souza et al. (2002) WO 2002/ 053166A1	Standardized extract compositions for the treatment of immunity-related disorders and method of treatment thereof	*Tinospora cordifolia*	Increase in phagocytosis by leukocytes Increase in WBC counts by induction of leukocytosis Induces release of granulocyte macrophage-colony-stimulating factor (GM-CSF) abating leukopenia
Lam (2003) WO 2003/ 068145A2	Dietary supplement stimulating the immune system without producing an allergic response	*Ganoderma lucidum*, *Dioscoreae oppositae*, *Chrysanthemum morifolium*, *Radix astragali*, *Folium isatidis*	Enhancement in production of interleukin-1 ß without causing an increase in the production of interleukin-4
Lu (2008) JP4980305B2	Herbal medicine extracts capable of inducing interferon production of immune cells and activating a Toll-like receptor and a method for producing the same	*Glycyrrhizae radix*, *Bupleuri radix*, *Scutellariae radix*, *Schisandrae fructus*, *Paeoniae radix rubra*	Significant interferon-α production induction in cell line (NK92) of T lymphocytes and NK cells Activation of various Toll-like receptors TLR2, TLR4, or TLR7
Lee-Huang et al. (2009) US 2009/ 0061031A1	Methods and pharmaceutical compositions for treating obesity or obesity-related disorders and for inhibiting the infectivity of HIV	Oleuropein, its analogue, derivatives, or metabolites	Modulation of adipogenesis, lipodystrophy, reduction of fat accumulation and weight gain. Prevention of HIV viral fusion/entry into a host cell and binding of the catalytic site of the HIV integrase
Wong et al. (2016) US 2016/ 0166661 A1	Carbohydrate-based immunogenic compositions containing fucose-enriched polysaccharides	*Ganoderma lucidum* (Reishi F3)	Antibodies generated specifically binds to tumor-associated antigens, stage-specific embryonic antigen, glycan antigen, α-L-fucose-specific lectin, *Ulex europaeus* agglutinin-I, trigger complement-dependent cytotoxicity in cancer cell

(continued)

Table 3.2 (continued)

References and patent number	Brief description of the invention	Patented herb/herbs and their parts or chemical constituents	Immunomodulatory effect claimed
Stamets (2018) US 9,931,316B2	Antiviral products useful in preventing the spread and proliferation of various viruses afflicting animals particularly viruses harming humans, pigs, birds, bats, and bees	Aqueous—ethanol extracts from the living mycelium of *Antrodia*, *Fomes*, *Fomitopsis*, *Ganoderma*, *Inonotus*, *Schizophyllum*, *Phellinus*, *Piptoporus*, *Trametes*, etc.	Activation of Th1 arm of lymphocyte activity Activation of the ER-mediated ROS pathway Induction of apoptosis

of biological activities. Resveratrol is effective anticancer, antioxidant, and anti-inflammatory (Pund et al. 2014; Yuan et al. 2012) having significant potential for the prevention or treatment of cardiovascular diseases, autoimmune disorders, age-related neurodegenerative diseases like Alzheimer's and Parkinson's disease, and other chronic disorders (Oliveira et al. 2017; Pund et al. 2020).

Resveratrol mediates its several biological effects through modulation of multiple signal inflammatory pathways which are generally dysregulated (Rieder et al. 2012). By suppressing inflammation, resveratrol alleviates inflammatory symptoms in cancer and several other autoimmune diseases such as colitis and rheumatoid arthritis (Pund et al. 2014; Gambini et al. 2015). A key mechanism of action of resveratrol is through inhibition of production of inflammatory factors through the activation of Sirt1 (Saqib et al. 2018). Sirt1(mammalian) is an important deacetylase member of the sirtuin family involved in a broad range of molecular events and physiological functions like control of gene expression, metabolism, cancer, embryonic development, and aging (Rahman and Islam 2011). Sirt1 maintains periphery T-cell tolerance and plays a major role in counteracting cellular stress and apoptosis (Cattelan et al. 2015). Sirt1 regulates two main pro-inflammatory pathways in the immune response, NF-κB and AP-1 pathways, and thus regulates the immune system (Manna et al. 2000). Sirt1 has a direct regulatory role in macrophages, the most abundant component of the innate immune system, and a main source of pro-inflammatory cytokines TNF-α, IL-6, and IL-1 secreted in response to infection and inflammation. The inhibitory effect of Sirt1 on the NF-κB pathway has an essential role in suppressing the pro-inflammatory phenotype of macrophages (Kong et al. 2012). Structural studies indicate that resveratrol binds to Sirt1, modulates the Sirt1 structure, and enhances binding activity to its substrates. RelA/p65 is an NF-κB member and principal substrate of Sirt1 that regulates leukocyte activation and inflammatory cytokine signaling (Malaguarnera 2019). Resveratrol activates Sirt1 and mediates RelA/p65 deacetylation, thus inhibiting TNF-α-induced NF-κB transcription and sensitizing cells to apoptosis (Signorelli and Ghidoni 2005). The antitumor effect of resveratrol was indirectly inhibiting its potential effect on nonspecific host immunomodulatory activity and direct inhibition of growth of implanted H22 cells tumor in Balb/c mice (Liu et al. 2003). Resveratrol by intraperitoneal route could insignificantly increase the host nonspecific immunological defense of mice with H22 tumor, by raising the level of serum IgG and TNF-α and the number of plaques forming cells.

Similar mechanisms were also observed in collagen-induced arthritis model in mice. Resveratrol inhibited T-cell activation by enhancing the expression and deacetylase activity of Sirt1, thereby decreasing the acetylation of c-Jun and blocking the translocation of c-Jun into the nucleus (Zou et al. 2013). Prophylactic or therapeutic administration of resveratrol attenuated clinical parameters and bone erosion in collagen-induced arthritis model in DBA1 mice model (Xuzhu et al. 2012). Researchers specifically assessed the immune mechanisms by which resveratrol prevented and suppressed ongoing arthritis and determined its role in the regulation of humoral immunity by measuring collagen-specific and total IgG levels in serum. Resveratrol significantly prevented the development of serum collagen-specific IgG2a and IgG1 without any influence on the levels of total IgG1 or IgG2a, suggesting that resveratrol selectively controls the collagen-specific B-cell response and is not a general B-cell-depleting/B-cell-suppressing factor (Xuzhu et al. 2012). Resveratrol prevented hyperplasia of synovial cells by disrupting mitochondrial membrane potentials, by decreasing B-cell lymphoma-extra-large expression, and by allowing cytochrome c from the mitochondria into the cytosol (Nakayama et al. 2012).Resveratrol was found to be very effective in suppressing the potent immune responses induced by inhalation of superantigen, staphylococcal enterotoxin B in C57BL/6 mice acute lung inflammation model. Upregulation of Sirt1, downregulation of NF-κB, and reduction in cytokine production were factors responsible for neutralization of toxicity of superantigen (Rieder et al. 2012). In case of inflammation in the colon induced with 2,4,6-trinitrobenzenesulfonic acid solution (TNBS) in BALB/C mice, resveratrol induced a shift from a pro-inflammatory to anti-inflammatory T-helper response in mesenteric lymph node. Resveratrol regulated several microRNAs that modulated inflammatory genes and factors. Decrease in the expression of several miRs (miR-31, Let7a, miR-132) that targets cytokines and transcription factors involved in anti-inflammatory T-cell responses (Foxp3 and TGF-β) was observed in resveratrol-treated animals (Alrafas et al. 2020). Earlier, Zhang et al. (2019a) confirmed the effectiveness of resveratrol treatment in dextran sulfate solution-induced colitis through upregulation of colonic tissue levels of Sirt1, along with downregulation of pro-inflammatory factors like autophagy-related 12, Beclin-1, and microtubule-associated protein light chain 3 II. Natural killer cells serve as an antitumor defense through their direct cytotoxic and indirect immunoregulatory capacities. Inactivation of natural killer cells causes immune escape of tumor cells leading to rapid progression of cancer and the poor efficacy of immunotherapy. Natural killer group 2 member D (NKG2D) is one of the most prominent activating receptors of natural killer cells, and resveratrol upregulates NKG2D ligands. Resveratrol promotes NKG2D recognition to induce the death of target cells by natural killer cells (Luis Espinoza et al. 2013). Resveratrol also increases the expression of major histocompatibility complex class I chain-related proteins A and B in cancer cell lines and xenograft models and thus increases cytolysis of breast cancer cells by natural killer cells (Pan et al. 2017).

Cardioprotective effect of resveratrol is also through the activation of Sirt1. It inhibits hypoxia-induced apoptosis via the Sirt1-FoxO1 pathway in H9c2 cells proving its potential in preventing cardiovascular disease, especially in coronary artery disease patients (Chen et al. 2009). Resveratrol reduces cardiomyocyte apoptosis by Sirt1 activity and NAD (+) level by an AMPK-dependent mechanism and attenuates caspase-3 cleavage via heat-shock factor 1 deacetylation and heat-shock protein expression upregulation (Cattelan et al. 2015). Peroxide-induced apoptosis in cardiomyocytes is arrested by resveratrol by activation of Sirt1, Sirt3, Sirt4, and Sirt7 (Yu et al. 2009). Activation of Sirt1 by resveratrol attenuates TNF-α-induced vascular adventitial fibroblast apoptosis, cleaves caspase-3 protein expression, and increases the Bcl-2/Bax ratio. The inhibitory effect of Sirt1 on vascular adventitial fibroblast apoptosis is partly mediated by the deacetylation of FoxO1 (Wang et al. 2013).

Resveratrol has shown to be effective for the treatment of organ-specific autoimmune diseases wherein a single organ is involved like type 1 diabetes mellitus, inflammatory bowel diseases (Al Azzaz et al. 2019; Nunes et al. 2018; Shi et al. 2017), and psoriasis (Lai et al. 2017; Oliveira et al. 2017). Beta cells in the islets of Langerhans are destroyed by the inflammatory environment resulting into insulin deficiency and hyperglycemia. Differentiation of T-helper 1-activating B lymphocytes produces autoantibodies against beta cells. Th1 will also activate macrophage and neutrophil migrations to the islet that will promote beta cell destruction by increasing ROS (Wallberg and Cooke 2013). Resveratrol inhibits Th1 cell migration by binding to chemokine receptor 6 and forming a complex with insulin for increased glucose intake (Lee et al. 2011). Leukocyte infiltrates in psoriatic lesions consist of dendritic cells, macrophages, neutrophils, and T cells. Dendritic cells synthesize various pro-inflammatory cytokines, like TNF-α, IL-1β, IL-6, and IL-23, that further promote IL-23 production in dendritic cells and IL-17 secretion from Th17 cells which are inhibited by resveratrol (Kjær et al. 2015; Lai et al. 2017). On similar lines, the adaptive immune system is generally considered as the primary contributor to IBD pathogenesis through increased pro-inflammatory cytokines driven by the T-helper cells or ineffective anti-inflammatory activity of regulatory T cells (Wallace et al. 2014). The innate immune response is mainly triggered by activation of the (NF-κB) pathway and macrophages, stimulating the secretion of pro-inflammatory cytokines such as IFN-γ, IL-1β, IL-6, IL-8, and TNF-α (Dou et al. 2014). Resveratrol has promising effects in bowel inflammation by modulation of inflammatory signaling pathways, inflammatory biomarkers, and intestinal microflora, thus improving clinical outcomes (Altamemi et al. 2014; Rahal et al. 2012; Shi et al. 2017; Yao et al. 2015; Yulug et al. 2015).

3.3.2 Epigallocatechin Gallate

The most popular global beverage, tea, is made from the dried leaves of the plant *Camellia sinensis*. Green tea encompasses 20% of the global tea consumption. It is usually prepared by steaming or panfrying the leaves; subsequently, these leaves are dried to inactivate enzymes to preserve the tea constituents from the oxidation. This process helps to preserve the distinctive polyphenols of the green tea, viz., catechins. Almost 250–375-mg catechins are present in a typical cup of brewed green tea. Majorly, these catechins include epicatechin (EC), epicatechin-3-gallate (ECG), epigallocatechin (EGC), and epigallocatechin-3-gallate (EGCG). EGCG is the most abundant and most biologically active catechin found in green tea. Peak plasma levels of EGCG reach a maximum concentration of 1 μM after 1.5–2 h of green tea consumption and have an elimination half-life of 5 h in humans. High doses of EGCG (525 mg) supplements on oral administration resulted in a peak plasma concentration of 4.4 μM, while 1600 mg of EGCG resulted in a peak concentration of 7.4 and 9 μM on administration of 2000-mg EGCG. EGCG is well distributed throughout the body, viz., the digestive tract, blood, brain, liver, kidney, and spleen, among others, and ~80% of EGCG occurs in free form in plasma. Interestingly, other catechins are found in highly conjugated forms like conjugation with glucuronic acid and/or sulfate groups. Tea catechins undergo intestinal and hepatic metabolism and undergo glucuronidation, salvation, methylation, and ring fission by the intestinal microflora. Excretion of EGCG occurs through the bile into the feces, while EGC and EC are eliminated through both bile and urine (Pae and Wu 2013). Japanese Zen monk Eisai published a book in 1211, entitled *Kissa Youjouki* (meaning "promotion of health by tea") and described the benefits of tea as "tea is a marvelous preventive medicine to maintain people's health and has an extraordinary power to prolong life." Undeniably, recent scientific evidence supports the benefits of green tea for health promotion (Hayakawa et al. 2016). Anti-inflammatory properties along with the ability to modulate immunity enables catechins to exert protective effects against various diseases such as cancer, obesity, diabetes, arteriosclerosis, neurodegenerative diseases, tooth decay, hepatitis, allergy, and bacterial and viral infections. EGCG has been found to alter immune function of the body predominantly T-cell-related activities such as T-cell activation, proliferation, and differentiation into different subsets of effector cells, both innate and adaptive (Pae and Wu 2013). In an in vitro test, EGCG supplementation reduced the neutrophil migration induced by chemokine IL-8 in a dose-dependent manner with an IC50 of >1 mM (Dona et al. 2003) and that of cytokine-induced neutrophil chemoattractant-1 at a dose of 15 mg ml^{-1} or higher (Takano et al. 2004). Also, suppression of neutrophil recruitment independent of chemokine level at the inflammation site in an ovalbumin-induced rat allergic inflammation model was observed on oral administration of EGCG (1 mg/rat). Oral gavage of 2% green tea was found to inhibit proteolytic enzymes such as neutrophil elastase both in vitro (Sartor et al. 2002) and in vivo (Chan et al. 2012) indicative of an anti-proteolytic activity of EGCG on neutrophils. All these effects of EGCG on the neutrophils suggest its protective effect in preventing inflammation and inflammation-induced tissue damage (Pae and Wu 2013). EGCG also prevents chemokine monocyte chemotactic protein-1 (MCP-1)-induced monocyte migration and its adhesion to fibronectin in the human monocyte cell line, THP-1, by obstructing β1 integrin activation. Monocyte chemotaxis is facilitated by two key mediators, viz., MCP-1 secretion and its receptor (CCR2) expression. FCG has been found to reduce the MCP-1 secretion (Melgarejo et al. 2009). ECGC has been found to modulate a number of inflammatory cytokines and cells involved in their production. EGCG has been reported to inhibit TNF-α, IL-1β, and IL-6 production in various cell lines in vitro, viz., human monocyte cell lines (U937,31 THP-1) (Singh et al. 2005; Choi et al. 2009; Wu et al. 2009a, b), murine macrophage cell lines (J774.1 (Ichikawa et al. 2004), RAW 264.7 (Yang et al. 1998), and primary mouse macrophages (Choi et al. 2009)). Further, EGCG has also been shown to inhibit pro-inflammatory gene expression in LPS-stimulated bone marrow-derived macrophages including IL-6, IL-12p40, MCP-1, ICAM-1, and VCAM-1 (Joo et al. 2012). Suppression of NF-κB by EGCG has been reported to be the underlying cause for all these effects (Joo et al. 2012). Another important immune cells that are involved in the presentation of antigens are dendritic cells. Dendritic cells are the cells bridging the innate and adaptive immunity. EGCG retards dendritic cell maturation which ultimately leads to impaired functionality as antigen-presenting cells (APC) to induce antigen-specific T-cell-mediated response (Ahn et al. 2004). Also, EGCG increases IL-10 production in mature dendritic cells. IL-10 is an inhibitor of dendritic cell differentiation and function. These effects of EGCG on dendritic cells suggest the protective effect of EGCG in autoimmune diseases. Mast cells are another cell involved in the innate immunity which is activated on exposure to allergen like bacteria, etc. These cells further produce cytokines to aid the adaptive immunity. It has been reported that EGCG inhibits the degranulation that results in the release of histamine upon stimulation by an IgE-antigen complex (Matsuo et al. 1997). Dietary supplementation of EGCG has been found to suppress the T-cell proliferation. The ability of T cells to proliferate reflects the immunocompetence of an individual. It has been shown that Con A-induced splenocyte T-cell proliferation is inhibited by EGCG at a dose of 25 mg ml^{-1} (55 mM) and higher via T-cell mitogen concanavalin A (Wilasrusmee et al. 2002). EGCG has been reported to greatly diminish T-cell proliferation by affecting both T cells and APC (Pae and Wu 2013). Survival, expansion, and differentiation of T cells are regulated via IL-15, IL-7, and IL-2. EGCG induces the downstream signaling of these receptors and thereby inhibits their expression. This eventually results in suppression of T-cell proliferation and its effector function (Shim et al. 2008). Further, EGCG has been reported to inhibit Th1, Th9, and Th17 differentiation while promoting the Treg development. This affects the differentiation of CD4+ T cells, thus further strengthening the fact that EGCG can be effectively used in autoimmune diseases (Pae and Wu 2013). Drinking green tea phenols protects against UV-induced immunosuppression by suppressing local and systemic contact hypersensitivity. Further, UV-induced DNA damage repairing was hastened by green tea phenols along with high levels of nucleotide excision repair genes in mice (Khan et al. 2013).

All these studies clearly suggest that ECGC modulates the immune function, both innate and adaptive immunity and inflammatory responses by altering various immune cell functions. Essentially, ECGC is an anti-inflammatory and T-cell proliferation inhibitor agent, thereby modulating immune functionality. Although most of the studies have been carried out in murine models, these findings need to be validated in human volunteers.

3.3.3 Gingerol

Ginger (*Zingiber officinale* Rosc.) belonging to the family Zingiberaceae is a native of Southeast Asia. It is widely used in Indian food as a spice and condiment. Apart from being used in food, the rhizome of ginger has therapeutic utility which is attributed to its rich phytochemistry. Ginger contains more than 60 compounds in that can be categorized into two categories, i.e., volatile and nonvolatile constituents. Volatile constituent includes sesquiterpene and monoterpenoid hydrocarbons that give distinct aroma and taste to ginger, while nonvolatile pungent constituents are gingerols, shogaols, paradols, and zingerone. Zingerones are converted to gingerols or shogaols by thermal degradation (Butt and Sultan 2011). Gingerols are of major importance and are homologous series of phenols that differ in their unbranched alkyl chain length. On drying ginger, gingerols are converted to shogaols. Of the different kinds of gingerols, 6- and 8-gingerol are of prime importance. Majority of the medicinal properties of ginger have been ascribed to 6-gingerol. 6-gingerol has antioxidant properties and suppresses the production of pro-inflammatory cytokines, namely, TNF-α, IL-1, and IL-12 from macrophages (Williams et al. 2007). Treatment with 6-gingerol in tumor-bearing mice has shown a massive increase in the infiltration of tumor-reactive lymphocytes (Ju et al. 2012). In various tumor cell lines, viz., OSCC and cervical HeLa, 6-gingerol has shown antiproliferative effect by induction (Kapoor et al. 2016). Further, 6-gingerol has also exhibited inhibitory effect on COX-2 expression by downregulation of p38 MAPK and NF-κB in vivo. Additionally, an antimetastasis effect on lung B16F10 melanoma in an animal model was shown by 6-gingerol (Kim et al. 2005). 8-gingerol has been found to be capable of simultaneously inhibiting the humoral as well as cellular immune responses in mice through B- and T-cell activation suppression (Lu et al. 2011). However, detailed studies are needed to reinforce the utility of gingerols as immunosuppressive agents.

3.3.4 Quercetin

Quercetin (3,3′,4′,5,7-pentahydroxyflavone) is a polyphenolic secondary plant metabolite commonly found in fruits and vegetables. It is the most abundant flavonoid and largely a part of diet. Onions, broccoli, apples, berry crops, and grapes, few herbs, tea, and wine are the major dietary sources (Mlcek et al. 2016). It is known to prevent cancer and cardiovascular and neurodegenerative disorders (Ay et al. 2016). Apart from being immunomodulator, it displays wide array of activities, viz., antioxidant, antidiabetic, anticancer, anti-inflammatory, antiplatelet, antiapoptotic, nephoprotective, gastroprotective, angioprotective, cardioprotective, and chondroprotective properties (Kawabata et al. 2015; Shebeko et al. 2018).

Major cellular signaling pathways regulated by quercetin to exert various therapeutic roles are apoptosis, cell cycle arrest, NF-κB pathway, MAPK pathway, PI3K-

AKT pathway, and mTOR (Sharma et al. 2018). It lowers Th1-lymphocyte differentiation, thereby reducing levels of cytokines and factors like IL-1β, IL-4, IL-6, IL-10, IL-12, IL-25, IL-33, MCP-1, NF-κB, VEGF-A, COX-2, 5-LOX, iNOS, NO, CRP, TNF-α, TNFγ, and IgE along with lowering expression of VCAM-1, ICAM-1, and MIP-2 (Burmańczuk et al. 2018).

Transcription factors of nuclear factor-kappa B family are mainly involved in the regulation of adaptive immune response, inflammation, lymphopoiesis, and osteogenesis. Quercetin helped improve conspicuously TNF-α, TRAF-2, TNFRSF1B, NF-κBp65, and IFN-γ while lowering expression when fed to healthy Arbor Acre broilers as a part of diet (Yang et al. 2020). During high-throughput screening of immunoregulatory activity of natural products in human umbilical cord blood mononuclear cells, quercetin displayed immunostimulatory effect. It reduced the CD3 expression of T cells and induced CD34 expression of stem cells (Chen et al. 2006).

Cellular immunity is promoted by cytokines IL-2, IFN gamma, and IL-12 derived by Th1, while humoral immunity is promoted by IL-4, IL-5, and IL-6 derived by Th-2. Quercetin interacts with both these pathways to exhibit antiviral and antitumor activities. Quercetin upregulates IFN gamma and downregulates IL-4 and acts as immunostimulant (Nair et al. 2002). Recently, quercetin was seen to significantly increase type I interferon-regulated genes and type I and II interferon and significantly decrease pro- and anti-inflammatory cytokine expressions in macrophages derived out of highly pathogenic porcine reproductive and respiratory syndrome virus infection (Ruansit and Charerntantanakul 2020).

Apart from being immunostimulant, it possesses strong adjuvant potential as evident by increased ovalbumin-specific serum IgG1 antibody titers indicative of boosted humoral immune response in ovalbumin immunized Balb/c mice. It enhanced infiltration of CD11c+ dendritic cells in the mouse peritoneum, and elevated levels of LPS activated IL-1b and nitric oxide by peritoneal macrophages. It also increased expression of Tbx21, GATA-3, and Oct-2 proteins in splenocytes. Unlike other plant-derived molecules, it does not show hemolysis in human RBCs dictating safety for in vivo administration (Singh et al. 2017). Quercetin exhibits leishmanicidal activity, and being potent, it is the most attractive lead among flavonoids. It induces reactive oxygen species generation in *L. amazonensis*-infected macrophages (Naman et al. 2018), upregulates inhibition of TNF-α in phagocyte migration, and suppresses *L. major*-induced apoptosis delay in vivo. It exerts weak control on NF-κB level and induces cell death by apoptosis and necrosis (Belkhelfa-Slimani and Djerdjouri 2017). Besides upregulation of Nrf2/HO-1 expression, it depleted labile iron store-infected macrophages hampering *L. braziliensis* replication (Cataneo et al. 2019).

Quercetin, however, had no effect on antibody production and did not alter white blood cell counts in the lambs as a consequence of road transport but protected the ruminal mucosa against the development of parakeratosis (Benavides et al. 2013). Quercetin along with melatonin is more effective in protecting lungs against hypoxic damage via downregulation of immuno-inflammatory mediators (TNF-a, IL-6, and CRP) and heat-shock protein expression vascular endothelial growth factor (VEGF,

HSP70) (Al-Rasheed et al. 2017). Suppression of c-met and VEGF by liposomal quercetin in the Eca109/9706 cells has been confirmed by using immunohistochemistry assay and Western blotting (Pei et al. 2009). Quercetin is known to stabilize hypoxia-inducible factor (HIF)-1α under normal oxygen pressure because of its ability to bind iron intracellularly leading to decreased iron stores in the cells and also downregulate HIF-1 transcriptional activity by impairing the phosphorylation and nuclear accumulation (Triantafyllou et al. 2008).

Quercetin reduced expression of all inflammation-related cytokines and transcription factors (IL-1b, IL-6, IL-8, IL-10 and TNF-a, NF-κB) of LPS-stimulated THP-1 macrophages, other than COX-2 (Chanput et al. 2010). Concentration of milk leukocytes, also known as somatic cell count (SCC), is a key indicator of mastitis in dairy animals (Rainard et al. 2018). In cows with clinical mastitis, quercetin was found to significantly reduce somatic cell count after treatment for a week. However, due to intramammary administration of quercetin, no general effect on the immune system was seen; particularly no effect was seen on TNF-α (Burmańczuk et al. 2018). Recently, the role of quercetin in treating allergic inflammation was demonstrated as it inhibited ovalbumin-induced allergic conjunctivitis in 16 C57BL/6 mice. Quercetin was found to inhibit DNP-HSA/IgE induced Ca2+ 3 influx, MC degranulation, 4 and chemokine release in LAD2 cells. Also, DNP-HSA/IgE induced Lyn/PLCγ/IP3R-Ca2+ 5 and Lyn/ERK1/2; Lyn/NF-κB activation was inhibited (Ding et al. 2020). Anti-allergic action of quercetin is attributed to inhibition of enzymes, inflammatory mediators, and human mast cell activation via inhibition of Ca2+ influx, histamine, leukotrienes, and prostaglandin release (Mlcek et al. 2016). Quercetin was found effective in allergic rhinitis, elevating the production of an endogenous antioxidant protein and immunomodulator "thioredoxin" and suppressing oxidative stress responses in nasal mucosa (Edo et al. 2018).

Potential beneficial allo-herbal combination effect was observed when quercetin was tested in combination with sitagliptin for streptozotocin-induced diabetes mellitus in rats. The duo-regulated serum C-peptide, malondialdehyde, significantly raised superoxide dismutase and glutathione and decreased nuclear factor kappa B more than standalone treatment (Eitah et al. 2019). Quercetin-3-*O*-β-D-glucuronide is a major metabolite of quercetin in the human body that might be responsible for beneficial effects of quercetin (Kawabata et al. 2015). Isoquercetin (quercetin-3-glucoside) which is a glycoside derivative of quercetin was found to reduce ultraviolet-A radiation-induced damage in living skin equivalent cultures via reduction in 8-hydroxy-2′-deoxyguanosine, TdT-mediated dUTP nick-end labeling, and IL-1 α that too are more effective than hesperetin (Dekker et al. 2005). Another glycosylated quercetin, i.e., quercetin 3-*O*-xyloside, was found more effective than quercetin in activating redox-dependent ASK1/MAPK/NF-κB signaling pathway in murine macrophage cell line RAW 264.7 resulting in secretion of TNF-α and IL-6 early innate immunity hinting its immunostimulator potential (Lee et al. 2016). A new quercetin glycoside (3′-methoxy-3-*O*-(4″-acetylrhamnoside)-7-*O*-α-rhamnoside) isolated from *Cleome droserifolia* showed significant antitumor activity in triple-negative breast cancer. It showed a dose-dependent decrease in MDA-MB-231 proliferation and viability through stimulation of TP53, repression of its downstream miR155, and sevenfold and twofold upregulation of MHC class I polypeptide sequence A and UL-16 binding protein 2, respectively (Abdel-Latif et al. 2019).

3.3.5 Curcumin

A potent antioxidant and anti-inflammatory polyphenol curcumin is most extensively studied for immunomodulatory activities. Curcumin is one of the constituents of curcuminoids, a fraction isolated from rhizomes of *Curcuma longa*. The other two compounds in curcuminoids are demethoxycurcumin and bisdemethoxycurcumin, and all have potential therapeutic activities (Upadhyaya et al. 2009a, b; Bose et al. 2015; Kesharwani et al. 2015). As an anticancer agent, curcumin recognizes and elicits specific immune response to eradicate neoplastic cells from the host body. It regulates several molecules in cell signal transduction and downregulates several invasions, cell adhesion, and extracellular matrix molecules which are essential for sustaining metastasis (Shehzad and Lee 2013). Various mechanistic pathways are proven for its immunomodulatory effect of curcumin. Curcumin activates proliferator-activated receptor-γ (PPAR-γ) and regulates various biochemical pathways in controlling the inflammatory responses (Fig. 3.3) (Mazidi et al. 2016; Upadhyaya et al. 2009a, b; Singh et al. 2015a, b, c; Kesharwani et al. 2018).

PPAR proteins function within the cell nuclei; control metabolism, development, and homeostasis; and play an important role in cell growth, differentiation, insulin sensitivity, inflammatory response, reproduction, and adipogenesis (Wu et al. 2009b). Inflammatory pathways of curcumin suggest that the effect is achieved mostly through the downregulation of NF-κB (Shishodia 2013). Narala et al. (2009) suggested that PPAR-γ-mediated effects of curcumin are indirect and mediated through endogenous ligands. By inhibiting PPAR-γ-mediated epithelial-mesenchymal transition, curcumin prevents intestinal fibrotic stricture which is a major complication of Crohn's disease (Xu et al. 2017) and also activates hepatic stellate cells to prevent fibrosis of the liver (Mazidi et al. 2016). Curcumin inhibited epithelial-mesenchymal transition, reduced expression of α-SMA and PAI-1, and increased E-cadherin in n renal tubular epithelial cells, thus preventing

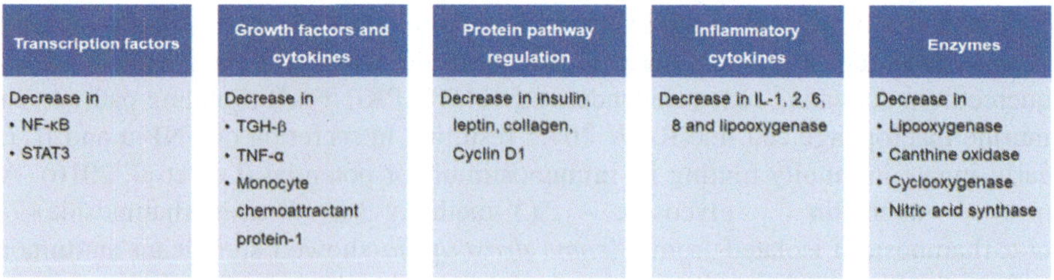

Fig. 3.3 Mechanism of immunomodulatory action of curcumin through PPAR-γ

tubulointerstitial fibrosis in renal disease (Li et al. 2013). In case of progressive and irreversible pulmonary fibrosis, a distinctive form of chronic interstitial pneumonia, curcumin triggered the expression of nuclear peroxisome PPAR-γ, upregulated cathepsin B and cathepsin L, and downregulated cystatin C (Saidi et al. 2019).

Recently, the role of curcumin on induction, expression, and production of anti-inflammatory cytokine IL-10 was studied (Mollazadeh et al. 2019). IL-10 plays an important role in the prevention of infection-related tissue damage and curcumin-alleviated coccidioidal infection in naturally infected lambs by downregulating pro-inflammatory cytokines, i.e., IFN-γ, and increasing IL-10 (Cervantes-Valencia et al. 2016).

By reducing the expression of iNOS, IFN-γ, and TNF-α, curcumin showed its effect on adaptive immunity in *Leishmania* infection model (Adapala and Chan 2008). It activated PPAR-γ and deactivated type 1 response.

Curcumin inhibits JAK/STAT signaling by reducing IL-10 levels and increases tumor apoptosis (Shiri et al. 2015). The anticancer activity of curcumin is by blocking the tumor-induced T-cell depletion by increasing the production of IFN-γ and IL-2. Several mechanisms, responsible for the potential antitumor activity of curcumin, proposed were restoration of activated/effector CD+4 and CD+8 T cells, secretion of IFN-γ, augmentation of tumor-infiltrating lymphocytes, and upregulation of IFN-γ mRNA expression (Bhattacharyya et al. 2010).

In a hamster model of accelerated atherosclerosis, curcuma oil, i.e., lipophilic fraction from turmeric, attenuated inflammation and arterial injury (Singh et al. 2015a, b, c). It also increased IL-10 and TGF-β mRNA expression in atherosclerotic arteries and decreased mRNA expression of pro-inflammatory cytokines, namely, TNF-α, IL-6, IL-1β, and IFN-γ. Curcuma oil-treated animals showed increased IL-10 expression in peritoneal macrophages. Curcumin also regulates the expression of genes implicated in tumor cell proliferation, metastasis, and angiogenesis. Physiological and pathological angiogenesis is governed by endothelial progenitor cells. Therefore, Vyas et al. (2015) studied the effect of curcumin on the growth properties of circulating endothelial progenitor cells and showed that curcumin inhibits cell growth by inhibiting the G1 to S phase transition in the cell cycle.

Low serum IL-10 and high serum TNF-α and IL-12 concentrations observed in intestinal colitis are very well normalized by curcumin, thus reducing the disease severity (Sreedhar et al. 2016). In small intestine inflammation, mice model of hyper-acute Th1-type ileitis following peroral infection with *Toxoplasma gondii*, curcumin, as well resveratrol downregulated Th1-type immune responses (Bereswill et al. 2010). Significant increase in anti-inflammatory IL-10 in the ileum, mesenteric lymph nodes, and spleen was observed, whereas pro-inflammatory cytokine expressions, namely, IL-23p19, IFN-γ, TNF-α, IL-6, and MCP-1, were significantly lowered in the ileum of treated animals. Curcumin, 50 mg/kg daily, by oral route decreased serum lactic acid dehydrogenase, IL-1β, and TNF-α levels, as well as colonic myeloperoxidase and lipid peroxide levels, and increased colonic prostaglandin E2 and IL-10 in acetic acid colitis model (Gopu et al. 2015). Lelli et al. (2017) reviewed the application of curcumin in pulmonary diseases showing abnormal inflammatory responses, like asthma or chronic obstructive pulmonary disease, acute respiratory distress syndrome, pulmonary fibrosis, and acute lung injury. Curcumin has pleiotropic effects, regulating transcription factor NF-κB, cytokine IL-6, TNF-alpha, adhesion molecule ICAM-1, and matrix metalloproteases, and thus is beneficial in inflammatory lung diseases and cancer (Singh et al. 2015a).

Non-neuronal cells such as glial cells play a critical role in the pathogenesis of chronic pain, and curcumin reduces chronic pain by blocking neuroinflammation (Chen et al. 2015). A new application of curcumin, i.e., intrathecal injection for the treatment of arthritic pain, was proven in complete Freund's adjuvant injection-induced arthritis model. Curcumin treatment efficiently reduced IL-1β, MCP-1, and monocyte inflammatory protein-1α in the spinal cord and showed similar effect in LPS-stimulated cultured astrocytes and microglia. In hind-paw incision model with C57BL/6 mice, curcumin, 50 mg/kg, significantly reduced the intensity of mechanical and heat sensitization (Sahbaie et al. 2014). In this experiment, the effectiveness of curcumin in reducing the pain and inflammation was due to the increase of TGF-β.

In progressive neurodegenerative inflammatory disorders like Parkinson's disease and Alzheimer's disease, curcumin exhibited neuroprotective actions. Wang et al. (2017) carried out systematic literature review of experiments performed on neuroprotective actions of curcumin in toxin-base animal models of Parkinson's disease (Singh et al. 2014). Curcumin showed protection of substantia nigra neurons and improvement in striatal dopamine levels through anti-inflammatory and antioxidant actions. The ability of curcumin to upregulate IL-10, a potent anti-inflammatory cytokine, limits the inducible nitric oxide synthase gene expression, thus reducing the production of neurotoxic nitric oxide. Even in chronic inflammatory autoimmune disease like systemic lupus erythematosus, curcumin showed protective effect by inhibiting maturation of immature dendritic cells, differentiation of naive CD4+ T cells into Th17 and Th1 cells via inhibition of IL-6 and IL-12 production, and also reduction of IFN-γ production (Momtazi-Borojeni et al. 2018).

3.3.6 Genistein and Daidezin

Isoflavones are plant-derived nonsteroidal polyphenolic compounds representing most common phytoestrogens, i.e., plant-based compounds resembling vertebrate steroids, e.g., 17-β-estradiol. Phytoestrogens are broadly classified as flavonoids and non-flavonoids. Flavonoids include isoflavones, coumestans, and prenylflavonoids, and non-flavonoids include lignans. Isoflavones are considered as phytoestrogens by virtue of their estrogen receptor-binding ability. Genistein, daidzein, glycitein, biochanin A, and formononetin are important isoflavone phytoestrogens (Krizova et al. 2019). Isoflavones have numerous health benefits such as prevention of hormone-related cancers, cardiovascular disease, osteoporosis, and hormone replacement therapy and improving cognitive function (Ko 2014). Specific isoflavones have also shown remarkable biological effects, viz., genistein as antioxidant, anticancer, cardiovascular, and antidiabetic; daidzein as antioxidant, antiosteoporotic, and immunostimulator; formononetin as cardiovascular, antioxidant, and estrogenic; biochanin A as anti-inflammatory, antiproliferative, antioxidant, antiviral, and anticancer; and glycitein as anti-inflammatory (Niaz and Khan 2020).

Genistein, 4′,5,7-trihydroxyisoflavone, is a tyrosine kinase inhibitor abundant in soy products such as soybeans. It is one of the promising and innovative phytoconstituents for the treatment of cancer, osteoporosis, cardiovascular diseases, and menopause exhibiting multifaceted action within the live cell and very low toxicity (Ganai and Farooqi 2015; Polkowski and Mazurek 2000).

Genistein suppresses both the cell-mediated and humoral components of the adaptive immune system as indicated by reduction in the number of developing CD4+ and CD8+ thymocytes (Ganai and Farooqi 2015). It also inhibits tyrosine kinase which is required for NK cell activation (Burkard et al. 2017). Masilamani et al. (2012) has extensively reviewed the role of isoflavones such as genistein, daidzein, glycitein, and their metabolites in regulating immune response. In pharmacological doses, isoflavones particularly genistein exert protective effects over tight junctions between intestinal epithelial cells and the mucosal barrier helping cure inflammation conditions. Being phytoestrogens, isoflavones regulate dendritic cells and downregulate TLR-dependent IL-6 expression. Genistein inhibited NF-κB-dependent gene expression by stimulating p53 expression. In addition, genistein suppressed MHC class II molecules. Inhibition of NK cell cytotoxicity and IgE signaling are other noticeable effects. These effects are varied depending on their cellular targets and also affected by the nature of the experimental setup, strain of the animal used, and dose and duration of isoflavone treatments. Readers are strongly urged to refer the text by Masilamani et al. (2012) for detailed account on the effect of isoflavones on various cells and tissues of the immune system.

Genistein exerts anticancer effects in different types of cancers via cell cycle arrest, apoptosis, antiangiogenesis, and antimetastasis. It acts via molecular targets such as caspases, B-cell lymphoma 2, Bcl-2-associated X protein, NF-κB, phosphoinositide 3-kinase/Akt, extracellular signal-regulated kinase 1/2, mitogen-activated protein kinase, and Wingless/integration 1/β-catenin signaling pathway (Tuli et al. 2019). Immunomodulatory effect of genistein has surfaced in a mouse model of human papillomavirus-associated cervical cancer with significant increase in lymphocyte proliferation, release of lactase dehydrogenase, and significant increment in IFN-γ (Ghaemi et al. 2012).

Antitumor activity of genistein is very well studied in prostate cancer, hepatoma cancer, and pancreatic and lung cancer. Anticancer effect is seen via induction of apoptosis through activation of caspase-3 and downregulation of antiapoptotic Bcl-2 and Bcl-XL family. Induction of G2/M cell cycle arrest by inhibition of AKT phosphorylation and Cdc2 kinase, antiangiogenesis, and antimetastasis are also observe. Genistein is also helpful as an adjuvant with other anticancer drugs such as 5-flurouracil, all-trans retinoic acid, and trichostatin A (Mohamed et al. 2017).

Genistein can also be combined with antiretroviral drugs for inhibiting HIV infection and HIV-related oxidative stress. It inhibits stromal cell-derived factor 1, HIV-mediated chemotactic signaling, and HIV infection of cultured blood T cells and macrophages (Guo and Wu 2018).

Genistein reduces inflammation and enhances Th1/Th2 cytokine-mediated responses mainly via IFN-γ and IL-10, IL-4, and IL-5 in rheumatoid arthritis and allergic asthma (Gandhi et al. 2018). Genistein being immunomodulator, phytoestrogen, and kinase inhibitor showed benefits as prophylactic antiviral drug against arenaviral hemorrhagic fever in the Syrian golden hamster (Vela et al. 2010).

Desmethylangolensin and equol are two major bioactive metabolites of isoflavones. Peripheral blood mononuclear cells pretreated with genistein and equol reduce IL-12/IL-18-induced IFN-γ production by natural killer cells without consistently affecting cytotoxicity (Mace et al. 2019).

Daidzein is known to enhance animal resistance to heat stress and regulate animal immunocompetence. It was found to raise serum IgG and IFN-α significantly upon supplementation through diet to the cows in late lactation (Liu et al. 2014). Daidzein significantly decreased the B-cell population (represented by $CD3^{-}IgM^{+}$) and increased the T-cell populations (represented by $CD3^{+}IgM$, $CD4^{+}CD8^{-}$, and $CD4^{-}CD8^{+}$) in hybrid B6C3F1 and inbred type 1 diabetes-prone nonobese diabetic mice (Huang et al. 2019). Daidzein supplementation to broiler breeder hens helps improve immune function in offsprings visa increase in B cells and CD4+T/CD8+T 386 ratio as well as the proliferative response of B cells, but no effect was seen on T lymphocytes of the offspring (Fan et al. 2018).

Recently, daidzein attenuated lipopolysaccharide-induced hepatic liver injury. It was found to reduce ALT and AST expressions, reactive oxygen species, IL-1β, IL-6, and TNF-α with no cytotoxicity. Inhibition of p-ERK1/2, p-IκBα, and p-p65 expression, in addition to downregulating Keap-1, and upregulating Nrf2 expression were revealed (Yu et al. 2020).

The duo genistein and daidzein were found to slow down aging processes by affecting immune processes. Genistein reduced vitellogenin expression and also resistance versus *P. luminescens*, whereas daidzein increased resistance versus the pathogen in a vitellogenin-dependent manner (Fischer et al. 2012).

Glucosylated derivatives of isoflavones such as glycitein 7-*O*-β-glucoside, glycitein 7-*O*-β-maltoside, and daidzein 7-*O*-β-glucoside inhibited the production of IgE antibody in rats against 7S-globulin as antigen (Shimoda and Hamada 2010). Ethyl acetate extract of *Cordyceps militaris* rich in isoflavones (genistein, daidzein) and their derivatives (genistein 7-*O*-β-D-glucoside 4″-*O*-methylate, genistein 4′-*O*-β-D-glucoside 4″-*O*-methylate, glycitein 7-*O*-β-D-glucoside 4″-*O*-methylate, daidzein 7-*O*-β-D-glucoside 4″-*O*-methylate and adenosine) showed significant antiallergic effect by inhibition of antigen-induced degranulation in RBL-2H3 cells and passive cutaneous anaphylaxis along with reduction of IL-4 and TNF-α . In mast cells, inhibition of phosphorylation of Syk, ERK, p38, and JNK expression was seen (Oh et al. 2011).

Other prominent isoflavones exerting immunomodulatory functions include isonigricin and isoirisolidone (Nazir 2013).

3.4 Concluding Remarks

The immune system is the most sensitive, however equally efficient integrated functioning of numerous cells and biochemicals developed by these cells to maintain the homeostasis and preserve the integrity of the organism against external harmful stress stimuli. The systematic functioning of the immune system and the balance between all chemicals are essential to the occurrence of any disease or disorder. To date, evidence from literature highlights an increase in immunological diseases because of change in lifestyle, and a great attention has been focused on the development of molecules to modulate the immune response. There is continuously increasing upward surge of demand for new effective therapies all over the world for safe and effective treatments, and researchers are always exploring new avenues for the same. One exciting strategy is the use of herbal medicines as their use in traditional medicine for their properties and health benefits are well recognized and established with safety, since ancient times. Natural products have a wide range of diversity of multidimensional chemical structures; in the meantime, the utility of natural products as biological function modifiers has also raised considerable attention and interest. Apart from versatility, safe and cost-effective therapy and abundant availability are the added advantage of phytochemicals. Scientists have successfully used isolated natural products and combinatorial chemistry in the discovery of new drugs. Despite developments in drug discovery technology, natural products from plants and other biological sources remain an undiminished source of new pharmaceuticals.

Even though great advancements are happening in the area of fundamental molecular biology and pathophysiology, the use of phytopharmaceuticals in modern medicine era is still in preclinical stage, with very few products successful in the market. Despite having beneficial health effects, therapeutic applications of bioactive phytoconstituents are not yet successful due to its poor solubility, short biological half-life, rapid metabolism, and elimination. In order to achieve a maximum benefit of phytochemicals as immunomodulators for antioxidant, anti-inflammatory, antifibrotic, chemo-preventive agents and also for metabolic and neurological disorders for human use, efforts should be focused to improve upon the solubility, stability, and targeted delivery for improving the bioavailability. Additionally, the toxicity associated with the high doses of phytochemicals required to achieve desirable response needs to be reduced. Nanotechnological strategies may be used to improve the bioavailability and targeting efficiency, thus reducing the dose requirements.

References

Abdel-Latif M, Afifi A, Soliman R, Elkhouly A, Abdelmotaal A, Youness RA (2019) 23P—A new quercetin glycoside enhances TNBC immunological profile through TP53/miR-155/MICA/ULBP2. Ann Oncol 30:7–8

Adapala N, Chan M (2008) Long-term use of an anti-inflammatory, curcumin, suppressed type 1 immunity and exacerbated visceral leishmaniasis in a chronic experimental model. Lab Investig 88:1329–1339

Ahn SC, Kim GY, Kim JH, Baik SW, Han MK, Lee HJ, Moon DO, Lee CM, Kang JH, Kim BH, Oh YH, Park YM (2004) Epigallocatechin-3-gallate, constituent of green tea, suppresses the LPS-induced phenotypic and functional maturation of murine dendritic cells through inhibition of mitogen-activated protein kinases and NF-kappaB. Biochem Biophys Res Commun 313(1): 148–155

Al Azzaz J, Rieu A, Aires V, Delmas D, Chluba J, Winckler P, Bringer MA, Lamarche J, Vervandier-Fasseur D, Dalle F, Lapaquette P, Guzzo J (2019) Resveratrol-induced xenophagy promotes intracellular bacteria clearance in intestinal epithelial cells and macrophages. Front Immunol 9:3149

Alrafas HR, Busbee PB, Nagarkatti M, Nagarkatti PS (2020) Resveratrol downregulates mir-31 to promote t regulatory cells during prevention of TNBS-induced colitis. Mol Nutr Food Res 64(1):e1900633

Al-Rasheed NM, Fadda L, Attia HA, Sharaf IA, Mohamed AM, Al-Rasheed NM (2017) Pulmonary prophylactic impact of melatonin and/or quercetin: a novel therapy for inflammatory hypoxic stress in rats. Acta Pharma 67:125–135

Altamemi I, Murphy EA, Catroppo JF, Zumbrun EE, Zhang J, McClellan JL, Singh UP, Nagarkatti PS, Nagarkatti M (2014) Role of microRNAs in resveratrol-mediated mitigation of colitis-associated tumorigenesis in Apc (Min/+) mice. J Pharmacol Exp Ther 350(1):99–109

Anonymous (1994) Randomised trial comparing tacrolimus (FK506) and cyclosporin in prevention of liver allograft rejection. European FK506 Multicentre Liver Study Group. Lancet 344(8920): 423–428

Ay M, Charli A, Jin H, Anantharam V, Kanthasamy A, Kanthasamy AG (2016) Quercetin. In: Gupta RC (ed) Nutraceuticals. Academic Press, Boston, pp 447–452

Banerjee G, Bhat J, Sinkar VP, Telkar A, Upadhyaya SA (2009) Composition for enhancing immunity. US Patent 2009/0175964 A1, 9 Jul 2009

Bascones-Martinez A, Mattila R, Gomez-Font R, Meurman JH (2014) Immunomodulatory drugs: oral and systemic adverse effects. Med Oral Patol Oral Cir Bucal 19:e24–e31

Belkhelfa-Slimani R, Djerdjouri B (2017) Caffeic acid and quercetin exert caspases-independent apoptotic effects on Leishmania major promastigotes, and reactivate the death of infected phagocytes derived from BALB/c mice. Asian Pac J Trop Biomed 7:321–331

Benavides J, Martínez-Valladares M, Tejido ML, Giráldez FJ, Bodas R, Prieto N, Perez V, Andres S (2013) Quercetin and flaxseed included in the diet of fattening lambs: effects on immune response, stress during road transport and ruminal acidosis. Livestock Prod Sci 158:84–90

Bereswill S, Munoz M, Fischer A, Plickert R, Haag LM, Otto B, Kuhl AA, Loddenkemper C, Gobel UB, Heimesaat MM (2010) Anti-inflammatory effects of resveratrol, curcumin and simvastatin in acute small intestinal inflammation. PLoS One 5(12):e15099

Besednova NN, Zvyagintseva TN, Kuznetsova TA, Makarenkova ID, Smolina TP, Fedyanina LN, Kryzhanovky SP, Zaporozhets TS (2019) Marine algae metabolites as promising therapeutics for the prevention and treatment of HIV/AIDS. Metabolites 9(5):87

Bhattacharyya S, Md Sakib Hossain D, Mohanty S, Sankar Sen G, Chattopadhyay S, Banerjee S, Chakraborty J, Das K, Sarkar D, Das T, Sa G (2010) Curcumin reverses T cell-mediated adaptive immune dysfunctions in tumor-bearing hosts. Cell Mol Immunol 7(4):306–315

Bose S, Panda AK, Mukherjee S, Sa G (2015) Curcumin and tumor immune-editing: resurrecting the immune system. Cell Div 10:6

Bule MH, Ahmed I, Maqbool F, Bilal M, Iqbal H (2018) Microalgae as a source of high-value bioactive compounds. Front Biosci 10:197–216

Burkard M, Leischner C, Lauer UM, Busch C, Venturelli S, Frank J (2017) Dietary flavonoids and modulation of natural killer cells: implications in malignant and viral diseases. J Nutr Biochem 46:1–12

Burmańczuk A, Hola P, Milczak A, Piech T, Kowalski C, Wojciechowska B, Grabowski T (2018) Quercetin decrease somatic cells count in mastitis of dairy cows. Res Vet Sci 117:255–259

Butt MS, Sultan MT (2011) Ginger and its health claims: molecular aspects. Crit Rev Food Sci Nutr 51:383–393

Cardwell G, Bornman JF, James AP, Black LJ (2018) A review of mushrooms as a potential source of dietary vitamin D. Nutrients 10(10):1498

Cataneo AHD, Tomiotto-Pellissier F, Miranda-Sapla MM, Assolini JP, Panis C, Kian D, Yamauchi LM, Simao ANC, Casagrande R, Pinge-Filho P, Costa IN, Verri WA, Conchon-Costa I, Pavanelli WR (2019) Quercetin promotes antipromastigote effect by increasing the ROS production and anti-amastigote by upregulating Nrf2/HO-1 expression, affecting iron availability. Biomed Pharmacother 113:108745

Cattelan A, Ceolotto G, Bova S, Albiero M, Kuppusamy M, De Martin S, Semplicini A, Fadini GP, de Kreutzenberg SV, Avogaro A (2015) NAD (+)-dependent SIRT1 deactivation has a key role on ischemia-reperfusion-induced apoptosis. Vascul Pharmacol 70:35–44

Cervantes-Valencia ME, Alcalá-Canto Y, Sumano-Lopez H, Ducoing-Watty AM, Gutierrez-Olvera L (2016) Effects of *Curcuma longa* dietary inclusion against *Eimeria* spp. in naturally-infected lambs. Small Rumin Res 136:27–35

Chan KH, Chan SC, Yeung SC, Man RY, Ip MS, Mak JC (2012) Inhibitory effect of Chinese green tea on cigarette smoke-induced up-regulation of airway neutrophil elastase and matrix metalloproteinase-12 via antioxidant activity. Free Radic Res 46(9):1123–1129

Chanput W, Mes J, Vreeburg RAM, Savelkoul HFJ, Wichers HJ (2010) Transcription profiles of LPS-stimulated THP-1 monocytes and macrophages: a tool to study inflammation modulating effects of food-derived compounds. Food Funct 1:254–261

Chen C-J, Tsai CC, Hsieh J-F, Chien CM, Wu T-H, Chen S-T (2006) A screening platform for compounds with potential immuno-regulatory activities using human cord blood mononuclear cells. Comb Chem High Throughput Screen 9:777–784

Chen CJ, Yu W, Fu YC, Wang X, Li JL, Wang W (2009) Resveratrol protects cardiomyocytes from hypoxia-induced apoptosis through the SIRT1-FoxO1 pathway. Biochem Biophys Res Commun 378(3):389–393

Chen JJ, Dai L, Zhao LX, Zhu X, Cao S, Gao YJ (2015) Intrathecal curcumin attenuates pain hypersensitivity and decreases spinal neuroinflammation in rat model of monoarthritis. Sci Rep 5:10278

Chen X, Song L, Wang H, Liu S, Yu H, Wang X, Li R, Liu T, Li P (2019) Partial characterization, the immune modulation and anticancer activities of sulfated polysaccharides from filamentous microalgae *Tribonema* sp. Molecules 24(2):322

Choi KC, Jung MG, Lee YH, Yoon JC, Kwon SH, Kang HB, Kim MJ, Cha JH, Kim YJ, Jun WJ, Lee JM, Yoon HG (2009) Epigallocatechin-3-gallate, a histone acetyltransferase inhibitor, inhibits EBV-induced B lymphocyte transformation via suppression of RelA acetylation. Cancer Res 69(2):583–592

Cutignano A, Nuzzo G, Ianora A, Luongo E, Romano G, Gallo C, Sansone C, Aprea S, Mancini F, D'Oro U, Fontana A (2015) Development and application of a novel SPE-method for bioassay-guided fractionation of marine extracts. Mar Drugs 13:5736–5749

De Souza NJ, Yeole R, Jha R, Bagchi S (2002) Use of Tinospora extract in the treatment of immune system-modulated disorders. WO Patent 02/053166 A1, 11 Jul 2002

Dekker P, Parish WE, Green MR (2005) Protection by food-derived antioxidants from UV-A1-induced photodamage, measured using living skin equivalents. Photochem Photobiol 81:837–842

Ding Y, Li C, Zhang Y, Ma P, Zhao T, Che D, Cao J, Wang J, Liu R, Zhang T, He L (2020) Quercetin as a Lyn kinase inhibitor inhibits IgE-mediated allergic conjunctivitis. Food Chem Toxicol 135:110924

Dona M, DellAica I, Calabrese F, Benelli R, Morini M, Albini A, Garbisa S (2003) Neutrophil restraint by green tea: inhibition of inflammation, associated angiogenesis, and pulmonary fibrosis. J Immunol 170(8):4335–4341

Dou W, Zhang J, Ren G, Ding L, Sun A, Deng C, Wu X, Wei X, Mani S, Wang Z (2014) Mangiferin attenuates the symptoms of dextran sulfate sodium-induced colitis in mice via NF-κB and MAPK signaling inactivation. Int Immunopharmacol 23(1):170–178

Edo Y, Otaki A, Asano K (2018) Quercetin enhances the thioredoxin production of nasal epithelial cells in vitro and in vivo. Medicine 5:124

Eitah HE, Maklad YA, Abdelkader NF, Gamal el Din AA, Badawi MA, Kenawy SA (2019) Modulating impacts of quercetin/sitagliptin combination on streptozotocin-induced diabetes mellitus in rats. Toxicol Appl Pharmacol 365:30–40

Fan H, Lv Z, Gan L, Guo Y (2018) Transcriptomics-related mechanisms of supplementing laying broiler breeder hens with dietary daidzein to improve the immune function and growth performance of offspring. J Agric Food Chem 66:2049–2060

Fischer M, Regitz C, Kahl M, Werthebach M, Boll M, Wenzel U (2012) Phytoestrogens genistein and daidzein affect immunity in the nematode *Caenorhabditis elegans* via alterations of vitellogenin expression. Mol Nutr Food Res 56:957–965

Gambini J, Ingles M, Olaso G, Lopez-Grueso R, Bonet-Costa V, Gimeno-Mallench L, Mas-Bargues C, Abdelaziz KM, Gomez-Cabrera MC, Vina J, Borras C (2015) Properties of resveratrol: in vitro and in vivo studies about metabolism, bioavailability, and biological effects in animal models and humans. Oxid Med Cell Longev 2015:837042

Ganai AA, Farooqi H (2015) Bioactivity of genistein: a review of in vitro and in vivo studies. Biomed Pharmacother 76:30–38

Gandhi GR, Neta MTSL, Sathiyabama RG, de Quintans JSS, de Olive e Sirailva AM, de Araújo AAS, Narain N, LSJ Q, Gurgel RQ (2018) Flavonoids as Th1/Th2 cytokines immunomodulators: a systematic review of studies on animal models. Phytomedicine 44:74–84

Ghaemi A, Soleimanjahi H, Razeghi S, Gorji A, Tabaraei A, Moradi A, Alizadeh A, Vakili M (2012) Genistein induces a protective immunomodulatory effect in a mouse model of cervical cancer. Iran J Immunol 9:119–127

Gibbons S (2003) An overview of plant extracts as potential therapeutics. Expert Opin Ther Patents 13(4):489–497

Gopu B, Dileep R, Rani MU, Kumar CS, Kumar MV, Reddy AG (2015) Protective role of curcumin and flunixin against acetic acid-induced inflammatory bowel disease via modulating inflammatory mediators and cytokine profile in rats. J Environ Pathol Toxicol Oncol 34(4): 309–320

Guo J, Wu Y (2018) Genistein and HIV infection. In: Preedy VR, Watson RR (eds) HIV/AIDS. Academic Press, Boston, pp 125–134

Haraoui B (2015) Leflunomide. In: Hochberg MC, Silman AJ, Smolen JS, Weinblatt ME, Weisman MH (eds) Rheumatology, 6th edn. Elsevier, pp 451–458

Hayakawa S, Saito K, Miyoshi N, Ohishi T, Oishi Y, Miyoshi M, Nakamura Y (2016) Anti-cancer effects of green tea by either anti- or pro- oxidative mechanisms. Asian Pac Cancer Prev 17: 1649–1654

Hirahashi T, Matsumoto M, Hazeki K, Saeki Y, Ui M, Seya T (2002) Activation of the human innate immune system by Spirulina: augmentation of interferon production and NK cytotoxicity by oral administration of hot water extract of Spirulina platensis. Int Immunopharmacol 2:423–434

Holt S (2007) Herbal compositions, methods of stimulating immunomodulation and enhancement of immunomodulating agents using the herbal compositions US Patent 2007/0196381 A1, 23 Aug 2007

Huang G, Xu J, Guo TL (2019) Isoflavone daidzein regulates immune responses in the B6C3F1 and non-obese diabetic (NOD) mice. Int Immunopharmacol 71:277–284

Ichikawa D, Matsui A, Imai M, Sonoda Y, Kasahara T (2004) Effect of various catechins on the IL-12p40 production by murine peritoneal macrophages and a macrophage cell line, J774.1. Biol Pharm Bull 27:1353–1358

Jadhav SS, Patwardhan B, Gautam M (2010) Adjuvant composition for vaccine. US Patent 2010/0285064 A1, 11 Nov 2010

Jewett A, Tseng HC (2017) Immunotherapy. In: Dowd FJ, Johnson BS, Mariotti AJ (eds) Pharmacology and therapeutics for dentistry, 7th edn. Mosby, St. Louis, pp 504–529

Joo SY, Song YA, Park YL, Myung E, Chung CY, Park KJ, Cho SB, Lee WS, Kim HS, Rew JS, Kim NS, Joo YE (2012) Epigallocatechin-3-gallate inhibits LPS-induced NF-κB and MAPK signaling pathways in bone marrow-derived macrophages. Gut Liver 6(2):188–196

Ju SA, Park SM, Lee YS, Bae JH, Yu R, An WG, Suh JH, Kim BS (2012) Administration of 6-gingerol greatly enhances the number of tumor-infiltrating lymphocytes in murine tumors. Int J Cancer 130(11):2618–2628

Kapoor V, Aggarwal S, Das SN (2016) 6-gingerol mediates its anti tumor activities in human oral and cervical cancer cell lines through apoptosis and cell cycle arrest. Phytother Res 30(4): 588–595

Kawabata K, Mukai R, Ishisaka A (2015) Quercetin and related polyphenols: new insights and implications for their bioactivity and bioavailability. Food Funct 6:1399–1417

Keservani RK, Kesharwani RK, Sharma AK, Jarouliya U (2016a) Chapter 20—Dietary supplements, nutraceutical and functional foods in immune response (immunomodulators). In: Bagchi D, Preuss HG, Swaroop A (eds) Nutraceutical and functional foods in human life and disease prevention. CRC Press, Taylor and Francis, Boca Raton, FL, pp 343–358. isbn:9781482237214

Keservani RK, Sharma AK, Kesharwani RK (2016b) An overview and therapeutic applications of nutraceutical & functional foods. In: Keservani RK, Sharma AK, Kesharwani RK (eds) Handbook of research on novel approaches for drug delivery. IGI Global Publisher, USA

Keservani RK, Sharma AK, Kesharwani RK (2016c) Nutraceutical and functional foods for cardiovascular health. In: Meghwal M, Goyal MR (eds) Food process engineering: emerging technologies and innovations in food engineering, vol 5. Apple Academic Press, CRC Press, Taylor and Francis. isbn:9781771884020

Keservani RK, Sharma AK, Kesharwani RK (2018) Nutraceutical and functional foods in disease prevention. IGI Global. isbn13:9781522532675

Keservani RK, Sharma AK, Kesharwani RK (2020) Nutraceuticals and dietary supplements: applications in health improvement and disease management. isbn:9781771888738

Kesharwani RK, Misra K (2010) Prediction of binding site for Curcuminoids at human Topoisomerase II α protein; an in silico approach. Curr Sci 101(8):1060–1064

Kesharwani RK, Srivastava V, Singh P, Rizvi SI, Adeppa K, Misra K (2015) A novel approach for overcoming drug resistance in breast cancer chemotherapy by targeting new synthetic curcumin analogues against Aldehyde Dehydrogenase 1 (ALDH1A1) and Glycogen Synthase Kinase-3 β (GSK-3β). Appl Biochem Biotechnol 2015. https://doi.org/10.1007/s12010-015-1696-x

Kesharwani RK, Singh DB, Singh DV, Misra K (2018) Computational study of curcumin analogues by targeting DNA topoisomerase II: a structure-based drug designing approach. Network Model Anal Health Infor Bioinformatics 7(1):1–7

Khan N, Monagas M, Urpi-sarda M, Llorach R, Andres-Lacueva C (2013) Contribution of bioactive foods and their emerging role in immunomodulation, inflammation, and arthritis. In: Bioactive food as dietary interventions for arthritis and related inflammatory diseases. Elsevier, Amsterdam, pp 43–65

Kim EC, Min JK, Kim TY, Lee SJ, Yang HO, Han S, Kim YM, Kwon YG (2005) [6]-Gingerol, a pungent ingredient of ginger, inhibits angiogenesis in vitro and in vivo. Biochem Biophys Res Commun 335(2):300–308

Kim BR, Yang S, Doh EJ, Choi CW, Youn SW (2018) Risk factors affecting adverse effects of cyclosporine a in a real-world psoriasis treatment. Ann Dermatol 30(2):143–149

Kjær TN, Thorsen K, Jessen N, Stenderup K, Pedersen SB (2015) Resveratrol ameliorates imiquimod-induced psoriasis-like skin inflammation in mice. PLoS One 10(5):e0126599

Ko K-P (2014) Isoflavones: chemistry, analysis, functions and effects on health and cancer. Asia Pac Cancer Prev 15:7001–7010

Kong S, McBurney MW, Fang D (2012) Sirtuin 1 in immune regulation and autoimmunity. Immunol Cell Biol 90(1):6–13

Krizova L, Dadakova K, Kasparovska J, Kasparovsky T (2019) Isoflavones. Molecules 24:1076

Kumar D, Arya V, Kaur R, Bhat ZA, Gupta VK, Kumar V (2012) A review of immunomodulators in the Indian traditional health care system. J Microbiol Immunol Infect 45:165–184

Lai CY, Su YW, Lin KI, Hsu LC, Chuang TH (2017) Natural modulators of endosomal toll-like receptor-mediated psoriatic skin inflammation. J Immunol Res 2017:7807313. https://doi.org/10.1155/2017/7807313

Lam PYS (2003) Herbal immune system stimulant. WO Patent 03/068145 A2, 21 Aug 2003

Lee SM, Yang H, Tartar DM, Gao B, Luo X, Ye SQ, Zaghouani H, Fang D (2011) Prevention and treatment of diabetes with resveratrol in a non-obese mouse model of type 1 diabetes. Diabetologia 54(5):1136–1146

Lee J, Choi JW, Sohng JK, Pandey RP, Il PY (2016) The immunostimulating activity of quercetin 3-O-xyloside in murine macrophages via activation of the ASK1/MAPK/NF-κB signaling pathway. Int Immunopharmacol 31:88–97

Lee-Huang S, Huang PL, Huang PL, Zhang D, Zhang J, Chang YT, Lee JW, Bao J, Sun Y (2009) Compositions and methods for treating obesity, obesity related disorders and for inhibiting the infectivity of human immunodeficiency virus. US Patent 2009/0061031A1, 5 Mar 2009

Lelli D, Sahebkar A, Johnston TP, Pedone C (2017) Curcumin use in pulmonary diseases: state of the art and future perspectives. Pharmacol Res Commun 115:133–148

Li R, Wang Y, Liu Y, Chen Q, Fu W, Wang H, Cai H, Peng W, Zhang X (2013) Curcumin inhibits transforming growth factor-β1-induced EMT via PPARγ pathway, not Smad pathway in renal tubular epithelial cells. PLoS One 8(3):e58848

Liu HS, Pan CE, Yang W, Liu XM (2003) Antitumor and immunomodulatory activity of resveratrol on experimentally implanted tumor of H22 in Balb/c mice. World J Gastroenterol 9(7): 1474–1476

Liu D-Y, He S-J, Liu S-Q, Tang Y-G, Jin E-H, Chen H-L, Li S-H, Zhong L-T (2014) Daidzein enhances immune function in late lactation cows under heat stress. J Anim Sci 85:85–89

Liu D, Chen YQ, Xiao XW, Zhong RT, Yang CF, Liu B, Zhao C (2019) Nutrient properties and nuclear magnetic resonance-based metabonomic analysis of macrofungi. Foods 8(9):397

Lu (2008) A herbal medicine extract that induces interferon production by immune cells and activates Toll-like receptors. JP 4980305B2, 18 Jul 2012

Lu J, Guan S, Shen X, Qian W, Huang G, Deng X, Xie G (2011) Immunosuppressive activity of 8-gingerol on immune responses in mice. Molecules 16:2636–2645

Luis Espinoza J, Takami A, Trung LQ, Nakao S (2013) Ataxia-telangiectasia mutated kinase-mediated upregulation of NKG2D ligands on leukemia cells by resveratrol results in enhanced natural killer cell susceptibility. Cancer Sci 104(6):657–662

Mace TA, Ware MB, King SA, Loftus S, Farren MR, McMichael E, Scoville S, Geraghty C, Young G, Carson WE III, Clinton C, Lesinski GB (2019) Soy isoflavones and their metabolites modulate cytokine-induced natural killer cell function. Sci Rep 9:5068

Malaguarnera L (2019) Influence of Resveratrol on the Immune Response. Nutrients 11(5):946

Mallard B, Leach DN, Wohlmuth H, Tiralongo J (2019) Synergistic immuno-modulatory activity in human macrophages of a medicinal mushroom formulation consisting of Reishi, Shiitake and Maitake. PLoS One 14(11):e0224740

Manna SK, Mukhopadhyay A, Aggarwal BB (2000) Resveratrol suppresses TNF-induced activation of nuclear transcription factors NF-kappa B, activator protein-1, and apoptosis: potential role of reactive oxygen intermediates and lipid peroxidation. J Immunol 164(12):6509–6519. https://doi.org/10.4049/jimmunol.164.12.6509

Manzo E, Cutignano A, Pagano D, Gallo C, Barra G, Nuzzo G, Sansone C, Ianora A, Urbanek K, Fenoglio D, Ferrera F, Bernardi C, Parodi A, Pasquale G, Leonardi A, Filaci G, De Palma R, Fontana A (2017) A new marine-derived sulfoglycolipid triggers dendritic cell activation and immune adjuvant response. Sci Rep 7(1):6286

Martínez Andrade KA, Lauritano C, Romano G, Ianora A (2018) Marine microalgae with anti-cancer properties. Mar Drugs 16(5):165

Masilamani M, Wei J, Sampson HA (2012) Regulation of the immune response by soybean isoflavones. Immunol Res 54:95–110

Matsuo N, Yamada K, Shoji K, Mori M, Sugano M (1997) Effect of tea polyphenols on histamine release from rat basophilic leukemia (RBL-2H3) cells: the structure-inhibitory activity relationship. Allergy 52(1):58–64

Mazidi M, Karimi E, Meydani M, Ghayour-Mobarhan M, Ferns GA (2016) Potential effects of curcumin on peroxisome proliferator-activated receptor-γ in vitro and in vivo. World J Methodol 6(1):112–117

Melgarejo E, Medina MA, Sanchez-Jimenez F, Urdiales JL (2009) Epigallocatechin gallate reduces humanmonocyte mobility and adhesion in vitro. Br J Pharmacol 158(7):1705–1712

Mishra H, Kesharwani RK, Singh DB, Tripathi S, Dubey SK, Misra K (2019) Computational simulation of inhibitory effects of curcumin, retinoic acid and their conjugates on GSK-3 beta. Netw Model Anal Health Inform Bioinform 8(1):3

Mitra SK, Saxena E, Dixit MN (2009) Natural immunostimulant compositions, methods for obtaining the same and pharmaceutical formulations thereof US Patent 2009/0136602 A1, 28 May 2009

Mlcek J, Jurikova T, Skrovankova S, Sochor J (2016) Quercetin and its anti-allergic immune response. Molecules 21(5):623

Mohamed SIA, Jantan I, Haque MA (2017) Naturally occurring immunomodulators with antitumor activity: an insight on their mechanisms of action. Int Immunopharmacol 50:291–304

Mollazadeh H, Cicero A, Blesso CN, Pirro M, Majeed M, Sahebkar A (2019) Immune modulation by curcumin: the role of interleukin-10. Crit Rev Food Sci Nutr 59(1):89–101

Momtazi-Borojeni AA, Haftcheshmeh SM, Esmaeili SA, Johnston TP, Abdollahi E, Sahebkar A (2018) Curcumin: a natural modulator of immune cells in systemic lupus erythematosus. Autoimmunol Rev 17(2):125–135

Morris PJ (2014) Azathioprine. In: Morris PJ, Knechtle SJ (eds) Kidney transplantation–principles and practice, 7th edn. Elsevier, Amsterdam, pp 216–220

Nair MPN, Kandaswami C, Mahajan S, Chadha KC, Chawda R, Nair H, Kumar N, Nair R, Schwartz S (2002) The flavonoid, quercetin, differentially regulates Th-1 (IFNgamma) and Th-2 (IL4) cytokine gene expression by normal peripheral blood mononuclear cells. Biochim Biophys Acta 1593:29–36

Nair A, Chattopadhyay D, Saha B (2019) Plant-derived immunomodulators. Elsevier, Amsterdam, pp 435–499

Nakayama H, Yaguchi T, Yoshiya S, Nishizaki T (2012) Resveratrol induces apoptosis MH7A human rheumatoid arthritis synovial cells in a sirtuin 1-dependent manner. Rheumatol Int 32(1): 151–157

Naman CB, Gomes CM, Gupta G (2018) Phytodrugs and immunomodulators for the therapy of leishmaniasis. In: Mandal SC, Mandal V, Konishi T (eds) Natural products and drug discovery. Elsevier, Amsterdam, pp 213–275

Narala VR, Smith MR, Adapala RK, Ranga R, Panati K, Moore BB, Leff T, Reddy VD, Kondapi AK, Reddy RC (2009) Curcumin is not a ligand for peroxisome proliferator-activated receptor-γ. Gene Ther Mol Biol 13(1):20–25

Nazir N (2013) Immunomodulatory activity of isoflavones isolated from Iris kashmiriana: effect on T-lymphocyte proliferation and cytokine production in Balb/c mice. Biom Prev Nutr 3:151–157

Niaz K, Khan F (2020) Analysis of polyphenolics. In: Sanches Silva A, Nabavi SF, Saeedi M, Nabavi SM (eds) Recent advances in natural products analysis. Elsevier, Amsterdam, pp 39–197

Nunes S, Danesi F, Del Rio D, Silva P (2018) Resveratrol and inflammatory bowel disease: the evidence so far. Nutr Res Rev 31(1):85–97

Ogino MH, Tadi P (2020) Cyclophosphamide. In: Stat Pearls. StatPearls Publishing, Treasure Island, FL

Oh JY, Choi W-S, Lee CH, Park HJ (2011) The ethyl acetate extract of Cordyceps militaris inhibits IgE-mediated allergic responses in mast cells and passive cutaneous anaphylaxis reaction in mice. J Ethnopharmacol 135:422–429

Oliveira A, Monteiro V, Navegantes-Lima KC, Reis JF, Gomes RS, Rodrigues D, Gaspar S, Monteiro MC (2017) Resveratrol role in autoimmune disease—a mini-review. Nutrients 9:1306

Oosthuizen CB (2015) Shoot extracts and compositions of *Euclea natalensis* for immune modulation and hepatoprotection. WO Patent 2015/049643 Al, 9 Apr 2015

Pae M, Wu D (2013) Immunomodulating effects of epigallocatechin-3-gallate from green tea: mechanisms and applications. Food Funct 4:1287–1303

Pan J, Shen J, Si W, Du C, Chen D, Xu L, Yao M, Fu P, Fan W (2017) Resveratrol promotes MICA/B expression and natural killer cell lysis of breast cancer cells by suppressing c-Myc/miR-17 pathway. Oncotarget 8:65743–65758

Patwardhan B, Kapadi AH (2004) Withasol and methods of use. US Patent 2004/0033273A1, Published on 19 Feb 2004

Pei Y, Du X, Li J, Gao H, Wang S (2009) Effect of proliferation, c-met and vascular endothelial growth factor expression in the Eca109/9706 cells by liposomal quercetin. Wei Sheng Yan Jiu 38:599–602. Abstract only

Pham PT, Pham PM, Pham SV, Pham PA, Pham PC (2011) New onset diabetes after transplantation (NODAT): an overview. Diabetes Metab Syndr Obes 4:175–186

Polkowski K, Mazurek AP (2000) Biological properties of genistein. A review of in vitro and in vivo data. Acta Pol Pharm 57:135–155

Pund S, Thakur R, More U, Joshi A (2014) Lipid based nanoemulsifying resveratrol for improved physicochemical characteristics, in vitro cytotoxicity and in vivo antiangiogenic efficacy. Colloids Surf B Biointerfaces 120:110–117

Pund S, Joshi A, Patravale V (2020) Exploration of nanonutraceuticals in neurodegenerative diseases. In: Keservani R (ed) Nutraceutical and dietary supplements. Applications in health improvement and disease management. Apple Academic Press

Pylypchuk V (2011) Composition and methods of use of an immunomodulator. US Patent 7.964.221 B2, Published on 21 Jun 2011

Rahal K, Schmiedlin-Ren P, Adler J, Dhanani M, Sultani V, Rittershaus AC, Reingold L, Zhu J, McKenna BJ, Christman GM, Zimmermann EM (2012) Resveratrol has antiinflammatory and antifibrotic effects in the peptidoglycan-polysaccharide rat model of Crohn's disease. Inflamm Bowel Dis 18:613–623

Rahman S, Islam R (2011) Mammalian Sirt1: insights on its biological functions. Cell Commun Signal 9:11

Rainard P, Foucras G, Boichard D, Rupp R (2018) Low milk somatic cell count and susceptibility to mastitis. J Dairy Sci 101:6703–6714

Rangel O (2008) Synergistic Inmune Phyto-Nutraceutical Composition. US Patent 2008/0118582 A1, Published on 22 May 2008

Riccio G, Lauritano C (2019) Microalgae with immunomodulatory activities. Mar Drugs 18(1):2

Rieder SA, Nagarkatti P, Nagarkatti M (2012) Multiple anti-inflammatory pathways triggered by resveratrol lead to amelioration of staphylococcal enterotoxin B-induced lung injury. Br J Pharmacol 167:1244–1258

Rosmarin DM, Lebwohl M, Elewski BE, Gottlieb AB (2010) National psoriasis foundation. cyclosporine and psoriasis: 2008 national psoriasis foundation consensus conference. J Am Acad Dermatol 62:838–853

Ruansit W, Charerntantanakul W (2020) Oral supplementation of quercetin in PRRSV-1 modified-live virus vaccinated pigs in response to HP-PRRSV-2 challenge. Vaccine 38:3570–3581

Ruiz R, Kirk AD (2015) Long-term toxicity of immunosuppressive therapy. Transplant Liver:1354–1363. https://doi.org/10.1016/B978-1-4557-0268-8.00097-X

Sahbaie P, Sun Y, Liang DY, Shi XY, Clark JD (2014) Curcumin treatment attenuates pain and enhances functional recovery after incision. Anesth Analg 118:1336–1344

Saidi A, Kasabova M, Vanderlynden L, Wartenberg M, Kara-Ali GH, Marc D, Lecaille F, Lalmanach G (2019) Curcumin inhibits the TGF-β1-dependent differentiation of lung fibroblasts via PPARγ-driven upregulation of cathepsins B and L. Sci Rep 9(1):491

Saqib U, Kelley TT, Panguluri SK, Liu D, Savai R, Baig MS, Schürer SC (2018) Polypharmacology or promiscuity? Structural interactions of resveratrol with its bandwagon of targets. Front Pharmacol 9:1201

Sartor L, Pezzato E, Garbisa S (2002) (-) Epigallocatechin-3-gallate inhibits leukocyte elastase: potential of the phyto-factor in hindering inflammation, emphysema, and invasion. J Leucoc Biol 71:73–79

Sgro C (1995) Side-effects of a monoclonal antibody, muromonab CD3/orthoclone OKT3: bibliographic review. Toxicology 105:23–29

Sharma A, Kashyap D, Sak K, Tuli HS, Sharma AK (2018) Therapeutic charm of quercetin and its derivatives: a review of research and patents. Pharm Pat Anal 7:15–32

Shebeko SK, Zupanets IA, Popov OS, Tarasenko OO, Shalamay AS (2018) Chapter 27—Effects of quercetin and its combinations on health. In: Watson RR, Preedy VR, Zibadi S (eds) BT-PM of A in HH and D, 2nd edn. Academic Press, pp 373–394

Shehzad A, Lee YS (2013) Molecular mechanisms of curcumin action: signal transduction. Biofactors 39:27–36

Shi Y, Zhou J, Jiang B, Miao M (2017) Resveratrol and inflammatory bowel disease. Ann N Y Acad Sci 1403:38–47

Shim JH, Choi HS, Pugliese A, Lee SY, Chae J, Choi BY, Bode AM, Dong Z (2008) (-)-Epigallocatechin gallate regulates CD3-mediated T cell receptor signaling in leukemia through the inhibition of ZAP-70 kinase. J Biol Chem 283:28370–28379

Shimoda K, Hamada H (2010) Synthesis of beta-maltooligosaccharides of glycitein and daidzein and their anti-oxidant and anti-allergic activities. Molecules 15:5153–5161

Shiri S, Alizadeh AM, Baradaran B, Farhanghi B, Shanehbandi D, Khodayari S, Khodayari H, Tavassoli A (2015) Dendrosomal curcumin suppresses metastatic breast cancer in mice by changing m1/m2 macrophage balance in the tumor microenvironment. Asian Pac J Cancer Prev 16:3917–3922

Shishodia S (2013) Molecular mechanisms of curcumin action: gene expression. BioFactors (Oxford, England) 39:37–55

Signorelli P, Ghidoni R (2005) Resveratrol as an anticancer nutrient: molecular basis, open questions and promises. J Nutr Biochem 16:449–466

Singh RS, Walia AK (2018) Lectins from red algae and their biomedical potential. J Appl Phycol 30:1833–1858

Singh U, Tabibian J, Venugopal SK, Devaraj S, Jialal I (2005) Development of an in vitro screening assay to test the antiinflammatory properties of dietary supplements and pharmacologic agents. Clin Chem 51:2252–2256

Singh DB, Gupta M, Kesharwani R, Sagar M, Dwivedi S, Misra K (2014) Molecular drug targets and therapies for Alzheimer's disease. Transl Neurosci *5*(3):203–217

Singh P, Kesharwani RK, Misra K, Rizvi SI (2015a) Modulation of erythrocyte plasma membrane redox system (PMRS) activity by curcumin, vol 2015. Biochem Res Int, p 702039

Singh P, Kesharwani RK, Misra K, Rizvi SI (2015b) The modulation of erythrocyte Na+/K+-ATPase activity by curcumin. J Adv Res 6(6):1023–1030

Singh V, Rana M, Jain M, Singh N, Naqvi A, Malasoni R, Dwivedi AK, Dikshit M, Barthwal MK (2015c) Curcuma oil attenuates accelerated atherosclerosis and macrophage foam-cell formation by modulating genes involved in plaque stability, lipid homeostasis and inflammation. Br J Nutr 113:100–113

Singh D, Tanwar H, Jayashankar B, Sharma J, Murthy S, Chanda S et al (2017) Quercetin exhibits adjuvant activity by enhancing Th2 immune response in ovalbumin immunized mice. Biomed Pharmacother 90:354–360

Sliva D (2004) Cellular and physiological effects of Ganoderma lucidum (Reishi). Mini Rev Med Chem 4:873–879

Sreedhar R, Arumugam S, Thandavarayan RA, Karuppagounder V, Watanabe K (2016) Curcumin as a therapeutic agent in the chemoprevention of inflammatory bowel disease. Drug Discov Today 21:843–849

Stamets PE (2018) Antiviral activity from medicinal mushrooms and their active constituents. US Patent 9931316B2 Published 3 Apr 2018

Takano K, Nakaima K, Nitta M, Shibata F, Nakagawa H (2004) Inhibitory effect of (-)-epigallocatechin 3-gallate, a polyphenol of green tea, on neutrophil chemotaxis in vitro and in vivo. J Agric Food Chem 52:4571–4576

Triantafyllou A, Mylonis I, Simos G, Bonanou S, Tsakalof A (2008) Flavonoids induce HIF-1alpha but impair its nuclear accumulation and activity. Free Radic Biol Med 44:657–670

Trung LQ, An D (2018) Is resveratrol a cancer immunomodulatory molecule? Front Pharmacol 9: 1255

Tuli HS, Tuorkey MJ, Thakral F, Sak K, Kumar M, Sharma AK, Sharma U, Jain A, Aggarwal V, Bishayee A (2019) Molecular mechanisms of action of genistein in cancer: recent advances. Front Pharmacol 10:1336

Upadhyaya J, Kesharwani RK, Misra K (2009a) Comparative study of antioxidants as cancer preventives through inhibition of HIF-1 alpha activity. Bioinformation *4*(6):233

Upadhyaya J, Kesharwani RK, Misra K (2009b) Metabolism, pharmacokinetics and bioavailability of ascorbic acid; synergistic effect with tocopherols and curcumin. J Comput Intell Bioinform *2*(1):77–84

Vela EM, Knostman KA, Mott JM, Warren RL, Garver JN, Vela LJ et al (2010) Genistein, a general kinase inhibitor, as a potential antiviral for arenaviral hemorrhagic fever as described in the Pirital virus-Syrian golden hamster model. Antivir Res 87:318–328

Vetchinkina E, Shirokov A, Bucharskaya A, Navolokin N, Prilepskii A, Burov A, Maslyakova G, Nikitina VE (2016) Antitumor activity of extracts from medicinal basidiomycetes mushrooms. Int J Med Mushrooms 18:955–964

Vyas D, Gupt S, Dixit V, Anita K, Kaur S (2015) To study the effect of curcumin on the growth properties of circulating endothelial progenitor cells. In vitro cellular & developmental biology. Animal 51:488–494

Wallace KL, Zheng LB, Kanazawa Y, Shih DQ (2014) Immunopathology of inflammatory bowel disease. World J Gastroenterol 20:6–21

Wallberg M, Cooke A (2013) Immune mechanisms in type 1 diabetes. Trends Immunol 34:583–591

Wang W, Yan C, Zhang J, Lin R, Lin Q, Yang L, Ren F, Zhang J, Ji M, Li Y (2013) SIRT1 inhibits TNF-α-induced apoptosis of vascular adventitial fibroblasts partly through the deacetylation of FoxO1. Apoptosis 18:689–701

Wang XS, Zhang ZR, Zhang MM, Sun MX, Wang WW, Xie CL (2017) Neuroprotective properties of curcumin in toxin-base animal models of Parkinson's disease: a systematic experiment literatures review. BMC Complement Altern Med 17:412

Wilasrusmee C, Kittur S, Siddiqui J, Bruch D, Wilasrusmee S, Kittur DS (2002) In vitro immunomodulatory effects of ten commonly used herbs on murine lymphocytes. J Altern Complement Med 8:467–475

Williams AS, Richards PJ, Thomas E, Carty S, Nowell MA, Goodfellow RM, Dent CM, Williams BD, Jones SA, Topley N (2007) Interferon-gamma protects against the development of structural damage in experimental arthritis by regulating polymorphonuclear neutrophil influx into diseased joints. Arthritis Rheumatol 56:2244–2254

Wong C-H, Wu C-Y, Hsu H-Y, Liao S-F, Liang C-H (2016) Antibody-mediated anti-tumor activity induced by reishi mushroom polysaccharides. US Patent 2016/0166661 A1, Published 16 Jun 2016

Wu CH, Wu CF, Huang HW, Jao YC, Yen GC (2009a) Naturally occurring flavonoids attenuate high glucose-induced expression of proinflammatory cytokines in human monocytic THP-1 cells. Mol Nutr Food Res 53:984–995

Wu J, Chen L, Zhang D, Huo M, Zhang X, Pu D, Guan Y (2009b) Peroxisome proliferator-activated receptors and renal diseases. Front Biosci (Landmark edition) 14:995–1009

Xu T, Beelman RB, Lambert JD (2012) The cancer preventive effects of edible mushrooms. Anticancer Agents Med Chem 12:1255–1263

Xu S, Jiang B, Wang H, Shen C, Chen H, Zeng L (2017) Curcumin suppresses intestinal fibrosis by inhibition of PPARγ-mediated epithelial-mesenchymal transition. Evid Based Complement Alternat Med 2017:7876064

Xuzhu G, Komai-Koma M, Leung BP, Howe HS, McSharry C, McInnes IB, Xu D (2012) Resveratrol modulates murine collagen-induced arthritis by inhibiting Th17 and B-cell function. Ann Rheum Dis 71:129–135

Yang F, de Villiers WJ, McClain CJ, Varilek GW (1998) Green tea polyphenols block endotoxin-induced tumor necrosis factor-production and lethality in a murine model. J Nutr 128:2334–2340

Yang JX, Maria TC, Zhou B, Xiao FL, Wang M, Mao YJ (2020) Quercetin improves immune function in Arbor Acre broilers through activation of NF-κB signaling pathway. Poult Sci 99: 906–913

Yao J, Wei C, Wang JY, Zhang R, Li YX, Wang LS (2015) Effect of resveratrol on Treg/Th17 signaling and ulcerative colitis treatment in mice. World J Gastroenterol 21:6572–6581

Yu W, Fu YC, Zhou XH, Chen CJ, Wang X, Lin RB, Wang W (2009) Effects of resveratrol on H2O2-induced apoptosis and expression of SIRTs in H9c2 cells. J Cell Biochem 107:741–747

Yu Z, Yang L, Deng S, Liang M (2020) Daidzein ameliorates LPS-induced hepatocyte injury by inhibiting inflammation and oxidative stress. Eur J Pharmacol 885:173399

Yuan J, Lu L, Zhang Z, Zhang S (2012) Dietary intake of resveratrol enhances the adaptive immunity of aged rats. Rejuvenation Res 15:507–515

Yulug E, Turedi S, Alver A, Turedi S, Kahraman C (2015) Effects of resveratrol on methotrexate-induced intestinal injury. Bratisl Med J 116:676–680

Zhang L, Xue H, Zhao G, Qiao C, Sun X, Pang C, Zhang D (2019a) Curcumin and resveratrol suppress dextran sulfate sodium-induced colitis in mice. Mol Med Rep 19:3053–3060

Zhang MY, Guo J, Hu XM, Zhao SQ, Li SL, Wang J (2019b) An in vivo anti-tumor effect of eckol from marine brown algae by improving the immune response. Food Funct 10:4361–4371

Zou T, Yang Y, Xia F, Huang A, Gao X, Fang D, Xiong S, Zhang J (2013) Resveratrol Inhibits CD4+ T cell activation by enhancing the expression and activity of Sirt1. PLoS One 8:e75139

Chapter 4
Effect of Plant-Derived Immunomodulators on the Immune System

Divya Vani Koraganji, Addanki Mounika, Pureti Sushanth, and Prameela Kandra

Abstract To regulate several immunological changes in the prevention of diseases, immunomodulators are helpful compounds which are derived naturally from plants. Since ancient times, people are using these natural compounds for treatment of several diseases. Immunomodulators are capable to fight against invaders by producing antibodies and cytotoxicity and maintain homeostasis of the immune system. Recent research focusing on investigations of new developments in immunomodulators has increased. Hence, this book chapter starts with introduction to immunomodulators and naturally existing plant-derived compounds and types. It also discusses about quercetin and its derived compounds with physicochemical properties and bioavailability. Furthermore, it highlights the mechanism of action of quercetin on the immune system with future challenges.

Keywords Immunomodulator · Quercetin · Bioavailability · Anticancer · Phytochemicals

4.1 Introduction

In recent times, therapeutic applications of medicinal plants and their derived products have played a major role in controlling and prevention of diseases. Immunomodulation is one of the approaches and homeostasis for host's defense system in impaired immune system (Singh et al. 2016). Even though natural plants have been used from ancient times, now, scientists isolated several innovative

D. V. Koraganji
Department of Biotechnology, GITAM School of Science, GITAM Deemed to be University, Visakhapatnam, Andhra Pradesh, India

A. Mounika · P. Sushanth · P. Kandra (✉)
Department of Biotechnology, GITAM School of Science, GITAM Deemed to be University, Visakhapatnam, Andhra Pradesh, India

Department of Biotechnology, GITAM School of Technology, GITAM Deemed to be University, Visakhapatnam, India
e-mail: pkandra@gitam.edu

R. K. Kesharwani et al. (eds.), *Nutraceuticals and Functional Foods as Immunomodulators*, https://doi.org/10.1007/978-981-19-2507-8_4

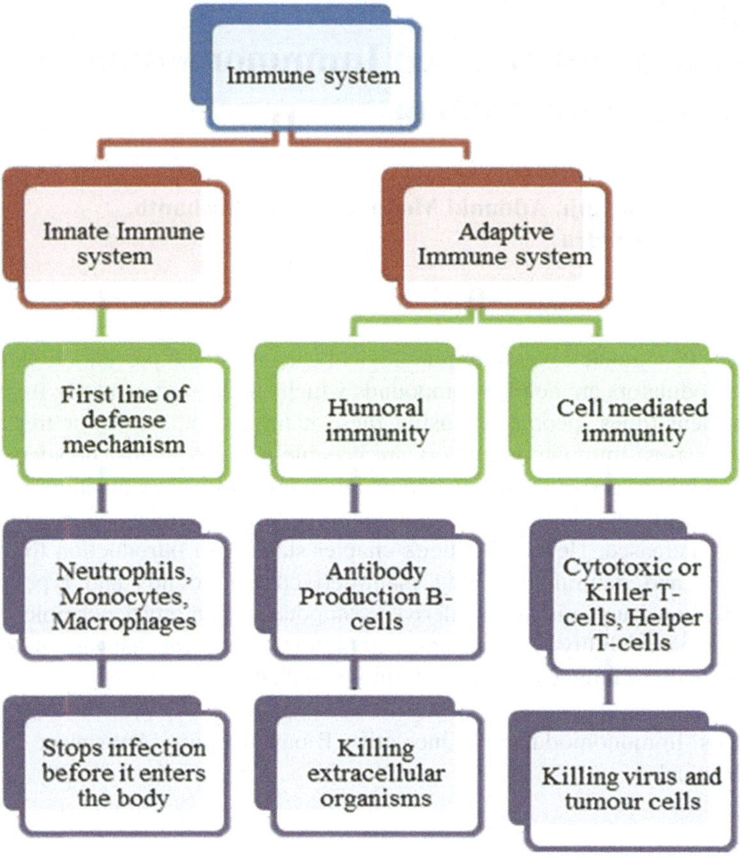

Fig. 4.1 Overview of the immune system

compounds which can be recognized as immune modulators. The primary function of the immune system is to protect the host against several disease-causing microorganisms. The collective action of immune cells in recognition of foreign pathogen and activation of cascade networks will help protect from nonself-antigens and self-antigens (Kharkar et al. 2017). Basically, innate and adaptive immune systems are primary defense mechanisms against pathogens. A combined effort of cells like macrophages, dendritic cells, granulocytes, and natural killer cells causes protection of an organism, and the acquired adaptive immune response is a slow process and is mediated by lymphocytes. It recruits different antigen receptors which give rise to DNA rearrangement in mammalian cells (Wen et al. 2012). Figure 4.1 shows the overview of both responses.

However, cell mediated immune responses are mediated by T-helper cells and cytotoxic T cells through the secretion of cytokines which stimulates various pathways to kill microorganisms in an efficient manner. Cytotoxic T cells are not only involved in killing of virally infected cells but also altered self-cells. In response

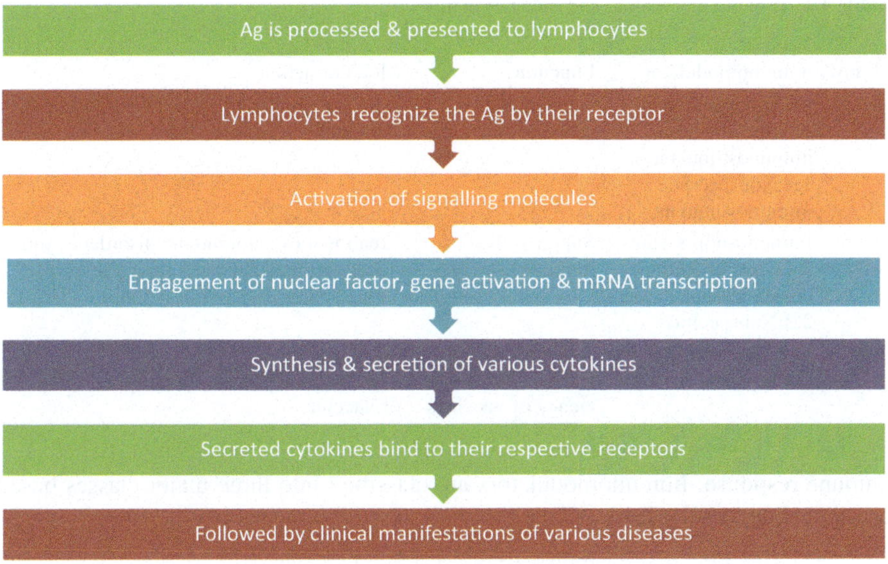

Fig. 4.2 Immune response in an orderly sequence of events

to the invaders, the immune system activates the B lymphocytes to proliferate into antibody-producing plasma cells. These secreted antibodies neutralize the antigen by forming cross-linked clusters that can be phagocytosed by phagocytic cells and eliminate the pathogen by several ways (Kharkar et al. 2017). Figure 4.2 shows the sequence of events that occur during immune response.

4.2 Immunomodulators

The compounds or agents involved to improve immunological response are called as immunomodulators. Most of immunomodulators are in developmental stage. They are used for transcription factor modulation, cytokine receptors and cell surface molecules, gene activation and signalling sequences. The immunomodulation approaches may be pathogen-specific and nonspecific to pathogens. Vaccines and antibody reagents are pathogen-specific type and antimicrobial agents, cytokines, and probiotics are nonspecific-type immunomodulators (Smit et al. 2009). Thus, immunomodulators function by stimulating or suppressing the natural and adaptive immune mechanism, which helps in tackling the body to various medical conditions. Immunomodulators can be derived from natural or synthetic sources. The IM compounds derived from synthetic sources have drawback of toxicity and other side effects. In order to reduce the toxic conditions, present research is focusing on natural immunomodulators. Natural IM derived from plant sources shows a potential

Table 4.1 Immunomodulators and mode of action

S. no	Type of immunomodulator	Function	Mode of action
1	Immunostimulants 1.1 Specific immunostimulant 1.2 Nonspecific immunostimulant	Activate the immune system	Enhance the defense mechanism. Used in the treatment of cancer and infection
2	Immunosuppressants 2.1 Specific immunosuppressant 2.2 Nonspecific immunosuppressant	Suppress the immune system	Treatment of autoimmune disorders. Anti-rejection medication in organ transplant
3	Immunoadjuvants	Enhance the efficiency of vaccines	Immune enhancer of antigenic component of vaccine

immune response. Immunomodulators are classified into three major classes based on clinical approach (Table 4.1).

4.2.1 *Natural Immunomodulators*

Plants contain secondary metabolite chemicals in the form of biomolecules such as carbohydrates, lipids, and proteins which humans are dependent and also contain several biologically active components such as alkaloids, glycosides, etc. These natural compounds have immunomodulating activity (Shukla et al. 2012). Among the secondary metabolites, flavonoids show strong antioxidant, anti-inflammatory, immunomodulatory, and antiproliferative activities. Immunosuppression can occur in aging, malnutrition, cancer and unwanted immunosuppresants may lead to increase in suseptibility to various infectious agents in chronic infections (Abbas et al. 2012; Kharkar et al. 2017).

4.2.2 *Plant-Derived Immunomodulators*

phytocompounds of plant extracts have immunomodulatory activity and they can increase the host's immune system against the pathogens by specific or nonspecific manner in an innate and adaptive immune responses. The mode of action could be immunostimulators, immunosuppressors, or immunoadjuvants to increase the antigen-specific immune response. Below, phytocompounds with immunomodulatory activity are elucidated (Table 4.2).

Immune system is responsible for maintaining homeostasis inside the body. Immune response is administered through various immune cells (lymphoid and myeloid cells) and their metabolic by-products (immunoglobulins, interleukins,

Table 4.2 Phytochemical compounds with immunological activity

S. no.	Phytochemical	Compound name	Source	Type of activity	References
1	Polyphenols	Resveratrol (belong to stilbenes)	Grapes (*Vitis vinifera* L.), *Polygonum cuspidatum*	Inflammation in pulmonary artery, cardiac, diabetes, cancer, and Alzheimer's diseases	Wen et al. (2012)
		EGCG (belongs to flavanol)	Tea (*Camellia sinensis*)	Inflammation in cardiovascular, hepatitis, aging process	Wen et al. (2012)
		Quercetin (belongs to flavonol)	Grapes, tea, onions, apples, berries	Inflammation in cancer, aging, and hepatitis	Wen et al. (2012)
2	Glycosides	Dendroside A (belong to sesquiterpene)	*Dendrobium nobile* (stems)	Stimulates T and B lymphocytes	Kumar et al. (2012)
		Esculetin (6,7-dihydroxycoumarin) (belong to coumarin)	*Citrus limonia*, *Euphorbia lathyris*	Suppress the radical-scavenging activity and anticancer activity	Kumar et al. (2012)
3	Terpenoids	Andrographolide (belong to diterpenoid lactone)	*Andrographis paniculata*	Interleukin-12, tumor necrosis factor, and COX-2 levels are reduced	Jantan et al. (2015)

interferons, colony-stimulating factors, etc.). While innate and adaptive immunities function complementarily to provide the human body with overall defense against pathogens, they emerged during different evolutionary periods. General mechanisms of innate immunity is present in vertebrates, invertebrates, and also plants, whereas adaptive immune response is unique to vertebrates. The immune system's function and efficacy are governed by several exogenous and endogenous factors that result in either in immunostimulation or in immunosuppression. Immunity is the mechanism employed by the body to protect itself from different infectious diseases. Immunisation prior to infection, and various external stimuli are the major factors that trigger the immunity. Furthermore, the immune system can differentiate between the self-cell and nonself-cells inside the body once a foreign particle is detected (Baxter 2007). Biomolecules of synthetic or biological origin have the ability to modulate, suppress, and stimulate any component of adaptive or innate immunity known as immunomodulators, immunorestoratives, immunoaugmentors, or modifiers of biological response.

Immunomodulators are usually divided into immunoadjuvants, immunostimulants, and immunosuppressants. Immunoadjuvants are unique immune stimulators that strengthen vaccine effectiveness. Immunostimulants improve resistance toward autoimmune diseases, cancer, asthma, and other infections. Immunosuppressants are the compounds that suppress the immune system; they can be used to regulate the pathological immune response which occurs after organ transplantation. In addition, these agents can also be used to treat infection-associated immunopathology, reactions to hypersensitivity, and autoimmune diseases.

Immunomodulatory compounds with high efficacy and safety standards are still in demand. Natural immunomodulators are likely to replace synthetic chemical drugs in medicinal therapies, due to adverse effects linked with chemical drugs. Most research and development are currently focused on biochemicals or compounds that target specific disease-related targets. Compounds with high potency and selectivity with low toxicity may target diseases. Therefore, the design and development of drug candidates from a wide range of traditional or alternative and complementary medicine are gathering interest (Jantan et al. 2015).

Flavonoids are secondary metabolites of phenolic-type compounds which are available in different forms like flavan, isoflavan, etc. This is rich in several foods most especially abundant in vegetables and different berries. The foods that are included are *Brassica* family vegetables, tomatoes, onions, green tea, berries, and nuts and some medicinal plant leaves and barks and flowers such as *Ginkgo biloba*, *Hypericum perforatum*, and *Sambucus canadensis*. The concentration of the availability of quercetin varies with variety of plants and their parts. Quercetin is a flavonoid that contains double bonds with ketone as functional group (apigenin) and hydroxyl group in the form of flavanols (quercetin, myricetin, etc.) and also consists of dihydrogenated flavanone (hesperetin, naringenin) and flavanols (taxifolin and dihydrokaempferol). Most predominant flavanol is quercetin (3,5,7,3′,4′-pentahydroxyflavone), rutin, and isoquercetin (Valentová et al. 2016; Istudor 1998; Chabane et al. 2009). Figure 4.3 shows the structure of quercetin.

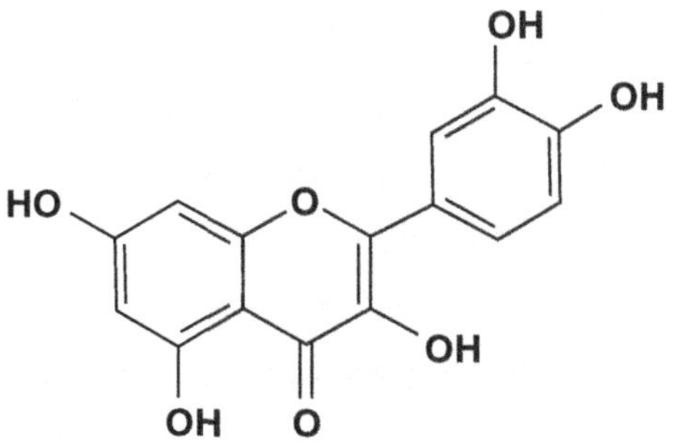

Fig. 4.3 Structure of quercetin (Source: Mlcek et al. 2016)

4.2.3 Physiochemical Properties of Quercetin

Quercetin is derived from the quercetun a oak forest since 1857. It is a flavonoid compound. It is a polar auxin transport inhibitor naturally. The International Union of Pure and Applied Chemistry (IUPAC) nomenclature for quercetin is 3, 31, 41, 5, 7-pentahydroxyflvanone (or its synonym 3, 31, 41, 5, 7-pentahydroxy-2-phenylchromen-4-one). It has five hydroxyl groups in its structure at different positions. It is an aglycone having molecular formula of $C_{15}H_{10}O_7$ (Chabane et al. 2009); hence, it is insoluble in cold water but soluble very slightly in hot water, alcohol, and lipids. It will form glycoside bond with sugar residues like rhamnose and glucose; hence, the solubility and absorption increase. Quercetin has brilliant citron yellow needle crystal. When a glycosyl group is attached to quercetin, it forms quercetin glycoside having high water solubility than quercetin aglycone (Li et al. 2016). Figure 4.4 shows the different forms of quercetin.

4.3 Bioavailability of Quercetin

The absorption of quercetin and its derivative are different from one another depending upon the type of sugar. Quercetin glycosides are better absorbed than its rutinosides which are found in tea. These glucosides are hydrolyzed to aglycone with the help of beta-glucosidases and then absorbed effectively (Ader et al. 2000). Quercetin sulfuric acid derivatives absorbed more quickly than quercetin. Hence, the absorption efficiency of quercetin is also depends on the source from which it is derived and also the dietary components such as fat and fiber (Guo et al. 2013). Quercetin and its derivatives are stable to gastric acid. Once the absorption of quercetin is completed, then it is metabolized in tissues like the liver, intestine, kidney, and colon where it is transformed to phenolic acids by bacteria and enzymes.

Aglycone in the small intestine and colon resulted in the breakdown of quercetin and formation of small phenolic compounds (Kim et al. 1999). These metabolites accumulate in the organs, which are involved in metabolism and also excretion, and also mitochondria (Konrad and Nieman 2014). The absorbed quercetin is

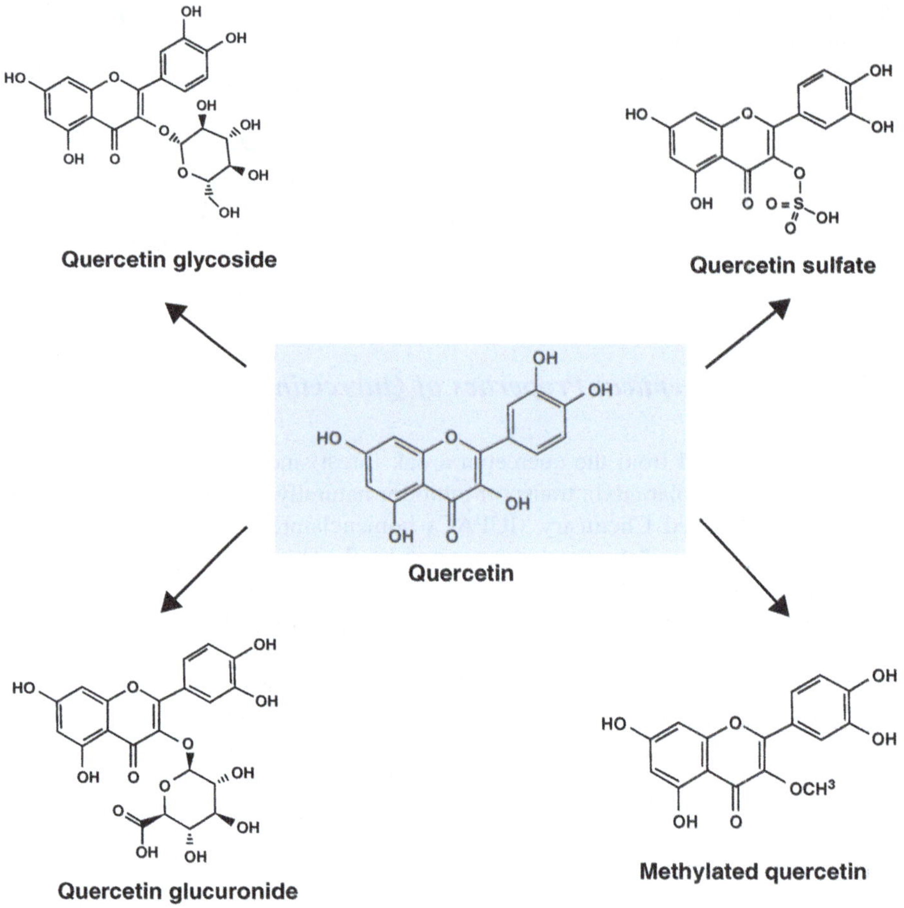

Fig. 4.4 Different forms of quercetin (Source: Li et al. 2016)

metabolized extensively and eventually eliminated by the lungs (Walle et al. 2001). Glycosylated quercetin is absorbed completely than quercetin in the aglycone form and that the simultaneous ingestion of quercetin with vitamin C, folate, and additional flavonoids improves bioavailability (Harwood et al. 2007). The enzymes of the small intestine may enhance absorption, and later, they are modified to methylated, sulfo-substituted, and glucuronidated by enzymes in the liver and are finally excreted in urine by the kidney.

4.4 Biological Functions of Quercetin

Quercetin is available in different forms and has shown unique properties which lead to several applications. The special biological functions of quercetin will reduce the risk of infections and also increase physical and mental ability. In the organism due to physical or chemical and biological factors, the production of free radicals is increasing which may damage the tissues. Quercetin will neutralize the free radicals especially smoke that damages erythrocyte membrane (Begum and Terao 2002). This property of quercetin will have potential effect on resistance to toxic chemicals. It is ubiquitous, having antiviral, anticarcinogenic, antioxidant, psychostimulant, and anti-inflammatory activities. Quercetin can inhibit lipid peroxidation and restorative mitochondrial biogenesis platelet and increases capillary permeability. It has antiallergic property and modulates EGFR-mediated pathways. Inhibition of cyclo-oxygenase (COX) and lipoxygenase pathways may prevent the production of pro-inflammatory molecules. It also inhibits xanthine oxidase and averts uric acid accumulation (Ahmad et al. 2008). One of the applications of quercetin is inhibit the viral replication and also prevent the formation of reactive oxygen and nitrogen species in MAPK and NF-kB pathways (Boots et al. 2008). Studies also refer to the reduction of LDL levels when they consume food rich in quercetin content. Oxidative damage of neurons can be inhibited, with the combination of ascorbic acid reducing leukocyte destruction and also oxidative stress in neurological disorders like Alzheimer's and other diseases (Lakhanpal and Rai 2007). Several other hypersensitivity reactions like asthma and bronchitis can be reduced by less production of histamine; hence, quercetin is considered as antihistamine. Quercetin is potent anticarcinogenic agent and reduces the growth of tumor in various tissues like the liver, brain, and colon and also will reduce the spread of tumor (Vásquez-Garzón et al. 2009).

4.5 Mechanism of Action of Quercetin

Lipopolysaccharide (LPS) is a bacterial endotoxin which is an inducer of inflammation and microglia activation. Treatment with quercetin inhibits the secretion of TNF-α production in macrophage cell. Furthermore, LPS treatment of A549 cells in the presence of quercetin results in the decrease of IL-8 production. The inhibitory effect of quercetin in LPS-induced mRNA-level TNF-α and interleukin (IL)-1α reduces the apoptotic neuronal death mediated by microglial activation. It inhibits the production of both cyclooxygenase (COX) and lipoxygenase (LOX), which are inflammation-producing enzymes. In-vitro treatment of quercetin on activated T cells blocks IL-12 induced tyrosine phosphorylation of Janus kinase, signal transducer and activator of transcription factors which resulted in reduced interleukin-12 induced proliferation and also differentiation of Th-1.

The effect of quercetin on immune system is by targeting numerous intracellular signaling kinases and phosphatases and membrane proteins. These are frequently crucial for a cellular function by either downregulation or suppression of the pathway and functions (Chirumbolo 2010). Few studies shows that LPS-induced inflammation is by several mediated kinases. The effect of quercetin on human cell lines is as follows: one of the studies showed the inhibitory effect of quercetin on human umbilical cord blood-derived cultured mast cells (hCBMCs) secretion of FcεRI-mediated release of pro-inflammatory cytokines, tryptase, and histamine, as well as on intracellular calcium ion levels and phosphoprotein kinase C (PKC). Another study shows that quercetin against H_2O_2-induced inflammation shows the protective effects of quercetin against inflammation in human umbilical vein endothelial cells (HUVECs) via the downregulation of vascular cell adhesion molecule 1 (VCAM-1) and CD80 expression (Yao Li et al. 2016).

The immunostimulatory effects of quercetin may be mediated through the induction of Th-1-derived cytokine, IFN-γ, and inhibition of Th-2-derived cytokine interleukin (IL-4) by normal peripheral blood mononuclear cells (PBMC). Furthermore, studies also explained that treatment of quercetin increased the phenotypic expression of interferon (IFN-γ) cells and decreased IL-4 positive cells by flow cytometry analysis, which give support to protein secretion and gene expression studies (Nair et al. 2002). In human dermal fibroblasts and matrix metalloproteinases, which are normally inhibited by plasminogen activator inhibitor 1 (PAI-1), studies show that quercetin is able to inhibit them (Lim and Hyun 2007). Another study explains that quercetin is able to inhibit the selective mast cell activation by inhibiting interleukin (IL-6) secretin induced by IL-1. IL-1 induce the stimulation of IL-6 production from human mast cells which is regulated by biochemical pathway distinguished from IgE-induced degranulation. Therefore, quercetin can block both IL-6 secretin and two key steps involved in signal transduction pathways (Kandere-Grzybowska et al. 2006; Penissi et al. 2003). It also showed in both effects basophils at nanomolar doses and therefore effect on allergic inflammation cells (Muthian and Bright 2004). The below flowchart explains the working model on ERK (extracellular signal-regulated kinase), JNK (c-Jun NH2-terminal kinase), NF-κB (nuclear factor-κB), and AP-1 (activator protein-1). ERK, JNK, c-Jun, and NF-κB are potent inducers of inflammatory gene expression and protein secretion and how quercetin block tumor necrosis factor (TNF-α) (Yao Li et al. 2016) (Fig.4.5).

4.6 Conclusions

Quercetin is a natural immunomodulator with several biological functions. As it is having poor solubility property, its use is restricted for oral administration. It is an interesting challenge to improve the solubility of quercetin to enhance the therapeutic usage. Although several modified quercetin forms are available, bioavailability is a future emerging challenge. However, literature to biological functions is available,

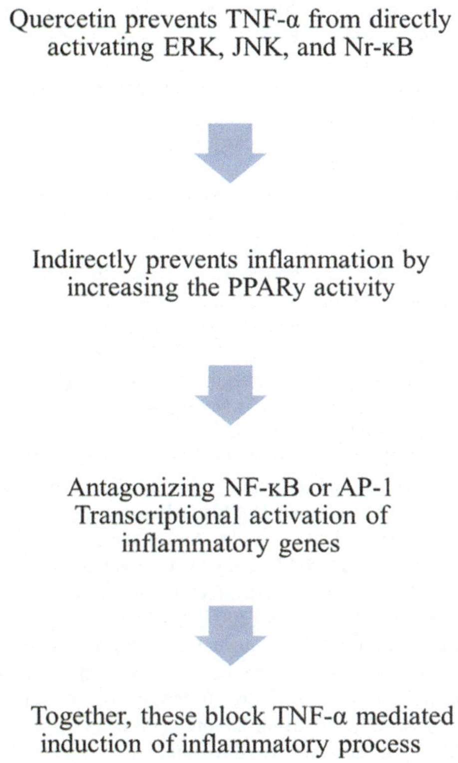

Fig. 4.5 Action of Quercetin

but research is needed to understand the mechanisms to evaluate the prognosis of immunomodulator research for therapeutic functions.

Competing Interest The authors declare that there are no competing interests.

References

Abbas AK, Lichtman AH, Pillai S (2012) Basic immunology: functions and disorders of the immune system. Elsevier Health Sciences, New York

Ader P, Wessmann A, Wolffram S (2000) Bioavailability and metabolism of the flavonol quercetin in the pig. Free Radic Biol Med 28(7):1056–1067

Ahmad NS et al (2008) Pharmacological basis for use of Pistacia integerrima leaves in hyperuricemia and gout. J Ethnopharmacol 117(3):478–482

Baxter D (2007) Active and passive immunity, vaccine types, excipients and licensing. Occup Med 57(8):552–556

Begum AN, Terao J (2002) Protective effect of quercetin against cigarette tar extract-induced impairment of erythrocyte deformability. J Nutr Biochem 13(5):265–272

Boots AW, Haenen GRMM, Bast A (2008) Health effects of quercetin: from antioxidant to nutraceutical. Eur J Pharmacol 585:325–337

Chabane MN, Ahmad AA, Peluso J, Muller CD, Ubeaud-Séquier G (2009) Quercetin and naringenin transport across human intestinal Caco-2 cells. J Pharm Pharmacol 61(11): 1473–1483
Chirumbolo S (2010) The role of quercetin, flavonols and flavones in modulating inflammatory cell function. Inflamm Allergy Drug Targets 9(4):263–285
Guo Y et al (2013) Dietary fat increases quercetin bioavailability in overweight adults. Mol Nutr Food Res 00:1–10
Harwood M et al (2007) A critical review of the data related to the safety of quercetin and lack of evidence of in vivo toxicity, including lack of genotoxic/carcinogenic properties. Food Chem Toxicol 45:2179–2205
Istudor V (1998) Pharmacognosy, phytochemistry, phytotherapy. In: Glycosides and lipids, 1
Jantan I, Ahmad W, Bukhari SNA (2015) Plant-derived immunomodulators: an insight on their preclinical evaluation and clinical trials. Front Plant Sci 6(655):1–18
Kandere-Grzybowska K et al (2006) Regulation of IL-1-induced selective IL-6 release from human mast cells and inhibition by quercetin. Br J Pharmacol 148:208–215
Kharkar PB et al (2017) Nanosystems for oral delivery of immunomodulators, nanostructures for oral medicine. Elsevier
Kim DH, Kim SY, Park SY, Han MJ (1999) Metabolism of quercitrin by human intestinal bacteria and its relation to some biological activities. Biol Pharm Bull 22(7):749–751
Konrad M, Nieman D (2014) Evaluation of quercetin as a countermeasure to exercise-induced physiological stress. In: Antioxidants in sport nutrition
Kumar D et al (2012) A review of immunomodulators in the Indian traditional health care system. J Microbiol Immunol Infect 45:165–184
Lakhanpal P, Rai DK (2007) Quercetin: a versatile flavonoid. Internet J Med Update 2(2):22–37
Li Y, Yao J, Han C, Yang J, Chaudhry MT, Wang S, Liu H, Yin Y (2016) Quercetin, inflammation and immunity. Nutrients 8(3):167. https://doi.org/10.3390/nu8030167
Lim H, Hyun PK (2007) Inhibition of mammalian collagenase, matrix metalloproteinase-1, by naturally-occurring flavonoids. Planta Med 73:1267–1274
Mlcek J, Jurikova T, Skrovankova S, Sochor J (2016) Quercetin and its anti-allergic immune response. Molecules (Basel, Switzerland) 21(5):623. https://doi.org/10.3390/molecules21050623
Muthian G, Bright JJ (2004) Quercetin, a flavonoid phytoestrogen, ameliorates experimental allergic encephalomyelitis by blocking IL-12 signaling through JAK-STAT pathway in T lymphocyte. J Clin Immunol 24(5):542–552
Nair MPN et al (2002) The flavonoid, quercetin, differentially regulates Th-1 (IFNγ) and Th-2 (IL4) cytokine gene expression by normal peripheral blood mononuclear cells. Biochim Biophys Acta 1593:29–36
Penissi AB, Rudolph MI, Piezzi RS (2003) Role of mast cells in gastrointestinal mucosal defense. Biocell 27(2):163–172
Shukla S, Bajpai VK, Kim M (2012) Plants as potential sources of natural immunomodulators. Rev Environ Sci Biotechnol 13:17–33
Singh N, Tailang M, Mehta SC (2016) A review on herbal plants as immunomodulators. Int J Pharm Sci Res 7(9):3602–3610
Smit E, Oberholzer HM, Pretorius E (2009) A review of immunomodulators with reference to Canova®. Homeopathy 98:169–176
Valentová K et al (2016) (Anti)mutagenic and immunomodulatory properties of quercetin glycosides. J Sci Food Agric 96:1492–1499
Vásquez-Garzón V et al (2009) Inhibition of reactive oxygen species and pre-neoplastic lesions by quercetin through an antioxidant defense mechanism. Free Radic Res 43(2):128–137
Walle T, Walle UK, Halushka PV (2001) Carbon dioxide is the major metabolite of quercetin in humans. J Nutr 131(10):2648–2652
Wen CC, Chen HM, Yang NS (2012) Developing phytocompounds from medicinal plants as immunomodulators. In: Advances in botanical research, 1st edn. Elsevier

Chapter 5
Polyphenols and Its Effect on the Immune System

Kanchan Gairola, Shriya Gururani, and Shiv Kumar Dubey

Abstract The mainstay of good health is well functioning and strong immune system which involves interaction between acquired and innate components of the immune system that leads to immune responses. In the recent few years, polyphenols are extensively studied for their immunomodulatory properties to achieve the preferable and desirable effects. Polyphenolic compounds exert many biological activities and are known to considerably affect the efficiency of the immune system either in a positive or in a negative manner by either stimulating or suppressing the immune response; hence, they are regarded as important immunomodulatory agents. As these bioactive leads generally show minimal toxicity and have better efficacy, a lot of scientific studies are focused on their clinical uses. The aim of this chapter is to focus on the adequacy and effect of the plant's polyphenol, such as tannins, lectins, flavonoids, glycosides, anthocyanins, and other phenolic compounds on the immune system at the molecular level and their mechanism of immunomodulation. Keeping the importance of polyphenols in medicinal field and immunomodulation in the view, this chapter mainly highlights the effect of polyphenol on modulation of the immune system.

Keywords Immune system · Polyphenols · Flavonoid · Cell signaling pathway · Antioxidant · Anti-inflammation

5.1 Introduction

The extensive study of immune mechanism is generally correlated to pathological context as immune function is mainly affected by pathogen leading to immune imbalances and dysfunction. Other causes that affect the immune system are physiological stress. Nowadays, intensive and prolonged exertion results into perturbation of the immune system that leads to malfunctioning of multiple compartments of

K. Gairola · S. Gururani · S. K. Dubey (✉)
Department of Biochemistry, C.B.S.H., G. B. Pant University of Agriculture and Technology, Pantnagar, Uttarakhand, India

R. K. Kesharwani et al. (eds.), *Nutraceuticals and Functional Foods as Immunomodulators*, https://doi.org/10.1007/978-981-19-2507-8_5

the body due to which the human body experiences oxidative stress, muscle microtrauma, inflammation, and suppressed function against pathogen (Nieman et al. 2007). The influence of certain nutritional countermeasure is found to be effective against these outcomes related to immune dysfunction (Nieman and Bishop 2006). Dietary interventions and functional food are considered to be potent immunomodulators as it is contemplated to be targeting multiple immune functional components.

Polyphenols, one of the many nutritional interventions, are plant metabolites, generally found in fruit and vegetables contributing their flavor and color and response to pathogens and related damages, are known for their pleiotropic pharmacological activities, and are well documented for its immunomodulatory activities. Polyphenols include flavonoids, tannins, stilbenes, phenolic acids, and their various chemically modified or polymerized derivatives. Polyphenols are secondary metabolites of plant that are produced ubiquitously in plants and have a defensive mechanism to protect the plant through hazardous environment. Numerous studies and prevailing evidence show that polyphenols not only can potentially prevent several immune diseases but also can directly target the pathogen in the living system affecting the immune processes (Ding et al. 2018).

Polyphenol present in red wine can induce interleukin-(IL-) 21 levels and reduce the IL-1β and IL-6 level significantly (Magrone et al. 2017). Polyphenols have the ability to inactivate prooxidant, scavenge free radicals, and regulate systemic inflammation by inhibiting various cell signaling pathways that trigger inflammation in the body. They are considered to be very effective in regulating innate and adaptive immunity as they can modulate different type of cytokines involved in inflammation process (Mao et al. 2017). Modulations of certain inflammatory process by polyphenols involve either individual mechanism or synergistic effects. Certain mechanisms modulated by polyphenol include altering the signaling and enzymatic processes involved in inflammation such as tyrosine and serine-threonine protein kinases which involve B-lymphocyte activation and T-cell proliferation. They are also involved in inhibiting the key inflammatory mediator, NF-κB, inducible nitric oxide synthase (iNOS), pro-inflammatory enzymes such as COX-2, MAPK, and protein kinase-C. Polyphenols exhibit blunt effects on inflammatory cell secretions, scavenge free radicals and inflammatory prooxidant such as superoxide anions and hydrogen peroxide, and protect cells from oxidative stress. Polyphenols modulate inflammatory mediators such as cytokines, peptides, and arachidonic acid (Hussain et al. 2016).

5.2 Structure and Function of Polyphenols

Polyphenols are generally characterized as aromatic rings with hydroxyl groups or multiple phenol structural units (Cheynier 2005). Polyphenols are generally classified into flavonoids and non-flavonoids (Fig. 5.1) based on their chemical structure, or it can be subdivided into many subclasses depending on the presence of phenol

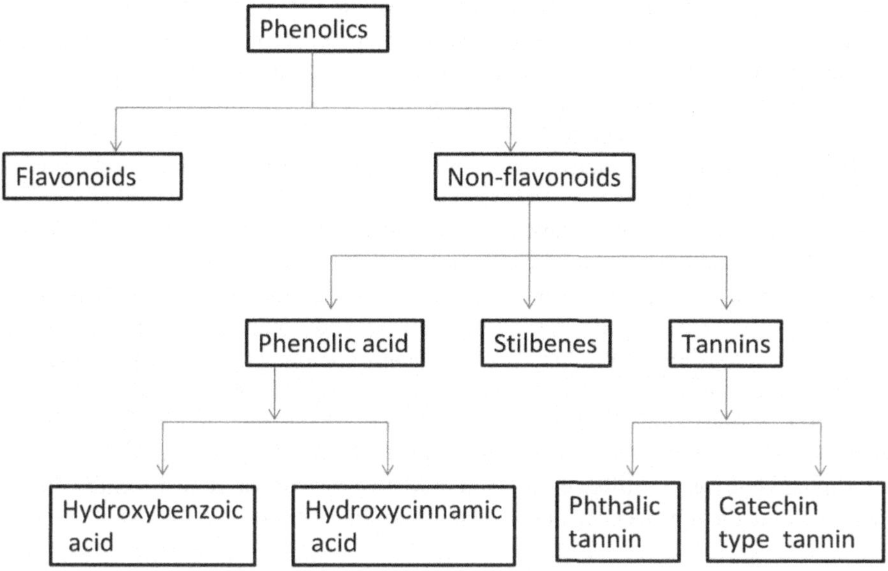

Fig. 5.1 Classification of polyphenols

units, substituent groups, or the linkage type between phenol units (Cardona et al. 2013). Flavonoid comprises largest group of polyphenols and are most structurally diversified compound among polyphenols because of the variation in their hydroxylation pattern and in oxidation state of central pyran ring. Flavonoids include flavones, flavanols, anthocyanins, isoflavones, anthocyanidins, flavanones, and flavanonols. Non-flavonoid includes phenolic acids, tannins, and stilbenes. Basic structure of non-flavonoids is single aromatic ring. This group mainly consists of phenolic acid predominantly benzoic acid and cinnamic acid derivatives (Singla et al. 2019).

Polyphenols are generally present as a complex mixture in food, and they do not get absorbed easily in a natural form as they exist in the form of esters, glycosides, or polymers. Some of the polyphenols may be absorbed in small intestine, and some have pro-host effect, but those polyphenols which tend to be not absorbable are hydrolyzed enzymatically by the intestine to facilitate the absorption. Polyphenols with different structure have different function. In previous studies, it is reported that polyphenols have various activities including antioxidant, anti-inflammatory, and antitumor properties (Kim et al. 2014; Romier et al. 2009; Szliszka and Krol 2013). Flavonoids generally consist of benzophenone structure with more than one aromatic ring; each has one or more hydroxyl group that is connected by Carbon Bridge (Szliszka and Krol 2011). Phenolic acid which is a non-flavonoid secondary metabolite averts damage caused by ultraviolet rays in plants. It also protects plant from insects, viruses, bacteria, and other microorganisms. Some plant produces phenolic acids which are known to inhibit the growth of their plant competitors (Heleno et al. 2015). Subgroup of phenolic acid is hydroxybenzoic acid and hydroxycinnamic acid

(Taofiq et al. 2015). In a study, it is demonstrated that in cell RAW264.7, the production of TNF-α which is stimulated by lipopolysaccharide (LPS) inhibited by ferulic acid (Park et al. 2011). In an immunological-related study, it was reported that the activity of interferon gamma-induced protein in response to TNF is significantly decreased by phenyl ethyl caffeate. The study also reports the strong effect of phenyl ethyl caffeate against the activation of nuclear factor-(NF-) kB and production of lymphoid factor (Wang et al. 2009). Tannins are non-flavonoid phenolic compounds, generally present in bark and fruits of many plants; they are well known for its anticarcinogenic, antioxidant, and antimutagenic activity. The structure of tannin has central carbohydrate and ten galloyl group (Lopes et al. 1999). In a report, the effect of total 67 tannins and structurally related compounds is tested against *Leishmania* parasite RAW 264.7 cells. Tannins show effective antileishmanial activity, and it also shows its effect on macrophage activation for the release of TNF, NO (nitric oxide), and interferon activities (Kolodziej and Kiderlen 2005). Stilbenes are that class of compound, which is characterized by the 1,2-diphenylethylene. Stilbenes are reported to have remarkable potential in the field of biomedical especially in the human immune system.

5.3 Immunomodulatory Mechanism of Polyphenols

After decades of studies, researchers develop the insight on the effect of polyphenols on immune system functions. Earlier, it was reported that polyphenols regulate host immune response by interfering cell signaling pathways. It binds to one or more than one receptor, triggering cell signaling that ultimately leads to the regulation of the immune system. Polyphenols interfere cell regulation, pro-inflammatory cytokine synthesis, and gene expression. Polyphenols are known to modulate epigenetic mechanisms and have effect on enzymes too. It is reported that polyphenols are involved in inhibiting certain enzyme that is responsible for ROS (reactive oxygen species) production such as NADPH oxidase (NOX) and xanthine oxidase. It is also involved in upregulating the endogenous antioxidant, modulation of mitogen-activated protein kinase, and arachidonic acid pathways. Polyphenol immunomodulatory mechanism involves inhibition of NF-κB, phosphatidylinositide 3-kinases/protein kinase B, kappa kinase/c-Jun amino-terminal kinases (IKK/JNK), and mammalian target of rapamycin complex 1 (mTORC1) which is a protein complex that controls protein synthesis and inhibition of JAK/STAT pathway. Pro-inflammatory genes and Toll-like receptor are also suppressed by polyphenols (Yahfoufi et al. 2018). Polyphenols have their effect on different types of immune cell. In a study, it was reported that red wine, having polyphenols, can induce the production of NO from human monocytes that leads to the vasodilatory action and could prevent atherosclerosis. Dihydroxyl phenolic acid is known for exhibiting anti-inflammatory properties. It reduces the secretion of IL-6, TNF-α, and IL-1β from PBMCs (peripheral blood mononuclear cells) (Monagas et al. 2009). Their experimental data suggest that polyphenols can be used for the treatment of controlling tumor growth

(Oršolić et al. 2004). Researchers while working with CBA mouse on metastatic effect of transplantable mammary carcinoma (MCa) reported that polyphenolic compound such as caffeic acid (CA), caffeic acid phenethyl ester (CAPE), quercetin (QU), and water-soluble derivative of propolis (WSDP) significantly decreases the tumor nodules present in the lungs.

Flavonoids are responsible for downregulating the expression of cyclooxygenase (COX)-2 which is demonstrated in lipopolysaccharide (LPS)-stimulated J774A.1 cells. Similar effect was shown by citrus polymethoxy flavones and quercetin in human synovial fibroblast and in mouse macrophage, respectively. Polyphenols present in green tea is reported to have the ability of suppressing mRNA and protein expression of COX-2 in the RAW 264.7 cells. Similarly, genistein shows its effect in colon cancer cell by downregulating the COX-2 promoter (González-Gallego et al. 2010). Polyphenols are involved in the production of eicosanoid that contributes to the anti-inflammatory property of polyphenols. It also inhibits enzymes that are involved in the process of ROS production such as NADPH oxidase (NOX) and xanthine oxidase and enhances the upregulation of enzymes such as catalase, peroxidase, and superoxide dismutase (Bengmark 2004). Polyphenols are associated with the regulation of T regulatory (T-reg)/T-helper 17 (Th17) balance and the level of plasma and intestinal mucosal cytokines including interleukin 10 (IL-10), transforming growth factor-beta1 (TGF-β1), interleukin 6 (IL-6), and interleukin 17 (IL-17). Polyphenols also have an activity against NF-κB activation; they inhibit its activation and suppress the growth of dextran sulfate sodium (DSS)-induced colitis (Mileo et al. 2019).

5.4 Polyphenol's Regulatory Effects on Different Immune Responses

Polyphenols have immense application in insulin resistance, inflammation, neurological disorders, allergic responses and in carcinogenesis involving modulation of many signaling pathways (Table 5.1).

5.4.1 Polyphenol and Insulin Resistance

Polyphenols regulate insulin by reducing its resistance. It generally promotes glycolysis by inhibiting mTORC1 and PI3K/AkT and activating the AMP-activated protein kinase. Polyphenols by activating the AMPKs effectively increase the uptake of glucose. Activation of AMPK positively affects eNOS imitating muscle contraction and activity of insulin which in turn increases the glucose uptake (Fryer et al. 2000). It was demonstrated in a study that polyphenols inhibit p/AkT and JNK activation of the AMPK-SIRT1-PGC1α axis by which it lowers the insulin

Table 5.1 Cell signaling pathway that is mainly targeted by polyphenols (Khan et al. 2019)

Polyphenols	Signaling pathway	Actions	References
Quercetin, kaempferol, galangin, catechol (1,2-dihydroxybenzen), resveratrol, p-coumarin, myricetin, morin	Arachidonic acid-dependent pathway	Inhibiting LOX enzyme Inhibiting PLA2 Inhibiting COX enzyme	Baumann et al. (1980) Laughton et al. (1991) Makanjuola et al. (2018)
Resveratrol, quercetin, EGCG, ellagic acid,	Phosphatidylinositol 3-kinase/protein kinase B (PI3K/Akt) signaling pathway	Downregulating the expression of pro-inflammatory cytokines by alteration of PI3K/Akt signaling pathway Protect HUVECs from apoptosis mediated by PI3K/Akt activation Neuroprotection in rats undergoing ischemia induced cerebral damage	Abdel-Aleem et al. (2016) Venkatachalam et al. (2008) Yang et al. (2018a, b)
Resveratrol Fisetin Curcumin Quercetin, myricetin EGCG	Epigenetic modulation (DNA methylation, histone modifications, and posttranscriptional regulation by miRNAs)	Activation of SIRT1 resulting in the inhibition of NF-κB and its downstream genes such as COX-2 and iNOS Increase SIRT1 and SIRT3 expression Inhibiting DNMTs Regulation of HATs and HDACs Regulate miRNA expression (miR-181b) Reactivate the neprilysin gene leading to inhibition of NF-κB Downregulate c-Myc, PHD2, and β-catenin expressions via SIRT1 activation in a manner that mimics hypoxic preconditioning Hypoacetylation of p65 by inhibiting the activity of HAT which results in reduction of the activity of NF-κB	Boyanapalli and Kong (2015) Choi et al. (2009) Cordero-Herrera et al. (2017) Hong et al. (2012) Li et al. (2013) Song et al. (2016)
Quercetin, resveratrol, EGCG, apigenin, luteolin	MAPK pathway	Suppressed the phosphorylation of NF-κB, ERK1/2, JNK, and p38 pathway proteins which alter the production and expression of pro-inflammatory Cytokines	Chen et al. 2004, Yang et al. 2018a, b

(continued)

Table 5.1 (continued)

Polyphenols	Signaling pathway	Actions	References
Resveratrol, EGCG, quercetin, genistein	NF-κB signaling pathway	Modulating NF-κB signaling pathway which alters the genetic regulation of COX enzyme, pro-inflammatory cytokines, and chemokines	Chen et al. 2018, Wheeler et al. 2004

resistance and prevents diabetes and reduction of insulin. Polyphenols such as flavonoids effectively reduce apoptosis of pancreatic β-cells and improve insulin secretion (Chu 2014).

5.4.2 *Polyphenols and Inflammatory Neurological Diseases*

Neurological diseases include epilepsy, Parkinson's disease, Alzheimer disease, and other dementia. In earlier studies, polyphenols show protective effect in neurological disorder. Polyphenol such as flavonoid significantly reduces the aging and dementia by 50%. It is also helpful in flaying the onsets of diseases like Alzheimer's and Parkinson's diseases (Dai et al. 2006). Due to the activities like reduction in reactive oxygen species (ROS) and presence of cellular GSH contents, EGCG has neuroprotective properties. Polyphenols can modulate glial signaling pathway related to signaling in the brain and also neuronal signaling pathways. Polyphenols can downregulate NF-κB related with iNOS generation in glial cells (Vafeiadou et al. 2009; Singh et al. 2014).

5.4.3 *Polyphenols and Allergic Diseases*

Polyphenols significantly work against allergic reaction. It effectively regulates the pathways that lead to allergic reaction in the human body. Allergic disease can affect elder one as well as a newborn child. It can occur at any stage of life. Sometimes, it occurs due to the genetic predisposition. Polyphenols exhibit anti-allergic effects including inhibition of histamine release, leukocyte production, and reduction of pro-inflammatory cytokines (Di Meo et al. 2013). Polyphenols inhibit IgE antibody formation and regulate Th1/Th2 balance by which polyphenol might affect the binding of allergen-IgE complex to its receptor (FceRI) on mast cells and basophils and can affect the formation of allergen-IgE complex (Choi and Yan 2009).

5.4.4 *Polyphenol and Tumor*

Polyphenols have a very crucial role in regulatory activities and suppressing tumor formation by altering the cell signaling pathways and regulating enzyme kinases. Polyphenols induce tumor cell death (apoptosis) as determined by the activation of caspase-3 and caspase-9, mobilization of cytochrome C, and externalization of annexin V on the cell surface (Gomez-Cadena et al. 2016). It significantly inhibits the chymotrypsin like activity in proteasome and proliferation of multiple myeloma cells. It was reported that polyphenols bind to 20S proteasome and inhibit it to accompany the suppression of tumor cell proliferation (Mujtaba and Dou 2012). In earlier study, it was reported that dietary polyphenols inhibit cancer cell growth by inducing apoptosis of tumor cell. They mainly target the tumor necrosis factor-related apoptosis inducing ligand (TRAIL)-mediated apoptotic pathway. Most of the tumor cell has anti-apoptotic property in it, and polyphenols significantly target that tumor cell and induce apoptotic activity in it by targeting TRAIL cell (Szliszka and Krol 2011).

5.5 Different Types of Polyphenolic Compound and Their Effect on the Immune System

Polyphenols are generally secondary metabolites of plants and are reported to have more than 8000 structural variants in their family. Important polyphenols (Fig. 5.2) and their important immune responses are discussed in this section.

5.5.1 *Quercetin*

Quercetin is an aglycone categorized as flavanol, basically obtained from fruits and vegetable, known for its unique and exceptional biological properties for improving mental or physical performance, and reduces the risk for infection (Davis et al. 2009). It shows potential benefits toward improvement in overall health and acts as potential resistant toward carcinogens, inflammation, free radicals, and viral infection. Quercetin also has the ability to inhibit capillary permeability, platelet aggregation, and lipid peroxidation and is known for stimulating mitochondrial biogenesis (Aguirre et al. 2011). Quercetin is reposted to be responsible for the enhancement in neutrophil chemotaxis, respiratory burst activity, NK cell lytic activity, macrophage phagocytosis, and mitogen-stimulated lymphocyte proliferation (Gleeson 2006). Quercetin regulates the expression of some genes including transcription factor nuclear factor-kappa B (NFκB). It shows antipathogenic activities against a wide variety of viruses and bacteria (Cushnie and Lamb 2005) and also shows activity against HSV-1 and HSV-2 at early stages; adenoviruses 3, 8, and 11; and

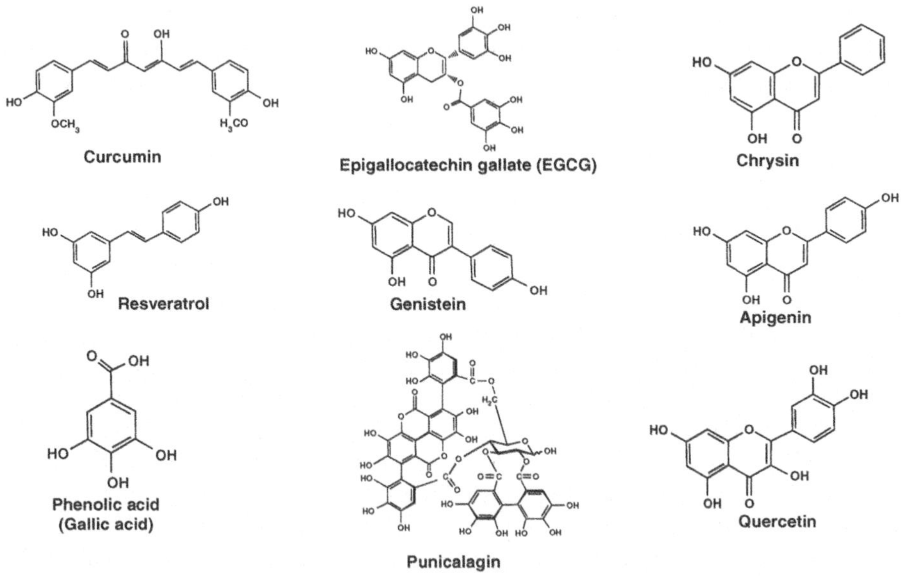

Fig. 5.2 Structure of immunomodulatory polyphenols

coronaviruses (Dimova et al. 2003). Another study shows that conjugates of quercetin reduce replication of various viruses such as HIV, SARS, and rhinovirus through different pathways such as reducing lipase activity, inhibition of DNA gyrase and proteases, and also binding of viral capsid (Nieman et al. 2007). It induces the gene expression and production of helper T-lymphocyte-1 (Th-1)-derived interferon gamma (IFNγ), and it downregulates Th-2-derived IL-4 (Nair et al. 2002).

5.5.2 *Resveratrol*

Resveratrol, chemically known as 3,5,49-trihydroxystilbene, is found usually in grapes and is an important constituent of red wine. Resveratrol exerts chemopreventive properties and anti-inflammatory activity in macrophage. It is reported to inhibit the development of DMBA-induced preneoplastic lesions in mammary glands of mouse (Jang and Pezzuto 1999). Resveratrol also shows its effect on TLR-mediated inflammatory responses and chronic diseases that are associated with TLR activation such as Crohn's disease, type 2 diabetes mellitus (T2DM), rheumatoid arthritis, and neurodegenerative and cardiovascular disorders (Narayanankutty 2019). It was reported that COX-2 expression and NF-κB activation in LPS were decreased by resveratrol. It also actively reduces the production of granulocyte-macrophage colony-stimulating factor (Schwager et al. 2017). Resveratrol also influences the activity of NK cells and simultaneously has an effect on other cells like $CD8^+$and $CD4^+$ T cells (Falchetti et al. 2001). It was observed in a study that resveratrol inhibits active single transducers and STAT3 signaling activators, in turn of which it helps in augmenting NK activity against cell leukemia (Malaguarnera 2019).

5.5.3 *Punicalagin*

Punicalagin is an ellagitannin which is isolated from the pomegranate (*Punica granatum*) and also present in abundant amount in fruit husk and juices. In China, it is widely used in the form of medicine for the treatment of various diseases including inflammatory bowel disease, ulcer, and diarrhea. It significantly reduces LPS -induced formation of nitric oxide (NO), TNF- α, (IL)-1β, IL-6, and prostaglandin E2 in RAW264.7 cells in a dose-dependent manner. It could also suppress the phosphorylation of extracellular signal-regulated kinase, mitogen-activated protein kinase including p38, and c-Jun N-terminal kinase (Xu et al. 2014). It was reported that punicalagin works as a potent immunosuppressant and inhibits the activation of NFAT (nuclear factor of activated T cells). Interleukin-2 which is the expression of mRNA and soluble protein was downregulated by punicalagin from anti-CD3-/anti-CD28-stimulated murine splenic $CD4^+$ T cells. It is also documented to reduce and suppress MLR (mixed leukocytes reaction) without exhibiting any kind of cytotoxicity of cell (Lee et al. 2008).

5.5.4 *Apigenin*

Apigenin is a very common dietary polyphenol classified as flavonoid. It is present in abundant amount in many vegetables and fruits. Apigenin potentially possesses anti-inflammatory, antioxidant, antibacterial, and antiviral activities and is used in traditional medicine for centuries. It helps in blood pressure reduction significantly, is highly effective against cancer cells, and has low toxicity in comparison to other anticancerous medicine. It significantly suppresses the tumor cell by triggering cell apoptosis and autophagy, inducing cell cycle arrest, suppressing cell migration and invasion, and stimulating an immune response (Yan et al. 2017). Apigenin acts as an anti-allergen against airway inflammation. It significantly reduces the degree of the inflammatory cell infiltration, airway hyper-responsiveness, and total immunoglobulin E levels compared with the ovalbumin group. It also helps in triggering the switching of the immune response to allergen toward a T-helper (type 1) profile (Li et al. 2010). Among all the phenolic compounds, apigenin is considered to have countless nutritional and organoleptic characteristics. Apigenin exhibits potential therapeutic action such as it can induce apoptosis and cell cycle arrest. Other than that, it also shows anti-inflammatory and antioxidant activities. It has access to different proliferative stages and can have cell cycle arrest at G1/S and G2-M phase by modulating different CDKs and other factors required for their initiation. It also regulates change in the potential of mitochondrial membrane and is able to cause the release of cytochrome C in cytoplasm which in turn activates caspase-3 through mean of activation of APFA and turns on the apoptosis process in cells that grew uncontrollably. Apigenin is proved to be a very potent inducer of apoptosis by regulating intrinsic apoptotic as well as extrinsic apoptotic pathways and is involved in activation of caspase-8. Apigenin modulates Bcl-2, Bax, STAT-3, and Akt protein expression in cancer cells and activates apoptosis (Salehi et al. 2019).

5.5.5 Genistein

Genistein, an isoflavone generally present in soybean, is described to be an effective phytoestrogen in the field of therapeutics as a putative preventer of breast and prostate cancer. A study on genistein reported that it exerted beneficial effect on various diseases such as cardiovascular diseases, cancer, osteoporosis, and also postmenopausal symptoms (Verdrengh et al. 2003). Genistein plays an important role in signal transduction in cells related to the immune system, reduces inflammation, and shows antioxidant and anti-promotional effects. Genistein is also considered as a potent inhibitor of TPA-induced hydrogen peroxide (Wei et al. 1995). In a study with adult female B6C3F1 mice, it was demonstrated that genistein has a chemopreventive effect as it significantly increases the host resistance toward the tumor cell B16F10. It effectively decreased the number of tumor nodules in lungs (Guo et al. 2001). Genistein also has its effect against the cell that is responsible for allergy. Delayed-type hypersensitivity (DTH) comes under type IV allergy response which is mediated by macrophage and cell. Genistein suppresses DTH reaction to oxazolone- and granulocyte-mediated response. It is reported that genistein at high concentration shows its effect on inhibiting lymphocyte proliferation response that is induced by mitogen. Genistein not only shows cellular immune response but also suppresses antigen (Ag)-induced antibody (Ab) production. In a research, it was demonstrated that mice which was treated with genistein show less severe arthritis than mice which are taken as control concluding that genistein has its effect on collagen-induced arthritis (CIA). It is also reported that genistein-treated mice show less frequent synovial hyperplasia and bone or cartilage destruction in joints (Sakai and Kogiso 2008).

5.5.6 Chrysin

Chrysin is present in honey, propolis, and numerous plant extracts. It is also found in passion flowers including *Passiflora incarnata*, *Passiflora caerulea*, and *Oroxylum indicum* (Su et al. 2013). Chrysin exerts antitumor, antioxidant, antibacterial, antiallergic, anti-inflammatory, antiaging, antihypertensive, antiviral, antiangiogenic, antiatherogenic, antidiabetic, hepatoprotective, and neuroprotective activities (Zeinali et al. 2017). In a study, it was demonstrated that chrysin shows its effect on M1 and M2 macrophages. It increases the upregulation of M2 marker genes such as Arg1, Ym1, and CD206 and reduces the mRNA level in M1 macrophage marker genes such as IL-12b and CCL3. It also suppresses the M1 phenotype and reduces anti-inflammatory effect by inducing M2 phenotype (Feng et al. 2014). Chrysin have response against allergic reaction. It suppresses the OVA-induced airway hyper-responsiveness and inhibits IgE level in serum. It inhibits TH2 cytokines including IL-4 and IL-13 production that is responsible for allergic reactions. It also decreases the production of gene expression of TNF-a, IL-1b, IL-4, and IL-6 in mast cells which are responsible for inflammatory diseases (Vickers 2017).

5.5.7 Curcumin

Curcumin is generally present in rhizomes of the plant *Curcuma longa*, i.e., turmeric. It is commonly used as spice and flavoring agent in food product. It is used as traditional medicine in India from centuries because of its pleiotropic therapeutic properties (Dubey et al. 2008; Upadhyaya et al. 2009a). Curcumin shows antioxidant, anti-inflammatory, antiaging, antidiabetic, antimicrobial, antidepressant, antiarthritic, and neuroprotective properties. It inhibits the ROS enzymes such as COX, LOX, and xanthine dehydrogenase to prevent inflammation. It also induces NOS associated with inflammation (Aggarwal et al. 2007; Mishra et al. 2019; Kesharwani et al. 2015). Curcumin blocks signaling pathways such as activator protein (AP)-1, nuclear factor (NF)-B transcription factors, and MAPKs in particular chondrocytes which are responsible for activation of the inflammatory cytokinesis and matrix metalloproteinase. By blocking these signaling pathways, curcumin inhibits inflammatory cytokinesis and matrix metalloproteinase which is associated with rheumatoid arthritis. It also protects human chondrocytes from IL-l-induced inhibition of collagen type II and β1 integrin expression and activation of caspase-3. Curcumin synergistically potentiates the growth inhibitory and proapoptotic effects of celecoxib in osteoarthritis synovial adherent cells (Momtazi-Borojeni et al. 2018). Curcumin has been shown to have antiproliferative properties against many cancers including breast and colon cancer. Curcumin is known to reduce self-renewal capacity of breast cancer stem cells and induce differentiation in it with least toxicity to normal cells (Upadhyaya et al. 2009b; Kakarala et al. 2010; Kesharwani et al. 2018). It is also known to modulate cancer stem cell signaling pathways, i.e., Wnt, Notch, and Hedgehog pathways, dysregulation of which leads to modulation of cancer stem cells (Gairola et al. 2021; Norris et al. 2013). However, poor bioavailability of curcumin is one of the main barriers in exploring it for therapeutic use because of its poor absorption, rapid metabolism, and rapid systemic elimination (Kesharwani and Misra 2010; Singh et al. 2016a, b). Glucuronidation and sulfatation at free phenolics of curcumin lead to systemic elimination of curcumin leading to poor bioavailability. Therefore, an efficient drug delivery system is anticipated to be a breakthrough technology for the successful medical application of curcumin (Singh et al. 2016a, b).

5.5.8 Epigallocatechin Gallate (EGCG)

EGCG is a flavanol and one of the major polyphenols present in tea. It was demonstrated that EGCG is very effective in reduction of autoimmune disease and very effective antioxidant, anti-inflammatory, anti-allergic, and anti-tumorigenic agent. EGCG reduces neutrophil migration by inducing chemokine interleukin-8 and cytokine-induced neutrophil chemoattractant. It represses the recruitment of neutrophils by reducing the chemokine and lipopolysaccharide (LPS) and prevents inflammation. It effectively reduces the proteolytic enzymes such as neutrophil elastase (Donà et al. 2003). In a recent report, it was demonstrated that EGCG shows potent activity against rheumatoid arthritis. EGCG test on a male Sprague Dawley rats by gavage shows that the EGCG help in administrating the elevated expression of BAFF receptor, P110d, pAKT, mTORC1, and anti-apoptotic Bcl-xL and normalized their effect by which it becomes a protection against rheumatoid arthritis (Liu et al. 2012). EGCG inhibits MAPK pathway, NF-κB activity, and activator protein-1 (AP-1) activity. It was demonstrated that EGCG inhibit DNA synthesis in human epidermoid carcinoma cell line, A431. The protein tyrosine kinase activity of fibroblast, platelet-derived, and epidermal growth factor receptor was inhibited by EGCG. It also inhibited the phosphorylation of EGF receptor by EGF and blocked the binding of EGF to its receptor (Kanwar et al. 2012).

5.5.9 Phenolic Acid

Phenolic acid that includes ellagic, gallic, ferulic, caffeic, p-coumaric, syringic, and catechol shows antioxidant, anti-inflammatory, and other therapeutic activities such as chemoprevention and efficacy against cardiovascular disease. Caffeic acid works as a selective blocker for the biosynthesis of leukotrienes by which it is involved in the regulation of various diseases, allergic reactions, and asthma. Phenolic acid effectively acts on the transcriptional activity of AP-1 and regulates cell differentiation, cell proliferation, and inflammation. p-Coumaric is one of the polyphenol compounds that have its effects against autoimmune diseases such as rheumatoid arthritis. p-Coumaric acid efficiently works as an immunosuppressor against autoimmune diseases. Recently, the blocking activity of p-coumaric acid on MAPKs and NF-κB pathways was significantly observed (Kilani-Jaziri et al. 2017). A remarkable antioxidant effect of various phenolic acids such as gallic acid, caffeic acid, ferulic acid, and p-coumaric acid is observed in different literatures. The multifaceted effect of phenolics was also observed against the cancerous cells, and they are also known to modulate the cell mechanism related to tumor cell. Phenolics alter the cell signaling pathway and regulated the growth factor receptor interaction including transcription factor and kinases in cancer cell that eventually lead to cell cycle arrest and apoptosis. Phenolics enhance the immune system of the body to recognize and destroy the cancer cells and inhibit the development of new blood vessels, thus preventing angiogenesis that is essential and vital for tumor growth (Wahle et al. 2010).

5.6 Cross Talk

Recent findings with modern medicine are that multi-targeted therapy works well than monotherapy. The role of polyphenol on the immune system is interconnected and converges at some regulatory point. Most of the polyphenol have synergistic effect on each other for boosting the immune system. The advantage of using a combination of these interventions establishes synergistic/additive effects of these bioactive components and would corroborate the idea if the combination of these nutritional interventions improves bioavailability of each other. Polyphenols, quercetin, and apigenin show synergistic effect on reducing the number of tumor nodules in the lungs. In combination, quercetin and apigenin show an effective inhibition of melanoma growth and metastatic potential (Caltagirone et al. 2000). The activity of quercetin with EGCG, iso-quercetin, and n-3-PUFA was effectively improved against inflammation. The report supports the concept that the effect of quercetin was amplified as an anti-inflammatory agent with other polyphenols (Nieman et al. 2009). It was demonstrated that polyphenols such as EGCG, along with lycopene, oleanolic acid, and kaempferol-3-O-sophoroside, play important role in anti-inflammatory activity. They show inhibition of high mobility group box1 protein which is a very important chromatin interacting with transcription factors, histones regulating transcription, and nucleosomes (Bae 2012). Earlier, it was reported that the polyphenols present in green tea mainly EGCG are highly effective in protecting lipid peroxidation when intervened with vitamins E and C. It was revealed that the antioxidant synergism in EGCG was due to the regeneration of vitamin E by green tea (Dai et al. 2008). Study by molecular docking reveals that three polyphenols, i.e., quercetin, cinnamic acid, and ferulic acid, might have synergistic role toward the inhibition of tyrosinase. Curcumin with quercetin shows increased therapeutic efficacy. Curcumin with quercetin reduce adenomas in patient with adenomatous polyposis. Curcumin with piperine effectively reverse the lipid peroxidation to prevent tropical pancreatitis (Kunnumakkara et al. 2017).

5.7 Conclusion

Polyphenols are significantly effective on the immune system as reported in earlier in vitro studies. It shows effective result against malfunctioning of the immune system. Discoveries in the field of dietary polyphenol are a significant step toward enhancing immune response against factors that are involved in immune imbalances ultimately leading to various diseases. Although therapeutic efficacy of polyphenols is reported in earlier studies, there are some aspects that we have to resolve for further enhancement of polyphenol's effect on the human body using advanced clinical trials. Polyphenol might have different expression in different population of the world since different population have different microbial population according to their age, environment, life schedule, and other related things. By knowing the effect of polyphenol on different microbiota, the immune response could be increased. The intervention result of polyphenol in previous studies differ vigorously, so it is important to study the mechanism to increase the understanding about the doses formulated to be given for increasing immunity, the interaction with food matrix, and the bioavailability of the polyphenols. The improvements in the bioavailability of polyphenol prove to be beneficial in enhancing the immune response in host against pathogens. The immunoregulatory studies of polyphenol help us to understand the capacity of these magic molecules toward the immune system and also their therapeutic management toward the inflammatory activities, cell apoptosis, and other cell signaling pathway regulatory activities which lead to enhancement of health.

References

Abdel-Aleem GA, Khaleel EF, Mostafa DG, Elberier LK (2016) Neuroprotective effect of resveratrol against brain ischemia reperfusion injury in rats entails reduction of DJ-1 protein expression and activation of PI3K/Akt/GSK3b survival pathway. Arch Physiol Biochem 122(4): 200–213

Aggarwal BB, Surh YJ, Shishodia S (eds) (2007) The molecular targets and therapeutic uses of curcumin in health and disease, vol 595. Springer Science & Business Media, pp 425–451

Aguirre L, Arias N, Teresa Macarulla M, Gracia A, Portillo MP (2011) Beneficial effects of quercetin on obesity and diabetes. Open Nutraceuticals J 4(1):189–198

Bae JS (2012) Role of high mobility group box 1 in inflammatory disease: focus on sepsis. Arch Pharm Res 35(9):1511–1523

Baumann J, Bruchhausen FV, Wurm G (1980) Flavonoids and related compounds as inhibitors of arachidonic acid peroxidation. Prostaglandins 20(4):627–639

Bengmark S (2004) Acute and "chronic" phase reaction—a mother of disease. Clin Nutr 23(6): 1256–1266

Boyanapalli SS, Kong ANT (2015) "Curcumin, the king of spices": epigenetic regulatory mechanisms in the prevention of cancer, neurological, and inflammatory diseases. Curr Pharmacol Rep 1(2):129–139

Caltagirone S, Rossi C, Poggi A, Ranelletti FO, Natali PG, Brunetti M, Aiello FB, Piantelli M (2000) Flavonoids apigenin and quercetin inhibit melanoma growth and metastatic potential. Int J Cancer 87(4):595–600

Cardona F, Andrés-Lacueva C, Tulipani S, Tinahones FJ, Queipo-Ortuño MI (2013) Benefits of polyphenols on gut microbiota and implications in human health. J Nutr Biochem 24(8): 1415–1422

Chen CC, Chow MP, Huang WC, Lin YC, Chang YJ (2004) Flavonoids inhibit tumor necrosis factor-α-induced up-regulation of intercellular adhesion molecule-1 (ICAM-1) in respiratory epithelial cells through activator protein-1 and nuclear factor-κB: structure-activity relationships. Mol Pharmacol 66(3):683–693

Chen L, Teng H, Jia Z, Battino M, Miron A, Yu Z, Cao H, Xiao J (2018) Intracellular signaling pathways of inflammation modulated by dietary flavonoids: the most recent evidence. Crit Rev Food Sci Nutr 58(17):2908–2924

Cheynier V (2005) Polyphenols in foods are more complex than often thought. Am J Clin Nutr 81(1):223S–229S

Choi KC, Jung MG, Lee YH, Yoon JC, Kwon SH, Kang HB, Kim MJ, Cha JH, Kim YJ, Jun WJ, Lee JM (2009) Epigallocatechin-3-gallate, a histone acetyltransferase inhibitor, inhibits EBV-induced B lymphocyte transformation via suppression of RelA acetylation. Cancer Res 69(2):583–592

Choi YH, Yan GH (2009) Silibinin attenuates mast cell-mediated anaphylaxis-like reactions. Biol Pharm Bull 32(5):868–875

Chu AJ (2014) Antagonism by bioactive polyphenols against inflammation: a systematic view. Curr Drug Targets-Inflamm Allergy 13(1):34–64

Cordero-Herrera I, Chen X, Ramos S, Devaraj S (2017) (−)-Epicatechin attenuates high-glucose-induced inflammation by epigenetic modulation in human monocytes. Eur J Nutr 56(3):1369–1373

Cushnie TT, Lamb AJ (2005) Antimicrobial activity of flavonoids. Int J Antimicrob Agents 26(5):343–356

Dai F, Chen WF, Zhou B (2008) Antioxidant synergism of green tea polyphenols with α-tocopherol and l-ascorbic acid in SDS micelles. Biochimie 90(10):1499–1505

Dai Q, Borenstein AR, Wu Y, Jackson JC, Larson EB (2006) Fruit and vegetable juices and Alzheimer's disease: the kame project. Am J Med 119(9):751–759

Davis JM, Murphy EA, Carmichael MD (2009) Effects of the dietary flavonoid quercetin upon performance and health. Curr Sports Med Rep 8(4):206–213

Di Meo F, Lemaur V, Cornil J, Lazzaroni R, Duroux JL, Olivier Y, Trouillas P (2013) Free radical scavenging by natural polyphenols: atom versus electron transfer. Chem A Eur J 117(10):2082–2092

Dimova S, Mugabowindekwe R, Willems T, Brewster ME, Noppe M, Ludwig A, Jorissen M, Augustijns P (2003) Safety-assessment of 3-methoxyquercetin as an antirhinoviral compound for nasal application: effect on ciliary beat frequency. Int J Pharm 263(1–2):95–103

Ding S, Jiang H, Fang J (2018) Regulation of immune function by polyphenols. J Immunol Res 2018:1–9

Donà M, Dell'Aica I, Calabrese F, Benelli R, Morini M, Albini A, Garbisa S (2003) Neutrophil restraint by green tea: inhibition of inflammation, associated angiogenesis, and pulmonary fibrosis. J Immunol 170(8):4335–4341

Dubey SK, Sharma AK, Narain U, Misra K, Pati U (2008) Design, synthesis and characterization of some bioactive conjugates of curcumin with glycine, glutamic acid, valine and demethylenated piperic acid and study of their antimicrobial and antiproliferative properties. Eur J Med Chem 43(9):1837–1846

Falchetti R, Fuggetta MP, Lanzilli G, Tricarico M, Ravagnan G (2001) Effects of resveratrol on human immune cell function. Life Sci 70(1):81–96

Feng X, Qin H, Shi Q, Zhang Y, Zhou F, Wu H, Ding S, Niu Z, Lu Y, Shen P (2014) Chrysin attenuates inflammation by regulating M1/M2 status via activating PPARγ. Biochem Pharmacol 89(4):503–514

Fryer LG, Hajduch E, Rencurel F, Salt IP, Hundal HS, Hardie DG, Carling D (2000) Activation of glucose transport by AMP-activated protein kinase via stimulation of nitric oxide synthase. Diabetes 49(12):1978–1985

Gairola K, Gururani S, Bahuguna A, Garia V, Pujari R, Dubey SK (2021) Natural products targeting cancer stem cells: implications for cancer chemoprevention and therapeutics. J Food Biochem 45(7):e13772–e13772

Gleeson M (2006) Immune system adaptation in elite athletes. Curr Opin Clin Nutr Metabol Care 9(6):659–665

Gomez-Cadena A, Urueña C, Prieto K, Martinez-Usatorre A, Donda A, Barreto A, Romero P, Fiorentino S (2016) Immunesystem-dependent anti-tumor activity of a plant-derived polyphenol rich fraction in a melanoma mouse model. Cell Death Dis 7(6):e2243–e2243

González-Gallego J, García-Mediavilla MV, Sánchez-Campos S, Tuñón MJ (2010) Fruit polyphenols, immunity, and inflammation. Br J Nutr 104(S3):S15–S27

Guo TL, McCay JA, Zhang LX, Brown RD, You L, Karrow NA, Germolec DR, White KL Jr (2001) Genistein modulates immune responses and increases host resistance to B16F10 tumor in adult female B6C3F1 mice. J Nutr 131(12):3251–3258

Heleno SA, Martins A, Queiroz MJR, Ferreira IC (2015) Bioactivity of phenolic acids: metabolites versus parent compounds: A review. Food Chem 173:501–513

Hong KS, Park JI, Kim MJ, Kim HB, Lee JW, Dao TT, Oh WK, Kang CD, Kim SH (2012) Involvement of SIRT1 in hypoxic down-regulation of c-Myc and β-catenin and hypoxic preconditioning effect of polyphenols. Toxicol Appl Pharmacol 259(2):210–218

Hussain T, Tan B, Yin Y, Blachier F, Tossou MC, Rahu N (2016) Oxidative stress and inflammation: what polyphenols can do for us? Oxidative Med Cell Longev 2016:1–9

Jang M, Pezzuto JM (1999) Cancer chemopreventive activity of resveratrol. Drugs Exp Clin Res 25(2–3):65–77

Kakarala M, Brenner DE, Korkaya H, Cheng C, Tazi K, Ginestier C, Liu S, Dontu G, Wicha MS (2010) Targeting breast stem cells with the cancer preventive compounds curcumin and piperine. Breast Cancer Res Treat 122(3):777–785

Kanwar J, Taskeen M, Mohammad I, Huo C, Chan TH, Dou QP (2012) Recent advances on tea polyphenols. Front Biosci (Elite Ed) 4:111–131

Kesharwani RK, Misra K (2010) Prediction of binding site for Curcuminoids at human topoisomerase II α protein; an in silico approach. Curr Sci 101(8):1060–1064

Kesharwani RK, Singh DB, Singh DV, Misra K (2018) Computational study of curcumin analogues by targeting DNA topoisomerase II: a structure-based drug designing approach. Network Modeling Analysis in Health Informatics and Bioinformatics 7(1):1–7

Kesharwani RK, Srivastava V, Singh P, Rizvi SI, Adeppa K, Misra K (2015) A novel approach for overcoming drug resistance in breast cancer chemotherapy by targeting new synthetic curcumin analogues against aldehyde dehydrogenase 1 (ALDH1A1) and glycogen synthase Kinase-3 β (GSK-3β). Appl Biochem Biotechnol 176(7):1996–2017. https://doi.org/10.1007/s12010-015-1696-x

Khan H, Sureda A, Belwal T, Çetinkaya S, Süntar İ, Tejada S, Devkota HP, Ullah H, Aschner M (2019) Polyphenols in the treatment of autoimmune diseases. Autoimmun Rev 18(7):647–657

Kilani-Jaziri S, Mokdad-Bzeouich I, Krifa M, Nasr N, Ghedira K, Chekir-Ghedira L (2017) Immunomodulatory and cellular anti-oxidant activities of caffeic, ferulic, and p-coumaric phenolic acids: a structure–activity relationship study. Drug Chem Toxicol 40(4):416–424

Kim HS, Quon MJ, Kim JA (2014) New insights into the mechanisms of polyphenols beyond antioxidant properties; lessons from the green tea polyphenol, epigallocatechin 3-gallate. Redox Biol 2:187–195

Kolodziej H, Kiderlen AF (2005) Antileishmanial activity and immune modulatory effects of tannins and related compounds on Leishmania parasitised RAW 264.7 cells. Phytochemistry 66(17):2056–2071

Kunnumakkara AB, Bordoloi D, Padmavathi G, Monisha J, Roy NK, Prasad S, Aggarwal BB (2017) Curcumin, the golden nutraceutical: multitargeting for multiple chronic diseases. Br J Pharmacol 174(11):1325–1348

Laughton MJ, Evans PJ, Moroney MA, Hoult JRS, Halliwell B (1991) Inhibition of mammalian 5-lipoxygenase and cyclo-oxygenase by flavonoids and phenolic dietary additives: relationship to antioxidant activity and to iron ion-reducing ability. Biochem Pharmacol 42(9):1673–1681

Lee SI, Kim BS, Kim KS, Lee S, Shin KS, Lim JS (2008) Immune-suppressive activity of punicalagin via inhibition of NFAT activation. Biochem Biophys Res Commun 371(4): 799–803

Li H, Jia Z, Li A, Jenkins G, Yang X, Hu J, Guo W (2013) Resveratrol repressed viability of U251 cells by miR-21 inhibiting of NF-κB pathway. Mol Cell Biochem 382(1):137–143

Li RR, Pang LL, Du Q, Shi Y, Dai WJ, Yin KS (2010) Apigenin inhibits allergen-induced airway inflammation and switches immune response in a murine model of asthma. Immunopharmacol Immunotoxicol 32(3):364–370

Liu D, Li P, Song S, Liu Y, Wang Q, Chang Y, Wu Y, Chen J, Zhao W, Zhang L, Wei W (2012) Pro-apoptotic effect of epigallo-catechin-3-gallate on B lymphocytes through regulating BAFF/PI3K/Akt/mTOR signaling in rats with collagen-induced arthritis. Eur J Pharmacol 690(1–3): 214–225

Lopes GK, Schulman HM, Hermes-Lima M (1999) Polyphenol tannic acid inhibits hydroxyl radical formation from Fenton reaction by complexing ferrous ions. Biochimica et Biophysica Acta (BBA)-General Subjects 1472(1–2):142–152

Magrone T, Jirillo E, Spagnoletta A, Magrone M, Russo MA, Fontana S, Laforgia F, Donvito I, Campanella A, Silvestris F, De Pergola G (2017) Immune profile of obese people and in vitro effects of red grape polyphenols on peripheral blood mononuclear cells. Oxidative Med Cell Longev 2017:1–11

Makanjuola SB, Ogundaini AO, Ajonuma LC, Dosunmu A (2018) Apigenin and apigeninidin isolates from the Sorghum bicolor leaf targets inflammation via cyclo-oxygenase-2 and prostaglandin-E2 blockade. Int J Rheum Dis 21(8):1487–1495

Malaguarnera L (2019) Influence of resveratrol on the immune response. Nutrients 11(5):946

Mao X, Gu C, Chen D, Yu B, He J (2017) Oxidative stress-induced diseases and tea polyphenols. Oncotarget 8(46):81649

Mileo AM, Nisticò P, Miccadei S (2019) Polyphenols: immunomodulatory and therapeutic implication in colorectal cancer. Front Immunol 10:729

Mishra H, Kesharwani RK, Singh DB, Tripathi S, Dubey SK, Misra K (2019) Computational study of curcumin, retinoic acid and its conjugates against GSK-3 beta. Netw Model Anal Health Inform Bioinform 8(1):1–7. https://doi.org/10.1007/s13721-018-0177-x

Momtazi-Borojeni AA, Haftcheshmeh SM, Esmaeili SA, Johnston TP, Abdollahi E, Sahebkar A (2018) Curcumin: A natural modulator of immune cells in systemic lupus erythematosus. Autoimmun Rev 17(2):125–135

Monagas M, Khan N, Andrés-Lacueva C, Urpí-Sardá M, Vázquez-Agell M, Lamuela-Raventós RM, Estruch R (2009) Dihydroxylated phenolic acids derived from microbial metabolism reduce lipopolysaccharide-stimulated cytokine secretion by human peripheral blood mononuclear cells. Br J Nutr 102(2):201–206

Mujtaba T, Dou QP (2012) Black tea polyphenols inhibit tumor proteasome activity. In Vivo 26(2): 197–202

Nair MP, Kandaswami C, Mahajan S, Chadha KC, Chawda R, Nair H, Kumar N, Nair RE, Schwartz SA (2002) The flavonoid, quercetin, differentially regulates Th-1 (IFNγ) and Th-2 (IL4) cytokine gene expression by normal peripheral blood mononuclear cells. Biochimica et Biophysica Acta (BBA)-molecular. Cell Res 1593(1):29–36

Narayanankutty A (2019) Toll-like receptors as a novel therapeutic target for natural products against chronic diseases. Curr Drug Targets 20(10):1068–1080

Nieman DC, Bishop NC (2006) Nutritional strategies to counter stress to the immune system in athletes, with special reference to football. J Sports Sci 24(07):763–772

Nieman DC, Henson DA, Gross SJ, Jenkins DP, Davis JM, Murphy EA, Carmichael MD, Dumke CL, Utter AC, McAnulty SR, McAnulty LS (2007) Quercetin reduces illness but not immune perturbations after intensive exercise. Med Sci Sports Exerc 39(9):1561–1569

Nieman DC, Henson DA, Maxwell KR, Williams AS, McAnulty SR, Jin F, Shanely RA, Lines TC (2009) Effects of quercetin and EGCG on mitochondrial biogenesis and immunity. Med Sci Sports Exerc 41(7):1467–1475

Norris L, Karmokar A, Howells L, Steward WP, Gescher A, Brown K (2013) The role of cancer stem cells in the anti-carcinogenicity of curcumin. Mol Nutr Food Res 57(9):1630–1637

Oršolić N, Knežević AH, Šver L, Terzić S, Bašić I (2004) Immunomodulatory and antimetastatic action of propolis and related polyphenolic compounds. J Ethnopharmacol 94(2–3):307–315

Park JE, Cuong TD, Hung TM, Lee I, Na M, Kim JC, Ryoo S, Lee JH, Choi JS, Woo MH, Min BS (2011) Alkaloids from Chelidonium majus and their inhibitory effects on LPS-induced NO production in RAW264. 7 cells. Bioorg Med Chem Lett 21(23):6960–6963

Romier B, Schneider YJ, Larondelle Y, During A (2009) Dietary polyphenols can modulate the intestinal inflammatory response. Nutr Rev 67(7):363–378

Sakai T, Kogiso M (2008) Soy isoflavones and immunity. J Med Investig 55(3, 4):167–173

Salehi B, Venditti A, Sharifi-Rad M, Kręgiel D, Sharifi-Rad J, Durazzo A, Lucarini M, Santini A, Souto EB, Novellino E, Antolak H (2019) The therapeutic potential of apigenin. Int J Mol Sci 20(6):1305

Schwager J, Richard N, Widmer F, Raederstorff D (2017) Resveratrol distinctively modulates the inflammatory profiles of immune and endothelial cells. BMC Complement Altern Med 17(1): 1–12

Singh DB, Gupta M, Kesharwani R, Sagar M, Dwivedi S, Misra K (2014) Molecular drug targets and therapies for Alzheimer's disease. Transl Neurosci 5(3):203–217

Singh P, Kesharwani RK, Misra K, Rizvi SI (2016a) Modulation of erythrocyte plasma membrane redox system activity by curcumin. Biochem Res Int, Article ID, p 702039. https://doi.org/10.1155/2016/6025245

Singh P, Kesharwani RK, Misra K, Rizvi SI (2016b) The modulation of erythrocyte Na+/K+-ATPase activity by curcumin. J Adv Res 6(6):1023–1030

Singla RK, Dubey AK, Garg A, Sharma RK, Fiorino M, Ameen SM, Haddad MA, Al-Hiary M (2019) Natural polyphenols: chemical classification, definition of classes, subcategories, and structures. J AOAC Int 102(5):1397–1400

Song J, Jun M, Ahn MR, Kim OY (2016) Involvement of miR-Let7A in inflammatory response and cell survival/apoptosis regulated by resveratrol in THP-1 macrophage. Nutr Res Pract 10(4): 377–384

Su Y, Qadri SM, Hossain M, Wu L, Liu L (2013) Uncoupling of eNOS contributes to redox-sensitive leukocyte recruitment and microvascular leakage elicited by methylglyoxal. Biochem Pharmacol 86(12):1762–1774

Szliszka E, Krol W (2011) The role of dietary polyphenols in tumor necrosis factor-related apoptosis inducing ligand (TRAIL)-induced apoptosis for cancer chemoprevention. Eur J Cancer Prev 20(1):63–69

Szliszka E, Krol W (2013) Polyphenols isolated from propolis augment TRAIL-induced apoptosis in cancer cells. Evid Based Complement Alternat Med 2013:1–10

Taofiq O, Calhelha RC, Heleno S, Barros L, Martins A, Santos-Buelga C, Queiroz MJR, Ferreira IC (2015) The contribution of phenolic acids to the anti-inflammatory activity of mushrooms: screening in phenolic extracts, individual parent molecules and synthesized glucuronated and methylated derivatives. Food Res Int 76:821–827

Upadhyaya J, Kesharwani RK, Misra K (2009a) Comparative study of antioxidants as cancer preventives through inhibition of HIF-1 alpha activity. Bioinformation 4(6):233–236

Upadhyaya J, Kesharwani RK, Misra K (2009b) Metabolism, pharmacokinetics and bioavailability of ascorbic acid; synergistic effect with tocopherols and curcumin. J Comput Intell Bioinforma 2 (1):77–84

Vafeiadou K, Vauzour D, Lee HY, Rodriguez-Mateos A, Williams RJ, Spencer JP (2009) The citrus flavanone naringenin inhibits inflammatory signalling in glial cells and protects against neuroinflammatory injury. Arch Biochem Biophys 484(1):100–109

Venkatachalam K, Mummidi S, Cortez DM, Prabhu SD, Valente AJ, Chandrasekar B (2008) Resveratrol inhibits high glucose-induced PI3K/Akt/ERK-dependent interleukin-17 expression in primary mouse cardiac fibroblasts. Am J Phys Heart Circ Phys 294:H2078–H2087

Verdrengh M, Jonsson IM, Holmdahl R, Tarkowski A (2003) Genistein as an anti-inflammatory agent. Inflamm Res 52(8):341–346

Vickers NJ (2017) Animal communication: when i'm calling you, will you answer too? Curr Biol 27(14):R713–R715

Wahle KW, Brown I, Rotondo D, Heys SD (2010) Plant phenolics in the prevention and treatment of cancer. Bio-Farms Nutraceuticals 1:36–51

Wang LC, Lin YL, Liang YC, Yang YH, Lee JH, Yu HH, Wu WM, Chiang BL (2009) The effect of caffeic acid phenethyl ester on the functions of human monocyte-derived dendritic cells. BMC Immunol 10(1):1–13

Wei H, Bowen R, Cai Q, Barnes S, Wang Y (1995) Antioxidant and antipromotional effects of the soybean isoflavone genistein. Proc Soc Exp Biol Med 208(1):124–130

Wheeler DS, Catravas JD, Odoms K, Denenberg A, Malhotra V, Wong HR (2004) Epigallocatechin-3-gallate, a green tea–derived polyphenol, inhibits IL-1β-dependent proinflammatory signal transduction in cultured respiratory epithelial cells. J Nutr 134(5): 1039–1044

Xu X, Yin P, Wan C, Chong X, Liu M, Cheng P, Chen J, Liu F, Xu J (2014) Punicalagin inhibits inflammation in LPS-induced RAW264. 7 macrophages via the suppression of TLR4-mediated MAPKs and NF-κB activation. Inflammation 37(3):956–965

Yahfoufi N, Alsadi N, Jambi M, Matar C (2018) The immunomodulatory and anti-inflammatory role of polyphenols. Nutrients 10(11):1618

Yan X, Qi M, Li P, Zhan Y, Shao H (2017) Apigenin in cancer therapy: anti-cancer effects and mechanisms of action. Cell Biosci 7(1):1–16

Yang G, Chang CC, Yang Y, Yuan L, Xu L, Ho CT, Li S (2018a) Resveratrol alleviates rheumatoid arthritis via reducing ROS and inflammation, inhibiting MAPK signaling pathways, and suppressing angiogenesis. J Agric Food Chem 66(49):12953–12960

Yang N, Zhang H, Cai X, Shang Y (2018b) Epigallocatechin-3-gallate inhibits inflammation and epithelial-mesenchymal transition through the PI3K/AKT pathway via upregulation of PTEN in asthma. Int J Mol Med 41(2):818–828

Zeinali M, Rezaee SA, Hosseinzadeh H (2017) An overview on immunoregulatory and anti-inflammatory properties of chrysin and flavonoids substances. Biomed Pharmacother 92:998–1009

Chapter 6
Mineral Nutraceuticals and Immunity Enhancement

Manoj Kumar Mishra, Jitendra Gupta, and Reena Gupta

Abstract The health and wellness of human beings is largely dictated by the consumption of nutritious foods. Various studies have linked foods as helpful in combating a number of degenerative diseases, as such a lot of research on functional attributes linked directly to the health benefits. Explosive growth, research developments, lack of standards, marketing zeal, quality assurance, and regulation will play a vital role in its success or failure. The importance of a proper functioning and well-balanced immune system for maintaining health has become strikingly evident over the past decades. With a concomitant increase in immune-related disorders, both diet and nutrition can affect the functioning of various immune parameters. Functional food not only enters the concept of considering food as a source of mental and physical well-being but also contributes in the reduction of several diseases or certain physiological functions. Some micronutrients incorporated into functional foods contribute to an enhancement of immunocompetence. The ultimate goal of scientific community and food industry is to develop functional foods for improving quality of life. Malnutrition and deficient intake of quality foods facing up a progressive increase in immune-mediated health problems. When functional food aids in the prevention or treatment of disease or disorder other than anaemia known as nutraceutical. The present review emphasized on probiotics, essential amino acids, vitamins, dietary antioxidants, β-glucans, and other immunomodulators. In the global marketplace, nutraceuticals and functional foods have become a multibillion-dollar industry, but the optimum intake level and recommended amounts of functional food have not yet been established which can be overcome by both experimental animals and humans.

Keywords Nutrients · Immunomodulators · Probiotics · Food industry · Immune system · Immunity

---

M. K. Mishra
Shambhunath Institute of Pharmacy, Jhalwa, Prayagraj, Uttar Pradesh, India

J. Gupta (✉) · R. Gupta
Institute of Pharmaceutical Research, GLA University, Mathura, Uttar Pradesh, India

R. K. Kesharwani et al. (eds.), *Nutraceuticals and Functional Foods as Immunomodulators*, https://doi.org/10.1007/978-981-19-2507-8_6

6.1 Introduction

A healthy immune system is the only weapon currently to combat the COVID-19 pandemic. This immunity system defends the body against attack by pathogens and maintains the normalcy of health condition. Various factors like undesirable life-style, diseased state, aging, mental stress, malnutrition, etc. have an impact on it. The use of vitamins and trace elements in desired quantity leads to a successful functioning of the immune system (Wintergerst et al. 2007). The nutraceutical comprises a broad range of products which includes the complete diet or a nutrient imparting some certain health benefits by suitable prevention and treatment of diseases (Trottier et al. 2010; Braga et al. 1996). Reduced immunity is partially responsible for an increase in morbidity and mortality as an outcome of infectious disease. Hence, continuous ingestion of foods with immune-modulating activities is helpful to decline the risk of infection (Kaminogawa and Nanno 2004; Keservani et al. 2010a, b).

6.2 Overview of Nutraceutical and Its Classification

Application of nutraceuticals may be for several reasons like:

- To remain fit
- To remain young
- To be safe from diseases
- To enhance life span
- To improve proper functioning of body part or organs (Zhao 2007)

It consists of herbal products, food supplements, probiotics, and prebiotics which have medicinal health benefits (Artukoglu et al. 2016). Nutraceuticals are found and well established in many sectors like:

- Food industries
- Natural and dietary supplement industries
- Pharmaceutical industries
- New reformed pharmaceutical-agricultural-nutrition industries

Maintaining an optimal level of health condition and suppressing disease are major challenges with most of the nutraceuticals (Makrides and Gibson 2000).

For better application and understanding point of view, nutraceuticals can be organized as follows:

- In academics
- Clinical studies
- Functional food and dietary supplements

Classification of nutraceuticals is based on type of food, utilization, and chemical entities (Kokate et al. 2002). Many nutraceuticals have the potential to be established in the current and future market (Keservani et al. 2016a, b).

6.3 Role of Various Minerals as a Nutraceutical

Timely consuming of minerals enhances the nutritional quality of the diet and simultaneously prevents a number of chronic diseases like:

- Malignancy cases
- Heart-related disorders
- Parkinson's disease
- Early aging (Kersting et al. 2001)

Also, energy obtained from minerals could be assessed by bioavailability, assay, percentage of drug release, permeation coefficient, and other factors dependent on size, weight of the human body, age, sex, and health conditions (Dokkum 1995).

Various mineral elements like calcium, iodine, zinc, iron, manganese, magnesium, etc. are essential components obtained from plant and animal origins required in trace amounts for humans, and deficiencies of any or all may cause serious health problems among infants, children, and adults (Kaushik et al. 2003).

Many reasons creating the need for proper balanced nutrition are:

- Imbalanced and irregularities in diet practice
- Lipids and vitamin coenzymes
- Different absorption rate of minerals
- Health condition
- Lower drug absorption

6.3.1 Zinc

Zinc is required for growth, appetite, immunity (Read et al. 2019), and normal functioning of above 100 enzymes related to digestion, metabolism, and wound healing. Dietary sources of zinc include the liver, beef, and dark meat or poultry (Shrimpton et al. 2005). Zinc deficiency occurs in those who consume diets enriched with phytate and low protein (Tapiero and Tew 2003) (Table 6.1).

Continuous zinc supplementation exhibits a significant reduction of 25–30% in acute and chronic diarrhea and up to 50% in the incidence of pneumonia (Ruel et al. 1997). Intake of zinc on a daily basis improved immunity level in elderly patient. But excessive intake may have adverse effects like diarrhea, vomiting, and headache (Scherz and Kirchhoff 2006). Zinc affects immunity too (Keservani et al. 2020).

Table 6.1 Abnormalities occurred in the absence of various minerals

S. no.	Name of mineral	Major abnormalities occurred in the absence of the mineral	Reference
1	Zinc	Zinc deficiency has a bad impact on dermal, nervous system, immunity, gastrointestinal tract, skeletal, and reproductive systems	Verstraeten et al. (2004)
		In children, lack of zinc may reduce the numbers of lymphocytes (T cells), in the blood and peripheral lymphoid tissues	Prasad (2000)
		Major populations remain at high risk of HIV	Acevedo-Murillo et al. (2019)
		Viral infection cases enhance with zinc deficiencies	
2	Selenium	Less selenium use had increased risk of mortality, poor immune function	Rayman (2012)
		In the absence of selenium patients with AIDS, the health condition may worsen	Stone et al. (2010)
		Heart-damaging effects of the cytomegalovirus may occur	Du et al. (2018)
		The antioxidant activity of glutathione will not work properly to remove excess damaging radicals	Lobo et al. (2010)
3	Copper	A big loss of immunity in the absence of copper	Li et al. (2019)
		Patient may increase antiviral activities	
		Impaired intestinal absorption of nutrients	Institute of Medicine (2001)
4	Magnesium	Decreased immune function	Institute of Medicine (2001)
5	Iodine	Goiter and hypothyroidism	Institute of Medicine (2001)
		Delayed physical development and impairment of mental function in adults	Zimmermann and Crill (2010)
		Deficiency leads to brain damage	Laurberg et al. (2010)
		Thyroid hormone secretion highly influenced by iodine	Wolff (1989)
6	Iron	Physical built-up, immune strength, and behavior worsened apart from increased infection rate	Walker (1998)
7	Manganese	For normal immunity function, blood sugar regulation and development along with function of the brain	Takeda (2003)
		Skeletal defect, epilepsy, and abnormal absorption of carbohydrate and lipid	Nkwenkeu et al. (2002)

6.3.2 Selenium

These are found naturally in water, food, and soil, aid in cell apoptosis, and prevent oxidative damage and DNA synthesis and repair (Wu et al. 2017). Highest concentrations of selenium found in animal proteins are plenty available (Institute of Medicine 2000; Keservani et al. 2018).

6.3.3 Copper

It is a trace mineral found in Earth with many major activities like:

- Used as cofactor in enzymatic reactions during skin pigmentation, redox reaction, collagen cross-linking, and physiological pathways, such as connective tissue maturation, neurotransmission, energy production, and iron metabolism (Kanteev et al. 2015)

  Highest concentration of copper is found in seafood, meat, nuts, and grains.

6.3.4 Magnesium

Magnesium controls immune function by influencing:

- Immunoglobulin synthesis
- Macrophage response to lymphokines
- Attachment to immune cells
- Binding of IgM lymphocytes
- Attachment to T-helper B cell (Liang et al. 2012)

6.3.5 Iodine

Dietary intake of iodine produces:

- Thyroid hormones
- T_4
- T_3, regulating physiological processes (Haldimann et al. 2005)

Unawareness about proper utilization of iodine leads to more than 1.9 billion individuals with lowest iodine deficiency in America only (de Benoist et al. 2003). Excessive consumption of iodine may lead to thyrotoxicosis along with hyperthyroidism, euthyroidism, hypothyroidism, or autoimmune thyroid disease (Wolff 1989).

6.3.6 Iron

It utilizes oxygen and affects several metabolic pathways like:

- O_2 transport
- Synthesis of DNA
- Transportation of electron (Puntarulo 2005)

Around 60–70% of iron attached with hemoglobin present in RBCs and mostly preserves in the liver, spleen, and bone marrow. Excessive bleeding in menstrual cycles leads to iron deficiency worldwide (Deegan et al. 2005). Accumulation of excessive iron can be stopped by homeostatic mechanisms generating oxidative stress by catalysis number of chemical reactions, may lead to cell damage (Pietrangelo 2002) and cancer, and augments heart-related disorders (Keservani et al. 2016c). Iron overloading is seen in some cases like:

- Heavy dietary intake
- Hereditary diseases such as idiopathic hemochromatosis, congenital atransferrinemia, and thalassemia (Fontecave and Pierre 1993)

Iron develops immunity by cell proliferation and maturation by lymphocytes, responding to infections.

6.3.7 Manganese

It is an essential trace element highly concentrated in tissues like the brain, and the retina, dark skin, bone, liver, pancreas, and kidney are required for a variety of biological functions (Aschner and Aschner 2005). Only 3–5% of manganese in diet is absorbed, metabolized in the liver, and finally excreted in the bile (Mergler 1999). Overconsumption of manganese damages the ganglia and enhances neuropsychological and behavioral irregularities and parkinsonism (Ordonez-Librado et al. 2010).

6.4 Immune System and Effect of Nutrition

The immune system protects the most affected parts like the heart, thymus, pancreas, kidney, spleen, and lymphatic nodes. Regular intake of balanced diet or nutrition eradicates immunity-related complications (Basoglu and Turnagol 2004). Strain, traumas, and ambustions destroy protein, ultimately reduce body protection, and pave the way for infections and other related complications (Chandra 2003).

6.5 Effects of Nutrition on the Immune System

Annually, around six million children die in infections caused by malnutrition. So requirement of biological proteins is needed like:

- Milk
- Dairy products
- Eggs
- Vitamin C
- Vitamin E
- Beta-carotene

Antioxidants as a micronutrient can protect against free radicals (Palmer 2011). Repressive immunity cases have been increasing recently – a substantial loss in the immune system due to abnormal absorption of food energy.

Minerals in combination with vitamins eradicate many disorders like:

- Hypertension
- Heart arrhythmia
- Muscular disorders
- pH maintenance
- Antifertility

With aging, immune response of the body reduces, which contributes for more infections. An increased life expectancy in developed countries increases age-related conditions like decreased T cells, possibly from the thymus atrophying with age. Additional supplement of micronutrient, essential vitamins, and trace minerals enhanced immunity with food intake.

6.6 Nutrients and Food Components for Modulation of Immunity

For normal functioning, proper nutrients or diet is required. Nutritional immunology identifies dietary factors and an optimal requirement for development of immunological balance and stronger defense system against microbes.

6.6.1 Vitamin D

In the presence of sunlight, 7-dehydrocholesterol gets converted to vitamin D. Common function of vitamin D is the regulation of calcium homeostasis and bone health. New discoveries like vitamin D receptor (VDR) and vitamin D-activating enzymes (hydroxylases) are available in cell and tissue.

6.6.2 Vitamin E

Vitamin E is a generic name for all tocopherols, among which α-tocopherol is much more predominant compared to others in the blood due to the difference in the rate of ADME. Vitamin E is an antioxidant present in the cell membrane, and these cells have high PUFA content (Coquette et al. 1986).

6.6.3 Fish Oil

Many dietary lipids like PUFA regulate cell function. The marine animal-derived n-3 PUFA, composed of eicosapentaenoic acid (EPA) and docosahexaenoic acid (DHA), has a great impact on immune cell functions and has been studied extensively. N-6 PUFA, however, is less significant (Calder 2017; Whelan et al. 2016; Galli and Calder 2009).

6.6.4 Probiotics

These are living microorganisms when abundantly taken give health benefit (Hill et al. 2014). Various microorganisms included are *Lactobacillus* (*L.*), *Bifidobacterium* (*B.*), and *Streptococcus* (*S.*) found in dairy products and normalize the gut.

6.6.5 Green Tea

It contains majorly the following constituents:

- Catechins
- Epicatechin (EC)
- Epicatechin-3-gallate (ECG)
- Epigallocatechin (EGC)
- Epigallocatechin-3-gallate (EGCG)

EGCG is a biologically active, beneficial ingredient responsible for health benefit innate and adaptive immunity (Pae and Wu 2013).

6.6.6 Beta-Glucans

These are natural polysaccharides obtained from barley, oats, bacteria, algae, yeast, and mushrooms. With variations in their structure, beta-glucans are responsible for specific biological properties. β-glucans consist of (1,3)-β-linked backbone with small fraction (1,6)-β-linked-side chains and plays a major role as immunomodulators (Stier et al. 2014). Glucans has many activities like:

- Phagocytic activity
- Production of IL-2
- Release of antibody
- Superoxide forming
- IFNγ formulation
- Cancer model inhibition

Pleuran, a β-glucan obtained from mushroom, reduces upper respiratory tract infections and its related signs and increases circulated NK cell number (Bergendiova et al. 2011).

6.6.7 Polyphenols

Polyphenols are secondary metabolites of plants that protect them from photosynthetic stress. A large number of polyphenols are used like:

- Flavanols
- Flavones
- Flavanones
- Anthocyanins

Dietary polyphenols are gaining popularity as it controls many processes like:

- mRNA expression
- Apoptosis
- Aggregation of platelets
- Intercellular signal transduction (Duthie et al. 2003)

The bioavailability of polyphenol determines their therapeutic activity depending on the chemical's nature of polyphenol conjugation in the intestine (Yang et al. 2001).

6.6.8 Dietary Supplements

These are the products which supplement the diet with one or more dietary ingredients like:

- Vitamin
- Mineral
- Herb or other botanicals
- Amino acid
- Product concentrate
- Metabolite
- Extract

All of the above may be consumed in single or combinations in dosage forms like tablet, pill, capsule, soft gel, extract, gel cap, liquid, or powder (Regulation of nutraceuticals, 1999).

It is not a conventional food or meal or diet but intended to treat or cure disease (Ross 2000). The interest toward dietary supplements has increased for the following benefits:

- Improve health
- Fitness
- Delay aging
- Stamina
- Body-building

Vitamins and minerals are also dietary supplements that exist in single or multiple ingredients in a product (Pandey and Kumar 2011). The demand potential is increased in developing and developed countries such as China, Brazil, India, Russia, etc. (Norman et al. 2010).

6.7 Discussion and Future Scope

Nowadays, nutraceuticals are gaining popularity with its efficient nutrition, therapeutic, and safety profile. Consumer acceptance and demands made it a profitable business both for pharmaceutical and nutritional companies. Nutraceuticals come out with many names like functional foods, phytochemicals, herbal food, agri-foods, baby foods, dietary product, etc., although having marginal differences in all of them separately. Nutraceuticals prevent diseases, but its eye keeps on the current market demand, upcoming invented products in the name of research, and consumer's choice with warrant of its usage safe and effectiveness.

Various mechanisms like gene expression activation of signal transduction pathways, antioxidant protection, cell growth, differentiation, and mitochondrial preservation are useful for nutraceuticals.

6.7.1 Future Scope

As per the survey report, functional foods will be the fastest and largest growth market for future days to come. But the demand of herbal dietary supplements will not be diminished among the nutraceuticals (64% market value). Majority of the bulk demand from the pharma division in the name of vitamin and minerals. Developing countries are the major expansion centers of nutraceuticals. Current sedentary lifestyle, lack of exercise, and high intake of junk food lead to increased obesity and stepped-up cases of diabetes and cardiovascular diseases, consumer demand along with medication increasing. The long-term benefits of nutraceuticals depend on the purity, efficacy, and safety. The focus arena is the product evaluation compared to standards, various dosages allowed, product categories, country-based label claims, and advertisement. However, the future of nutraceuticals with meaningful doses with multiple therapies may combat cancer, immunity boosters, anti-inflammatory agents, arthritis, allergy, macular degeneration, vision impairment, and Parkinson's disease.

6.8 Conclusion

Any dietary deficiencies of iodine, iron, magnesium, manganese, copper, or zinc are major sources of increased death rate among children due to decreased immune systems and even lead to death. Due to the lack of knowledge regarding the usage, its outcome, and safety profile, many nutraceuticals and phytonutrients are not utilized by mankind. Modified biosynthetic pathways and better yield of minerals, nuts, long-chain PUFA, and crop of medicinal and aromatic plants produce sufficient phytonutrients. These technological grafted crops produce some magical properties. The easiest way of avoiding deficiencies and related disorders is to consume diets with high vegetables, fruits, and meats. In the current lifestyle, antioxidant defense mechanisms decrease appreciably with age prone to a variety of immunity-related health issues. Antioxidant acts intrinsically to scavenge free radicals. A detailed study on nutraceuticals is conducted in this present study, but individually acting response of nutraceuticals needs to be established although it has a proven health benefit.

References

Acevedo-Murillo JA, Leon MLG, Firo-Reyes V, Santiago-Cordova JL, Gonzalez-Rodriguez AP, Wong-Chew RM (2019) Zinc supplementation promotes a Th1 response and improves clinical symptoms in fewer hours in children with pneumonia younger than 5 years old. A randomized controlled clinical trial. Front Pediatr 14(7):431

Artukoglu BB, Beyer C, Zuloff-Shani A, Brener E, Bloch MH (2016) Efficacy of palmitoylethanolamide for pain: a meta-analysis. Pain Physician 20(5):353–362

Aschner JL, Aschner M (2005) Nutritional aspects of manganese homeostasis. Mol Asp Med 26(4–5):353–362

Basoglu S, Turnagol H (2004) Egzersiz ve immün sistem: Karbonhidratlarin Etaisi. *Spor Bilimleri Dergisi* 15(2):100–123

Benoist BD, Andersson M, Takkouche B, Egli I (2003) Prevalence of iodine deficiency worldwide. Lancet 362(9398):1859–1860

Bergendiova K, Tibenska E, Majtan J (2011) Pleuran (β-glucan from Pleurotus ostreatus) supplementation, cellular immune response and respiratory tract infections in athletes. Eur J Appl Physiol 111(9):2033–2040

Braga M, Vignali A, Gianotti L, Cestari A, Profili M, Carlo VD (1996) Immune and nutritional effects of early enteral nutrition after major abdominal operations. Eur J Surg 162(2):105–112

Calder PC (2017) Omega-3 fatty acids and inflammatory processes: from molecules to man. Biochemical Society of Transactions 45(5):1105–1115

Chandra RK (2003) Nutrient regulation of immune functions. *Forum of Nutrition* 56(3):147–148

Coquette A, Vray B, Vanderpas J (1986) Role of vitamin E in the protection of the resident macrophage membrane against oxidative damage. Archives of International Physiology and Biochemistry 94(5):S29–S34

Deegan H, Bates HM, McCargar LJ (2005) Assessment of iron status in adolescents: dietary, biochemical and lifestyle determinants. J Adolesc Health 37(1):75

Dokkum WV (1995) The intake of selected minerals and trace elements in European countries. Nutr Res Rev 8(1):271–302

Du Y, Zhang G, Liu Z (2018) Human cytomegalovirus infection and coronary heart disease: a systematic review. Virol J 15(1):31

Duthie GG, Gardner PT, Kyle JAM (2003) Plant polyphenols: are they the new magic bullet? Proc Nutr Soc 62(3):599–603

Fontecave M, Pierre JL (1993) Iron: metabolism, toxicity and therapy. Biochimie 75(9):767–773

Galli C, Calder PC (2009) Effects of fat and fatty acid intake on inflammatory and immune responses: a critical review. Ann Nutr Metab 55(1–3, 123):–139

Haldimann M, Alt A, Blanc A, Blondeau K (2005) Iodine content of food groups. J Food Compos Anal 18(6):461–471

Hill C, Guarner F, Reid G, Gibson GR, Merenstein DJ, Pot B, Morelli L, Canani RB, Flint HJ, Salminen S, Calder PC, Sanders ME (2014) The international scientific association for probiotics and prebiotics consensus statement on the scope and appropriate use of the term probiotic. Nat Rev Gastroenterol Hepatol 11(8):506–514

Institute of Medicine (2000) Food and nutrition board. In: Dietary reference intakes: vitamin C, vitamin E, selenium, and carotenoids. National Academy Press, Washington, DC

Institute of Medicine (2001) Food and nutrition board. In: Dietary reference intakes for vitamin A, vitamin K, arsenic, boron, chromium, copper, iodine, iron, manganese, molybdenum, nickel, silicon, vanadium, and zinc. National Academy Press, Washington, DC

Kaminogawa S, Nanno M (2004) Modulation of immune functions by foods. Evid Based Complement Alternat Med 1(3):241–250

Kanteev M, Goldfeder M, Fishman A (2015) Structure-function correlations in tyrosinases. Protein Sci 24(9):1360–1369

Kaushik D, Dureja H, Kumar V (2003) Development in nutraceuticals. Indian Journal of Pharmacology 35(6):363–372

Kersting M, Alexy U, Sichert-Hellert W (2001) Dietary intake and food sources of minerals in 1 to 18-year-old German children and adolescents. Nutr Res 21(4):607–616

Keservani RK, Kesharwani RK, Sharma AK, Vyas N, Chadoker A (2010b) Nutritional supplements: An overview. International Journal of current pharmaceutical review and research 1(1): 59–75. ISSN: 0976-822X

Keservani RK, Sharma AK, Kesharwani RK (2016a) Nutracetical and Functional Foods for Cardiovascular Health. In: Meghwal M, Goyal MR (eds) Food Process Engineering: Emerging Technologies and Innovations in Food Engineering, vol 5. Apple Academic Press, CRC Press, Taylor and Francis. ISBN: 9781771884020

Keservani RK, Sharma AK, Kesharwani RK (2016c) An overview and therapeutic applications of nutraceutical & functional foods. In: Keservani RK, Sharma AK, Kesharwani RK (eds) Handbook of research on novel approaches for drug delivery. IGI Global Publisher

Keservani RK, Sharma AK, Kesharwani RK, Medicinal Effect of nutraceutical fruits for the cognition and brain health (2016b) Scientifica. https://doi.org/10.1155/2016/3109254. Article ID 3109254

Keservani RK, Sharma AK, Kesharwani RK (2018) Nutraceutical and functional foods in disease prevention. IGI Global. ISBN13: 9781522532675

Keservani RK, Sharma AK, Kesharwani RK (2020) Nutraceuticals and Dietary Supplements: Applications in Health Improvement and Disease Management. Apple Academic Press. ISBN: 9781771888738

Keservani RK, Vyas N, Jain S, Raghuvanshi R, Kesharwani RK, Sharma AK (2010a) Nutraceuticals and functional food as future food. Pharm Lett 2(1):106–116. ISSN: 0975-5071

Kokate CK, Purohit AP, Gokhale SB (2002) Nutraceutical and cosmaceutical. pharmacognosy, 21st edn. Nirali Prakashan

Laurberg P, Cerqueira C, Ovesen L, Rasmussen LB, Perrild H, Andersen S, Pedersen IB, Carle A (2010) Iodine intake as a determinant of thyroid disorders in populations. Best Pract Res Clin Endocrinol Metab 24(1):13–27

Li C, Li Y, Ding C (2019) The role of copper homeostasis at the host-pathogen Axis: from bacteria to fungi. Int J Mol Sci 20(1):175

Liang RY, Wu W, Huang J, Jiang SP, Lin Y (2012) Magnesium affects the cytokine secretion of CD4+T lymphocytes in acute asthma. J Asthma 49(10):1012–1015

Lobo V, Patil A, Phatak A, Chandra N (2010) Free radicals, antioxidants and functional foods: impact on human health. Pharmacogn Rev 4(8):118–126

Makrides M, Gibson RA (2000) Long-chain polyunsaturated fatty acid requirements during pregnancy and lactation. Am J Clin Nutr 71(1):307S–311S

Mergler D (1999) Neurotoxic effects of low level exposure to manganese in human populations. Environ Res 80(2):99–102

Nkwenkeu SF, Kennedy G, Philippe S, Zayed J (2002) Oral manganese intake estimated with dietary records and with direct chemical analysis. Sci Total Environ 287(1–2):147–153

Norman HA, Butrum RR, Feldman E, Heber D, Nixon D, Picciano MF, Rivlin R, Simopoulos A, Wargovich MJ, Weisburger EK, Zeisel SH (2010) The role of dietary supplements during cancer therapy. J Nutr 133(11):3794S–3799S

Ordonez-Librado JL, Anaya-Martinez V, Gutierrez-Valdez AL, Montiel-Flores E, Corona DR, Martinez-Fong D, Avila-Costa MR (2010) L-DOPA treatment reverses the motor alterations induced by manganese exposure as a Parkinson disease experimental model. Neurosci Lett 471(2):79–82

Pae M, Wu D (2013) Immunomodulating effects of epigallocatechin-3-gallate from green tea: mechanisms and applications. Food and Function 4(9):1287–1303

Palmer AC (2011) Nutritionally mediated programming of the developing immune system. Adv Nutr 2(5):377–395

Pandey M, Kumar V (2011) Nutraceutical supplementation for diabetes: a review. Int J Pharm Pharm Sci 3(4):33–40

Pietrangelo A (2002) Physiology of iron transport and the hemochromatosis gene, *American journal of physiology-gastrointestinal liver*. Physiology 282(3):G403–G414

Prasad AS (2000) Effects of zinc deficiency on immune functions. The Journal of Trace Elements in Experimental Medicine 13(1):1–20

Puntarulo S (2005) Iron, oxidative stress and human health. Mol Asp Med 26(4–5):299–312

Rayman MP (2012) Selenium and human health. Lancet 379(9822):1256–1268

Read SA, Obeid S, Ahlenstiel C, Ahlenstiel G (2019) The role of zinc in antiviral immunity. Adv Nutr 10(4):696–710
Ross S (2000) Functional foods: the food and drug administration perspective. Am J Clin Nutr 71(6):1735S–1738S
Ruel MT, Ribera JA, Santizo MC, Lonnerdal B, Brown KH (1997) Impact of zinc supplementation on morbidity from diarrhea and respiratory infections among rural Guatemalan children. Pediatrics 99(6):808–813
Scherz H, Kirchhoff E (2006) Trace elements in foods: zinc contents of raw foods-a comparison of data originating from different geographical regions of the world. J Food Compos Anal 19(5): 420–433
Shrimpton R, Gross R, Darnton-Hill J, Young M (2005) Zinc deficiency: what are the most appropriate interventions? Br Med J 330(7487):347–349
Stier H, Ebbeskotte V, Gruenwald J (2014) Immune-modulatory effects of dietary yeast beta-1, 3/1, 6-D-glucan. Nutr J 13(38):1–9
Stone CA, Kawai K, Kupka R, Fawzi WW (2010) The role of selenium in HIV infection. Nutr Rev 68(11):671–681
Takeda A (2003) Manganese action in brain function. Brain Res Rev 41(1):79–87
Tapiero H, Tew KD (2003) Trace elements in human physiology and pathology: zinc and metallothioneins. Biomed Pharmacother 57(9):399–411
Trottier G, Boström PJ, Lawrentschuk N, Fleshner NE (2010) Nutraceuticals and prostate cancer prevention: a current review. Nat Rev Urol 7(1):21–30
Verstraeten SV, Zago MP, Mackenzie GG, Keen CL, Oteiza PI (2004) Influence of zinc deficiency on cell-membrane fluidity in Jurkat, 3T3 and IMR-32 cells. Biochem J 378(2):579–587
Walker ARP (1998) The remedying of iron deficiency: what priority should it have? Br J Nutr 79(3):227–235
Whelan J, Gowdy KM, Shaikh SR (2016) N-3 polyunsaturated fatty acids modulate B cell activity in pre-clinical models: implications for the immune response to infections. Eur J Pharmacol 785: 10–17
Wintergerst ES, Maggini S, Hornig DH (2007) Contribution of selected vitamins and trace elements to immune function. Ann Nutr Metab 51(4):301–323
Wolff J (1989) Excess iodide inhibits the thyroid by multiple mechanisms. Adv Exp Med Biol 261: 211–244
Wu X, Cheng J, Wang X (2017) Dietary antioxidants: potential anticancer agents. Nutr Cancer 69(4):521–533
Yang CS, Landau JM, Huang MT, Newmark HL (2001) Inhibition of carcinogenesis by dietary polyphenolic compounds. Annu Rev Nutr 21:381–406
Zhao J (2007) Nutraceuticals, nutritional therapy, phytonutrients, and phytotherapy for improvement of human health: a perspective on plant biotechnology application. Recent Patent on Biotechnology 1(1):75–97
Zimmermann MB, Crill CM (2010) Iodine in enteral and parenteral nutrition. Best Pract Res Clin Endocrinol Metab 24(1):143–158

Chapter 7
Nutraceuticals as Disease Preventive Food and Immunity Boosters

Bhushan R. Rane, Rutuja R. Patil, Ashish S. Jain, Raj K. Keservani, and Rajesh K. Kesharwani

Abstract Nutraceuticals are medicinal foods that play a role in maintaining wellbeing, enhancing health, modulating immunity and thereby preventing as well as treating specific diseases. Thus, the field of nutraceutical can be envisioned as one of the missing blocks in the health benefit of an individual. Nutraceuticals are foods providing all the essential nutrients required for maintaining the optimal health. They are used as alternatives to modern medicines to promote quality of health, increase nutritive value of the diet, and prolong life expectancy. They have received considerable zest for their expected safety and therapeutic effects. Generally, consumers prefer food supplements to improve their health, as drugs show various side effects and adverse reactions. The principle reason for the growth of the nutraceutical market worldwide is current health status and lifestyle disorders. Nutraceutical market is seeing tidal growth mainly in the United States, India, and European countries.

Keywords Nutraceutical · Functional foods · Disease · Fortified foods · Health benefits

7.1 Introduction

The term nutraceutical was coined by Stephen DeFelice, founder and chairman of the foundation for Innovation in Medicine, located in Cranford, New Jersey (Lakshmana et al. 2012). It combines the word nutrient (a nourishing food or food

---

B. R. Rane (✉) · R. R. Patil · A. S. Jain
Department of Pharmaceutics, Shri D.D. Vispute College of Pharmacy and Research Center, New Panvel, Mumbai, Maharashtra, India

R. K. Keservani
Faculty of B. Pharmacy, CSM Group of Institutions, Prayagraj, Uttar Pradesh, India

R. K. Kesharwani
Department of Computer Application, Nehru Gram Bharati (Deemed to be University), Prayagraj, Uttar Pradesh, India

R. K. Kesharwani et al. (eds.), *Nutraceuticals and Functional Foods as Immunomodulators*, https://doi.org/10.1007/978-981-19-2507-8_7

component) and pharmaceutical (a medical drug) (Ross 2000). The word "nutraceutical" is defined as any substance which will be considered as a food or a part of a food and provides medical and health benefits, including the prevention and treatment of disease. Pharmaceuticals may be considered as the drugs used mainly to treat diseases, while nutraceuticals are those that are intended to prevent diseases. Both pharmaceuticals and nutraceuticals can cure and prevent diseases, but only pharmaceuticals have sanctions from the government. Nutraceuticals may range from isolated nutrients, dietary supplements, and diets to genetically engineered designer foods, herbal products, and processed foods like cereals, soups, and beverages. Hippocrates emphasized "Let food be your medicine and medicine be your food" (Rajasekaran et al. 2008, Brower 1998, Kalra 2003).

There is a slight difference between the terms functional foods, nutraceuticals, dietary supplement, and medical foods. When food is being cooked or prepared using "scientific intelligence" with or without the knowledge of how or why it is being used, then it is called "functional food." It provides the body with the required amount of vitamins, fats, carbohydrates, etc. for the healthy survival. When functional food aid in prevention and treatment of disease or disorder other than anemia, it is called a nutraceutical (Kalra 2003).

A dietary supplement is a product that is intended to supplement the diet that bears or contains one or more ingredients like vitamins, minerals, herbs, amino acid or a concentrate, metabolite, constituent, extract, or combinations of these. Medical foods are a specific category of therapeutic agents that are intended for nutritional management of a specific disease (Rajasekaran et al. 2008).

Phytonutrients are plant nutrients with particular biological activities in supporting human health. Compared to nutraceuticals, functional foods, and dietary supplements, phytonutrients are more natural bioactive compounds from plants. They have major role in many physiological functions like:

(a) Substrates for biochemical reactions
(b) Cofactors of enzymatic reactions
(c) Inhibitors of enzymatic reactions
(d) Absorbents that bind to and eliminate undesirable constituents from the intestine
(e) Enhancement of absorption and/or stability of essential nutrients
(f) Selective growth factors for beneficial bacteria
(g) Fermentation substrates for beneficial bacteria
(h) Selective inhibitors of deleterious intestinal bacteria
(i) Scavengers of reactive or toxic chemicals
(j) Ligands that agonize or antagonize cell surface or intercellular receptors (Padmavathi 2018)

Nutraceuticals create a new era of research to promote quality of life. They can reduce the risk of disease onset by retaining normal health condition and improving immunity (Dutta et al. 2018). Food is not only the source of energy and nutrients but also provides medicinal benefits, and nutritional therapy is based on complimentary therapy with nutraceuticals (Chauhan et al. 2013).

Large numbers of people are dependent on natural and alternative medicines in India due to imbalance in the diet and nutritional deficiencies. Nutraceuticals are now available as capsules, tablets, or powders in a prescribed dose. Vitamin D will see the fastest growth in demand due to its increasing clinical evidence in the treatment of swine flu, cancer and other preventive medicine benefits (Shinde et al. 2014). Some popular nutraceuticals include green tea (antioxidant), glucosamine (for arthritis), lutein (for macular degeneration), ginseng (for cold), *Echinacea* (anti-immune), folic acid, cod liver oil, etc. (Stephen 2012).

7.1.1 Advantages

Nutraceuticals play an important role in healthy eating and contributes in prevention and treatment of diseases, enable consumers to derive daily dose of vitamins and minerals and are less toxic when compared to conventional pharmaceuticals, cost effective, and easily available (Sayeed 2015).

7.1.2 Emerging Trends in Nutraceuticals

Nutraceuticals could also be divided into herbal/natural products, dietary supplements and functional foods. Out of those, most rapidly growing segment is herbal/natural products followed by dietary supplements. The generation of scientific research-linked foods of plant origin and health has resulted in understanding that plant bioactive compounds have antioxidant and other health-promoting properties.

High dietary intake of fibers within the sort of fruits, vegetables and whole grains is strongly linked to a reduced risk of chronic diseases like cancer and cardiovascular diseases. Cancer development is a chronic, stepwise complex process culminating into metastasis if not tackled in time. Epidemiological studies now provide convincing evidence that dietary factors may modify carcinogenesis. A number of phytochemicals also as some plant origin foods with yet unidentified components possess anticarcinogenic and antimutagenic properties. Thus, use of these bioactive compounds as chemopreventive substances, in future, cannot be overlooked (Cherdshewasart et al. 2009).

Similarly, isoflavonoids or soy products and flaxseed have the ability to decrease total and low density lipoprotein cholesterol (LDL-C) and increase high density lipoprotein cholesterol (HDL-C) resulting in reduced risk of cardiovascular diseases (CVDs). Phytoestrogens are also reported to be beneficial in prevention of CVDs. For CVD, important risk factors include obesity, hyperlipidemia, hypertension and diabetes which can be countered by phytochemicals. Phytochemicals help in reducing oxidative stress also which is implicated in process of atherosclerosis. Nutraceuticals help in boosting the antioxidant defense system of the body (Prakash et al. 2011; Prakash and Gupta 2011).

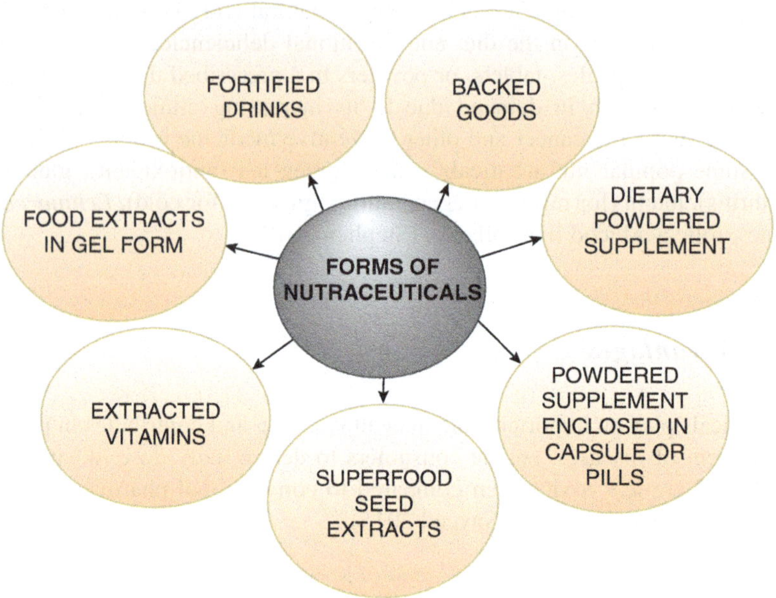

Fig. 7.1 Schematic representation of different forms of nutraceuticals (Kalra 2003)

7.1.3 Nutraceutical

The term "Nutraceutical" was coined from "Nutrition" and "Pharmaceutical" by Stephen De Felice, founder and chairman of the Foundation for Innovation in Medicine. According to De Felice, nutraceutical can be defined as, "a food (or a part of food) that gives medical or health benefits, including the prevention and or treatment of a disease" (Kalra 2003; Zhao 2007).

Nutraceuticals are found during a mosaic of products emerging from (a) the food industry, (b) the herbal and dietary supplement market, (c) pharmaceutical industry, and (d) the newly merged pharmaceutical/agribusiness/nutrition conglomerates (Fig. 7.1).

7.2 Classification of Nutraceuticals

(Singh and Sinha 2012) (Fig. 7.2):

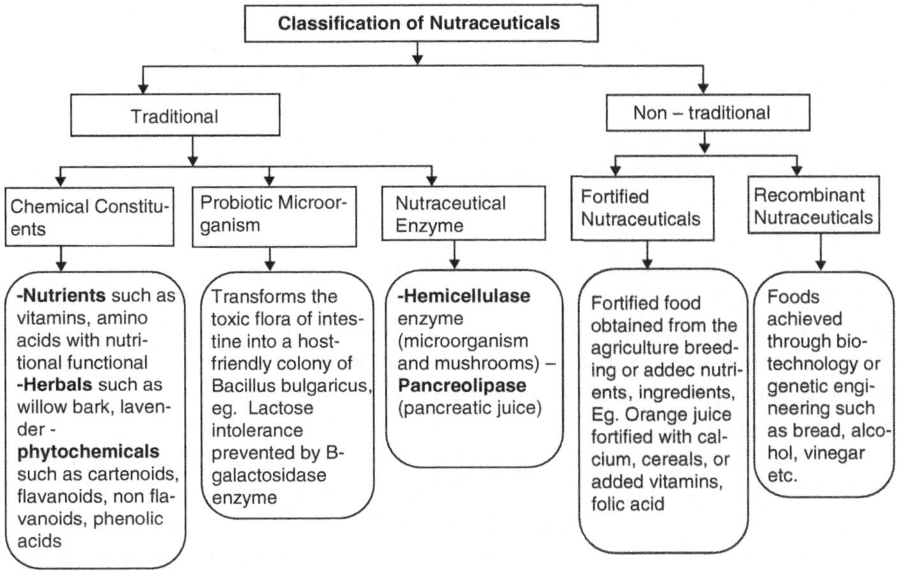

Fig. 7.2 Classification of nutraceuticals

7.2.1 Nutraceuticals Based on Food Availability

Traditional Nutraceuticals

These classes are generally sourced directly from nature, without any changes in the natural form. Various constituents such as lycopene in tomatoes, omega-3fatty acids in salmon, or saponins in soy are available and consumed for different health benefits. Further, various types of traditional nutraceuticals are as follows:

1. Chemical constituents
   (a) Nutrients
   (b) Herbals
   (c) Phytochemicals
2. Probiotic microorganisms
3. Nutraceutical enzymes

1. Chemical Constituents

(a) Nutrients

Primary metabolites such as amino acids, various vitamins and fatty acids had well-defined functions in various metabolic pathways. Plant and animal products alongside vitamin have many health benefits and are helpful in curing diseases associated with heart, kidney, lungs, etc.

Natural products obtained from plants are beneficial in treating various disorders like brittle bones and low hemoglobin count, and they provide strength to bones and muscles, help in neurotransmission, and maintain rhythm of heart muscles. Fatty acids, omega-3 PUFAs present in salmon, had influenced the general inflammatory response and brain function and reduced cholesterol within the arteries.

(b) Herbals

Nutraceuticals along with herbs had an excellent impact on prevention of varied chronic diseases to form life better. Salicin present within the willow bark (*Salix nigra*) had been proved for anti-inflammatory, analgesic, antipyretic, astringent, and antiarthritic response clinically. Flavonoids such as psoralen present in parsley (*Petroselinum crispum*) are useful in diuretic, carminative and antipyretic.

Peppermint (*Mentha piperita*) contains various terpenoids especially menthol, a bioactive constituent, and cures cold and flu. Tannin contents of lavender (*Lavandula angustifolia*) help release stress and blood pressure and are useful for lung disorders such as asthma (Ehrlich 2008).

(c) Phytochemicals

They are mainly classified on the based on phytochemicals. Carotenoids (isoprenoids) are present in vegetables, enhancing immune system, mainly killer cells accounting for an anticancer response. Legumes (chickpeas and soybeans), grains, and palm oil contain noncarotenoids, which remove cholesterol and are anticarcinogenic.

Flavonoids, a category of secondary metabolites, which are present in most of the plants, having quite 4000 varieties had been proven clinically for preventing various diseases like cancer, diabetes, heart diseases and kidney problem through its antioxidant properties and their bioactive components (Ehrlich 2009).

Phenolic acids are the most important class of secondary metabolites, mainly found in citrus fruits and wine, and have the antioxidant activity of scavenging the free radicals produced as a result of various metabolic pathways such as protein, carbohydrate, and fat. They also have anticancer and antitumor activity.

One of the classical examples is curcumin (turmeric), used as phytochemicals in most of the kitchen.

2. Probiotic Microorganisms

Metchniko coined the term "probiotic." Its application is well boosted in modern medicine due to its ability of making the intestine more friendly for processes like absorption and metabolism. Probiotics are vital to form life smoother by removing the toxic flora of the intestine and maintaining a friendly environment, e.g., useful consumption of *Bacillus bulgaricus* (Holzapfel et al. 2001). Currently, various probiotic products are available within the market with adequate nutrients to counter various pathogens in order that variety of ailments related to human body can be treated. The antimicrobial property usually had an altering impact on the microflora, making the epithelial tissues more grounded and making a situation for the supplements for better retention, which is required by the body. Moreover, probiotics is very useful in lactose intolerance by the production of related enzymes (ß-galactosidase) and hydrolyzing lactose into its sugar components (Pineiro and Stanton 2007).

3. Nutraceutical Enzymes

Enzymes are proteinous in structure, are produced by the cell, and act as a biocatalyst. It eases the metabolic rate and fastens the life process. The medical problem mainly associated with the GIT whether GERD (gastro esophageal reflux disease) or constipation or diarrhea or ulcerative colitis might be treated with enzyme supplements. The enzyme could be a better option for diabetic patients. Nowadays, enzyme therapies are used for several rare diseases such as Gaucher disease, Hunter syndrome, Fabry disease, and Pompe disease. Although enzymes are produced by their own cells, microbial sources are preferred more over plant and animal sources as they are more economical.

Nontraditional Nutraceuticals
They are foods enriched with supplements or biotechnologically designed crops to boost the nutrients; e.g., rice and broccoli are rich in β-carotene and vitamins, respectively. Food samples contain bioactive components which are engineered to supply products for human wellness. They are arranged as follows:

1. **Fortified Nutraceuticals**

   These types of nutraceuticals include breeding at the agriculture level or addition of compatible nutrients to the main ingredients like minerals added to cereals; flour fortified with calcium, iron, and folic acid (vitamin B); and milk fortified with cholecalciferol commonly used for vitamin D deficiency (Casey et al. 2010).

2. **Recombinant Nutraceuticals**

   Biotechnology tools have been well applied through a fermentation process in various food materials like cheese and bread to extract the enzyme useful for providing necessary nutrients at an optimum level.

7.2.2 Classification Based on Mechanism of Action

Nutraceuticals has been further classified in reference to specific therapeutic properties accounting for antimicrobial, anti-inflammatory, and antioxidant properties.

7.2.3 Classification Based on Chemical Nature

These types are classified depending upon their primary and secondary metabolite sources like is oprenoid derivatives, phenolic substances, fatty acids, carbohydrates, and amino acid-based substances.

7.3 Categorizing Nutraceuticals

Nutraceuticals can be organized in several ways depending upon its easier understanding and application, i.e., for academic instruction, clinical test design, functional food development, or dietary recommendations. Some of the foremost common ways of classifying nutraceuticals can be based on food sources, mechanism of action, chemical nature, etc. The food sources used as nutraceuticals are all natural and may be categorized as follows (Kalia 2005; Kokate et al. 2002):

1. Dietary fiber
2. Probiotics
3. Prebiotics
4. Polyunsaturated fatty acids
5. Antioxidant vitamins
6. Polyphenols
7. Spices

More broadly, nutraceuticals can be categorized in two groups (Pandey et al. 2010):

1. Potential nutraceuticals
2. Established nutraceuticals

A potential nutraceutical could become an established one only after efficient clinical data of its health and medical benefits are obtained. It is to be noted that much of the nutraceutical products still lay within the "potential" category.

7.3.1 Dietary Fiber

Dietary fiber is that the food material, more precisely the plant material that is not hydrolyzed by enzymes secreted by the digestive tract, but digested by microflora in the gut. Dietary fibers mostly include non-starch polysaccharides (NSP) like celluloses, hemicelluloses, gums and pectins, lignin, resistant dextrins, and resistant starches. Foods rich in soluble fiber include fruits, oats, barley, and beans. Chemically dietary fiber means carbohydrate polymers with a degree of polymerization not less than 3, which are neither digested nor absorbed in the small intestine. Based on their water solubility, dietary fibers could also be divided into two forms:

1. Insoluble dietary fiber (IDF), which incorporates celluloses, some hemicelluloses, and lignins which is fermented to a limited extend within the colon
2. Soluble dietary fiber (SDF), which incorporates β-glucans, pectins, gums, mucilages, and hemicelluloses that are fermented within the colon

The IDF and SDF compounds are collectively referred to as non-starch polysaccharides (NSP). The soluble components of dietary fiber by virtue of their bulking and viscosity producing capabilities retard the gastric emptying of the stomach. This affects the rate of digestion and the uptake of nutrients and creates a feeling of satiety. Soluble fiber has been shown to lower selectively serum LDL cholesterol and to enhance glucose tolerance (Glore et al. 1994). They also enhance insulin receptor binding and improve glycemic response. In colon, dietary fiber increases fecal bulking due to increased water retention, increased transit time, and increased fecal bacterial mass caused by soluble fiber fermentation. The fiber also promotes the growth of Bifidobacteria within the gut (especially fructooligosaccharides). Persons consuming generous amounts of dietary fiber, compared to those who have minimal fiber intake, are having low risk of CHR, stroke (Steffen et al. 2003), hypertension, diabetes (Montonen et al. 2003), obesity (Lairon et al. 2005), and certain gastrointestinal disorders (Petruzziello et al. 2006). Again, increase in the intake of high fiber food improves serum lipoprotein values (Brown et al. 1999), lowers blood pressure level, improves blood sugar control for diabetes (Anderson et al. 2004), aids weight loss, and promotes regularity (Cummings 2001). Research reveals that certain soluble fibers enhance the immunity in humans (Watzl et al. 2005). Some potential negative effects of dietary fiber include reduced absorption of vitamins, minerals, proteins, and calories. It is recommended that dietary fiber intake for adults generally falls in the range of 20–35 g/day. The recommended dietary fiber intake for youngsters and adults is estimated to be 14 g/1000 kCals. Several case histories have reported that consumption of excessive amounts of dietary fiber causes diarrhea (Saibil 1989).

7.3.2 Probiotics

The history of probiotics dates back as far as the first intake of fermented milks, over 2000 years ago. The scientific interest during this area boosted from the work of Metchinkoff (1907) to transform the toxic flora of the massive intestine into a host-friendly colony of *Bacillus bulgaricus* (Hord 2008). A probiotics are often defined as live microbial feed supplement, which when administered in adequate amounts beneficially affects the host animal by improving its intestinal microbial balance (Food and Agricultural Org World Health Org 2001). Probiotics generally include the following categories of bacteria:

1. Lactobacilli such as *L. acidophilus*, *L. casei*, *L. delbrueckii* subsp. *bulgaricus*, *L. brevis*, *L. cellobiosus*.
2. Gram-positive cocci such as *Lactococcus lactis*, *Streptococcus salivarius* subsp. *thermophilus*, *Enterococcus faecium*
3. Bifidobacteria such as *B. bifidum*, *B. adolescentis*, *B. infantis*, *B. longum*, and *B. thermophilum*. Probiotics are available in various forms as powder form, liquid form, gel or paste or granule forms, capsule forms, etc. (Suvarna and Boby 2005). Specific probiotics are generally wont to treat gastrointestinal (GI) conditions like lactose intolerance, acute diarrhea, and antibiotic-associated GI side effects. Probiotic agents possess the properties of non-pathogenic, non-toxic, resistance to gastric acid, and adherence to gut epithelial tissues producing antibacterial substances (Suvarna and Boby 2005). There are evidences that administration of probiotics decreases the risk of systemic conditions, like allergy, asthma, cancer, and number of other infections of the ear and urinary tract (Lenoir-Wijnkoop et al. 2007).

7.3.3 Prebiotics

Prebiotics are dietary ingredients that beneficially affect the host by selectively altering the composition or metabolism of the gut microbiota (Macfarlane et al. 2006). These are short-chain polysaccharides that have unique chemical structures that are not digested by humans, especially fructose-based oligosaccharides that exist naturally in food or are added within the food. The prebiotic consumption generally promotes the *Lactobacillus* and *Bifidobacterium* growth within the gut, thus helping in metabolism (Hord 2008). Vegetables like chicory roots, banana, tomato, and alliums are rich in fructooligosaccharides. Some other examples of these oligosaccharides are raffinose and stachyose, found in beans and peas. The health benefits of the prebiotics include improved lactose tolerance, antitumor properties, neutralization of toxins, and stimulation of intestinal immune system and reduction of constipation, blood lipids, and blood cholesterol levels (Isolauri et al. 1991). A daily intake of 5–20 g of insulin and oligosaccharides promote the growth of Bifidobacteria (Schrezenmeir and De Vrese 2001). Again, consumption of huge amounts of such oligosaccharides causes diarrhea, abdominal distension and flatulence (Guarner 2005).

7.3.4 Polyunsaturated Fatty Acids (PUFA)

PUFAs also are called "essential fatty acids" as these are crucial to the body's function and are introduced externally through the diet (Escott and Mahan 2000). PUFAs have two subdivisions: omega-3-(n-3) fatty acids and omega-6-(n-6) fatty acids. The major omega-3 fatty acids are α-linolenic acid (ALA), eicosapentaenoic acid (EPA), and docosahexaenoic acid (DHA). ALA is the precursor of EPA and DHA. EPA and DHA are found mainly in fatty fishes like mackerel, salmon, herring, trout, blue fin tuna, and fish oils. Principal sources of ALA are mainly flaxseed, soybeans, canola, some nuts (e.g., walnuts), and red/black currant seeds. Omega-6PUFAs mainly consist of linoleic acid (LA), γ-linolenic acid (GLA), and arachidonic acid (ARA). LA occurs mainly in vegetable oils, e.g., corn, safflower, soya bean, and sunflower. ARA is found in animal products like meat, poultry, and eggs. Studies suggest that omega-3 fatty acids have three major effects as cardiovascular diseases anti-arrhythmic (preventing or alleviating irregularities within the force or rhythm of the heart), hypolipidemic (promoting the reduction of lipid concentrations in the serum) (Bucher et al. 2002), and antithrombotic (decreased arteriosclerosis) (Bucher et al. 2002). Emerging research evidence shows the advantages of omega-3 oils in other areas of health including premature infant health (Carlson 1999), asthma (Hodge et al. 1996), bipolar and depressive disorders (Edwards et al. 1998), and dysmenorrhea and diabetes (Simopoulos 1991). Omega-3 fatty acids have been shown to be beneficial at various stages of life.

Infant formulas nowadays contain DHA alongside with ARA, which closely mimic the breast milk.

7.3.5 Antioxidant Vitamins

Vitamins like vitamin C, vitamin E, and carotenoids are collectively referred to as antioxidant vitamins. These vitamins act both singly also as synergistically for the prevention of oxidative reactions resulting in several degenerative diseases including cancer, cardiovascular diseases, cataracts, etc. (Elliott 1999). These vitamins are abundant in many fruits and vegetables and exert their protective action by free-radical scavenging mechanisms. Vitamin E which comprises of tocopherols together with tocotrienols transfers hydrogen atom and scavenge singlet oxygen and other reactive species thus protecting the peroxidation of PUFA within the biological membrane and LDL (Meydani 2000). Tocotrienols are more mobile within the biological membrane than tocopherols due to the presence of the unsaturated side chain and hence penetrate tissues with saturated fatty layers, i.e., in the brain and liver more efficiently. They have more recycling ability and are a far better inhibitor of liver oxidation. Vitamin E and selenium have a synergistic role against lipid peroxidation. Vitamin C, better referred to as ascorbic acid (vitamin c), donates hydrogen atom to lipid radicals, quenches singlet oxygen radical, and removes molecular oxygen. Scavenging of aqueous radicals by the synergistic effect of ascorbic acid along with tocopherol supplementation is a well-known antioxidant mechanism (Lee et al. 2004). Carotenoids like lycopene, βcarotene, lutein, and zeaxanthin are known to be the most efficient singlet oxygen quencher in the biological systems without the production of any oxidizing products. β-carotene traps peroxy free radicals in tissues at low-oxygen concentrations. Hence, β-carotene complements the antioxidant properties of vitamin E.

7.3.6 Polyphenols

Polyphenols form a huge group of phytochemicals, which are produced by plants as secondary metabolites to protect them from photosynthetic stress, reactive oxygen species. There are approximately 8000 different classes of polyphenols, the most important being flavonols, flavones, flavan-3-ols, flavanones, and anthocyanins. The highly branched phenylpropanoid pathway synthesizes majority of polyphenols. The most commonly occurring polyphenols in food include flavonoids and phenolic acids. Dietary polyphenols are of current interest because substantial evidence in vitro has suggested that they will affect numerous cellular processes like gene expression, apoptosis, platelet aggregation, and intercellular signaling, which can have anticarcinogenic and anti-atherogenic implications. These apart, polyphenols also possess antioxidant, anti-inflammatory, antimicrobial, and cardioprotective activities and play a role in the prevention of neurodegenerative diseases and diabetes mellitus (Scalbert et al. 2005).

7.3.7 Spices

Spices are esoteric food adjuncts that are used for thousands of years to enhance the sensory quality of foods. The quantity and therefore the sort of the spices consumed in the tropical countries are particularly extensive. These impart characteristic flavor, aroma, or piquancy and color to foods, stimulating our appetite also as to modify the texture of food. Recent research reveals that dietary spices in their minute quantities have an immense influence on the human health by their antioxidative, chemopreventive, antimutagenic, anti-inflammatory, and immunomodulatory effects on cells and a wide range of beneficial effects on human health by the action of gastrointestinal, cardiovascular, respiratory, metabolic, reproductive, neural, and other systems (Kochhar 2008; Lampe 2003).

7.4 Health Benefits (Chauhan et al. 2013; Baradaran et al. 2013)

- Avoid the side effect.
- May increase the health beneficial effect.
- May have naturally dietary supplement, so do not have unpleasant side effect.
- May increase the health value and our diet and improve medical condition of human. May easily be available and economically affordable.

Nutritional therapy may be a healing system using dietary therapeutics or nutraceuticals as a complementary therapy. This therapy is predicted on the assumption that foods cannot only be sources of nutrients and energy but could also provide medicinal benefits. According to nutraceutical and nutritional therapy theory, it achieves this goal by using efficacy of such nutraceuticals in detoxifying the body, avoiding vitamin and mineral deficiencies, and restoring healthy digestion and dietary habit. Phytonutrients basically is plant nutrients with particular biological activities in supporting human health.

The phytochemical works by the following ways:

1. Substrate for biochemical reactions
2. Cofactors of enzymatic reactions
3. Inhibitors of enzymatic reactions
4. Absorbents that bind to and eliminate undesirable constituent in the intestine
5. Enhance the absorption and/or stability of essential nutrients
6. Selective growth factor for beneficial bacteria

Table. 7.1 Health benefits of natural nutraceuticals

S. No	Nutraceuticals	Health benefits and functions
1	Green tea	Prevents allergy, possesses antibacterial activity. Helps in lowering blood cholesterol preventing bone loss
2	Aloe vera	Rich in antioxidant. Promotes immunity. Aids in weight loss
3	Garlic	Aids in digestion process. Strengthen immunity. Helps in weight loss. Possess antifungal and antibacterial properties
4	Ajwain	Contains thymol aids in digestion and prevents regulations. Prevents CVD (cardio vascular disease)
5	Cinnamon	Rich in antioxidants like polyphenols. Phenolic acid and flavonoid. Helps to control diabetes and prevent cancer

7. Fermentation substrate for beneficial bacteria
8. Selective inhibitors of deleterious intestinal bacteria
9. Scavengers of reactive or toxic chemicals
10. Ligands that agonize or antagonize cell surface or intracellular receptors (Tables 7.1 and 7.2)

7.5 Role of Nutraceuticals

7.5.1 Role of Bacterial Nutraceuticals

Bacteria like lactic acid bacteria (LAB) are used everywhere the world in a large variety of industrial food fermentations (Chelule et al. 2010). Some of the starter bacteria used in dairy fermentation are the yoghurt bacterium *Streptococcus thermophilus* and the cheese and butter (milk) bacterium *Lactococcus lactis*. This property offers the possibility to fortify fermented dairy products with folate by natural means, i.e., without the addition of food supplements. Some lactic acid bacteria, such as *Lactococcus lactis*, are able to produce and excrete riboflavin into the surrounding medium. Probiotics are live microbial food ingredients, which are beneficial to health. They are friendly bacteria that promote healthy digestion and absorption of some nutrients. They act to displace pathogens, like yeasts, other bacteria, and viruses which will otherwise cause disease and develop a mutually advantageous symbiosis with the human gastrointestinal tract. They have an antimicrobial effect through modifying the microflora, preventing adhesion of pathogens to the intestinal epithelium, competing for nutrients necessary for pathogen survival, producing an antitoxin effect, and reversing some of the consequences of infection on the intestinal epithelium (Singh and Sinha 2012). The various types of bacteria that are having the probiotics characteristics are *Lactobacillus*, *Lactobacillus rhamnosus*, *Lactobacillus reuteri*, *Lactobacillus case*, *Bifidobacterium*, *Bifidobacterium lactis*, *Bifidobacterium longum*, *Bifidobacterium breve*, *Bifidobacterium infantis*, *Streptococcus*, *Lactococcus*, *Lactococcus platinum*, *Lactococcus reuteri*, *Lactococcus agilis*, *Enterococcus*, *Saccharomyces*, *Bacillus*, and *Pediococcus*. There are large numbers of advantages of using probiotics as

Table. 7.2 Natural nutraceuticals along with mechanism

Nutraceuticals	Mechanism activity
Proanthocyanidin (chestnut fruits)	Inhibit IL-8 secretion by impairing NF-kappa-B signaling (Sangiovanni et al. 2018)
Fish-based diet	Severe osteoarthritis and hip and elbow dysplasia (Manfredi et al. 2018)
Curcuma extract	Decrease the level of PSA for prostate cancer (Fabiani et al. 2018)
Supplementation of live yeast fostered	Regulate inflammation and epithelial barrier in the rumen and express DFEB1 coding for an antimicrobial peptide (Bach et al. 2018)
Inulin-type friction dietary fiber	Immune responses against hepatitis-B (Vogt et al. 2017)
Bovine milk-derived oligosaccharide and B. Lactis	Modulate gut microbiota and immune system (Radke et al. 2017)
Lipid-based nutrient supplements	Prevent growth faltering in infants (Matsungo et al. 2017)
Partially hydrolyzed cow's milk proteins	Cow's milk allergy in children (Kiewiet et al. 2017)
Lactic acid bacteria (LAB) probiotic	Endometrial inflammation and infection (Genís et al. 2017)
Lipid-based nutrient supplement (LNS)	Moderate acute malnutrition (MAM) (Fabiansen et al. 2017)
Vitamin D supplementation	Extraskeletal benefits (Caprio et al. 2017)
Neutral amino acid supplements	Optimize neurocognitive function (Van Vliet et al. 2016)
Myo-inositol	Gestational diabetes (Santamaria et al. 2016)
Lactobacillus fermentum CRL1446	Enhances metabolism and oxidative parameters (Russo et al. 2016)
Dehydrozingerone and its dimer	Counteract the inflammation and oxidative stress (Profumo et al. 2016)

nutraceuticals. Some of these benefits include enhancing bowel function, preventing colon cancer, lowering cholesterol, lowering of blood pressure, improving immune function, reducing infections, reducing inflammation, improving mineral absorption, preventing growth of harmful bacteria, fighting off diseases like candida and eczema, and many more. As these "friendly bacteria" are beneficial for humans, similarly there are large numbers of soil bacteria which act as probiotics for plants. They are helpful in promoting the growth, health, and yield of crops (Pandey et al. 2011).

7.5.2 *Antioxidants in Disease Prevention*

Antioxidants are known to defuse free radicals leading to limited risk of oxidative stress (OS) and associated disorders. During the last few years, researches have confirmed that a lot of common disease (CVS, diabetes, cataracts, high blood pressure, infertility, respiratory tract infection, and rheumatoid arthritis) are related to with tissue deficiency and/or low dietary levels of compounds called antioxidants which make them an essential part of the nutraceutical market. Antioxidants are found in the vegetable oils, e.g., soybean oil, canola oil, corn oil, oat oil, wheat germ oil, palm oil, and evening prime rose oil. Antioxidants are very essential within the treatment of just about all diseases because most chronic diseases carry with them an excellent pact of oxidative stress. Oxidative stress plays a chief job in neurodegenerative diseases such as Alzheimer's disease (AD), Parkinson's disease (PD), and Huntington's disease (HD). Treatment with antioxidants is a hopeful loom for slowing disease progression (Klatte et al. 2003).

7.5.3 Phytochemicals as Potential Nutraceuticals

Phytochemicals (bioactive non-nutrient plant compounds), have raised interest in human nutrition because of their potential effects as antioxidants, antiestrogenics, anti-inflammatory, immunomodulatory, and anticarcinogenics (Laparra and Sanz 2010). Majority of foods, such as whole grains, beans, fruits, vegetables, and herbs, contain phytonutrients/phytochemicals. They are naturally occurring biochemicals that give plants their color, flavor, smell, and texture, which may help prevent diseases. They are biologically active natural products such as glucosinolates in cruciferous vegetables, limonoids in citrus fruits, lignans in flaxseed, lycopene in tomatoes, and catechins in tea. They all have specific actions and can be used variously, e.g., as antioxidants, and have a positive effect on health. Recently, much attention has been given to phytochemicals that possess cancer-preventive properties (Shukla and Singh 2007). Besides chemopreventive components in vegetables and fruits, some phytochemicals derived from herbs and spices even have potential anticarcinogenic and antimutagenic activities, among other beneficial health effect. A broad range of phytopharmaceuticals with a claimed hormonal activity, called "phytoestrogens," is recommended for prevention of prostate/breast cancer (Iriti and Faoro 2006). Tea contains catechin derivative and amino acid theanine (shown reduced blood pressure in hypertensive rats). Black and green tea both show equal activity in total plasma antioxidant status after single dose. But some study shows green tea to be more effective than black. Tea polyphenols have anti-inflammatory, antithrombotic, and antiplatelet properties and are effective in lowering risk of developing Coronary Heart Disease. Flavonoids have anticancerous properties by acting as antioxidants. They are found in citrus fruits, soyfoods which are unique dietary source of isoflavones, green tea rich in epigallocatechin gallate, and curcuma longa rich in curcumin. The main soybean isoflavones, genistein, daidzein, biochanin inhibits prostate cancer cell growth. Carotenoids and lycopenes are also important chemicals for human health (Pandey et al. 2011).

Flavonoids block the angiotensin-converting enzyme (ACE) that raises blood pressure by blocking the "suicide" enzyme cyclooxygenase that breaks down prostaglandins; they prevent platelet stickiness and hence platelet aggregation.

Flavonoids also protect the system and strengthen the small capillaries that carry oxygen and essential nutrients to all or any cells. Flavonoids block the enzymes that produce estrogen, thus reducing the risk of estrogen-induced cancers. In recent years, there is growing evidence that plant-food polyphenols, due to their biological properties, could also be unique nutraceuticals and supplementary treatments for various aspects of type 2 diabetes mellitus. Potential efficacies of polyphenols, including phenolic acids, flavonoids, stilbenes, lignans, and polymeric lignans, on metabolic disorders and complications induced by diabetes have been discovered (Bahadoran et al. 2013).

7.5.4 Polyunsaturated Fatty Acids as Nutraceuticals

The use of omega-3 oils/long-chain ω3 PUFA constitutes one among the most promising developments in human nutrition and disease risk reduction within the past three decades. Long-chain ω3 PUFAs are of great interest due of their effectiveness in prevention and treatment of coronary heart disease, hypertension, diabetes, arthritis and other inflammations, autoimmune disorders, and mental health (psychological state) and neural function as in depression and schizophrenia and cancers and are essential for maintenance and development of normal growth, especially for the brain and retina. Polyunsaturated fatty acids (PUFAs) (which include the omega-3 and omega-6 fatty acids) and phytochemicals also play a crucial role as healthy dietary bioactive compounds. A balanced PUFA composition of food influences diverse aspects of immunity and metabolism. Moreover, interactions between PUFAs and components of the gut microbiota can also influence their biological roles (Cencic and Chingwaru 2010).

7.6 Nutraceuticals in Diseases

7.6.1 Diabetes

Diabetes mellitus is characterized by abnormally high levels of blood glucose, either due to insufficient insulin production or due to its ineffectiveness. The most common forms of diabetes are type 1 diabetes (5%), an autoimmune disorder, and type 2 diabetes (95%), which is associated with obesity. Gestational diabetes occurs in pregnancy.

In diabetic patients, omega-3 fatty acids are suggested to reduce glucose tolerance and promote insulin sensitivity. Insulin is required for the synthesis of long chain n-3 fatty acids, which contain ethyl esters that may be potentially beneficial in diabetic patients (Sirtori and Galli 2002). A high isoflavone intake (20–100 mg/day) is associated with lower incidence and mortality rate of type 2 diabetes, heart diseases, osteoporosis, and certain cancers (Brouns 2002). Docosahexaenoic acid modulates insulin resistance, especially important in women with gestational diabetes mellitus which foster the recommendation for essential fatty acids during pregnancy (Thomas et al. 2006).

Lipoic acid is a universal antioxidant, may be more effective as a long-term dietary supplement aimed at the prophylactic protection of diabetic patients from complications. Dietary fibers from psyllium help in weight reduction and glucose control in diabetic patients and to reduce lipid levels in hyperlipidemia (Coleman et al. 2001).

Magnesium reduces diabetes risk and improves insulin sensitivity. Calcium and vitamin D also promote insulin sensitivity and improve glycemic control in some diabetics. Caffeic acid reduces elevated plasma glucose levels in insulin-resistant patients. Green tea and epicatechin-3-gallate reduce fasting and postprandial glucose levels and improve insulin resistance. Bitter melon and pomegranates are good for diabetes, as they regulate metabolism and transport glucose from the blood into cells (Stephen 2012).

7.6.2 Cardiovascular Diseases

Cardiovascular diseases are the name for the group of disorders of the heart and blood vessels and include hypertension (high blood pressure), coronary heart diseases (heart attack), cerebrovascular disease (stroke), heart failure, peripheral vascular diseases, etc. It was reported that low intake of fruits and vegetables is associated with a high mortality in cardiovascular diseases. Antioxidants, dietary fibers, omega-3 polyunsaturated fatty acids, vitamins, and minerals along with physical exercise are recommended for prevention and treatment of cardiovascular diseases (Mandala et al. 2010).

Nutraceuticals like phytosterols, policosanol, monacolin etc. tend to reduce circulating levels of LDL-cholesterol. This is achieved by modulating cholesterol production in liver, binding cholesterol within intestines, and/or increasing LDL-c receptor uptake in the liver. Polyphenols present in grapes alter cellular metabolism and signaling, thereby reducing arterial diseases. They also reduce the possibility of oxidation by neutralizing free radicals. Nutraceuticals that show antihypertensive activity by blocking calcium channels include alpha-lipoic acid, magnesium, vitamin B6 (pyridoxine), vitamin C, N-acetyl cysteine, celery, omega-3-fatty acids, etc. (Houston 2005).

Flavonoids in plants available as flavones, flavanones play a major role in curing cardiovascular diseases. They inhibit cyclooxygenase pathway and angiotensin-converting enzyme (ACE), which is responsible for high blood pressure, and also prevent platelet aggregation and stickiness. Because of high ability to transfer electrons, scavenge reactive oxygen species, i.e., decrease oxidative stress, flavonoids are considered as cardioprotectors in delaying the onset of atherosclerosis (Hu and Willett 2002).

Phytosterols compete with dietary cholesterol by blocking the uptake and facilitating cholesterol excretion from the body. Buckwheat proteins more importantly lower cholesterol and high blood pressure and are also beneficial in constipation and obesity. Dietary fibers from defatted rice bran have laxative and cholesterol lowering ability. Essential fatty acids are required for production and rebuilding of cells, to reduce blood pressure, lower cholesterol and triglycerides, reduce the risk of blood clots, and help in preventing many diseases including arthritis, arrhythmias, and other cardiovascular diseases. Octacosanol has gastroprotective and lipid-lowering effects. Allicin lowers blood pressure and cholesterol (Nasri et al. 2014).

7.6.3 Cancer

Nutraceutical rich bioactive dietary components have the ability to prevent cancer. Vitamins (folate) play a serious role in DNA methylation and cancer prevention. Ginseng is an anti-inflammatory compound that prevents chronic inflammation of cancer (Sabita and Trygve 2012). Flavonoids found in citrus fruits appear to supply protection against cancer by acting as antioxidants. Fruits and vegetables containing lycopene exert cancer-protective effect via decrease in oxidative and other damage to DNA in humans. Tannins and saponins are found to have anticarcinogenic effect (Stahl and Sies 2005).

Phenolics like ferulic, caffeic, gallic acids, and curcumin possess anticancer activity. Thiosulfonates present in garlic and onions possess anticarcinogenic properties. Astragalus membranaceus (Fabaceae) may be a traditional Chinese herb, used as anticancer, antidiabetic, immune enhancer, antioxidant, and hepatoprotective agent (Chanda et al. 2019).

7.6.4 Obesity

Obesity may be a medical condition characterized by accumulation of excess body fat and is related with various risk factors such as angina pectoris, congestive heart failure, hypertension, osteoarthritis, hyperlipidemia, etc. A tolerable and effective nutraceutical that can increase energy expenditure and/or decrease caloric intake is desirable for body weight reduction.

5-hydroxytryptophan decreases appetite, whereas green tea extract increases energy expenditure, thus promoting weight loss. Fenugreek, chitosan, vitamin C, curcumin, black gram, and bottle guard also reduce body weight (Nasri et al. 2014).

7.6.5 *Osteoarthritis*

Osteoarthritis is a disease with a multifactorial etiology affecting all joint tissues, involving both biochemical and mechanical factors that act in synergy to degrade cartilage. Nutraceuticals like glucosamine, chondroitin sulfate, ginger, green tea, pomegranate, curcumin, avocado, collagen hydrolysate, etc. are used to decrease the complications associated with osteoarthritis (Sacco et al. 2013). Nutraceutical antioxidant agents have considerable evidence for treating inflammation, pain, and joint destruction. Chondroitin sulfate and glucosamine supplementation prevent arthritic pain and narrowing of joint space. Application of olive oil reduces pain, stiffness, and swelling (Agarwal 2017).

7.6.6 *Alzheimer's Disease*

Oxidative stress is etiologically associated with variety of neurodegenerative disorders including Alzheimer's disease, which is characterized by progressive dementia with memory loss as the major clinical manifestation. Nutraceuticals like beta-carotene, lycopene, curcumin, lutein, etc. act as antioxidants and stop oxidative stress-induced neuronal damage (Klatte et al. 2003).

7.6.7 *Parkinson's Disease*

Parkinson's disease is a brain disorder that results from nerve damage in certain regions of the brain causing muscle rigidity, shaking, and difficulty in walking. Plant polyphenols, stilbenes, soybean and other phytoestrogens, vitamin C, vitamin D, vitamin E, coenzyme Q10, and unsaturated fatty acids revealed protective roles against progression of Parkinson's disease. Creatine seemed to modify Parkinson's disease features as measured by a decline within the clinical signs (Nasri et al. 2014).

7.6.8 *Allergy*

Allergy is a condition during which body has an exaggerated response to either a drug or food. Quercetin is a flavonoid, a polyphenolic substance, and a natural antihistamine. It inhibits some inflammatory enzymes such as lipid peroxidases and reduces leukotriene formation. It also exerts antioxidant properties by scavenging free radicals (Formica and Regelson 1995).

7.6.9 *Oral Diseases*

Odontonutraceuticals represent pleiotropic phytotherapeutic agents in dentistry as they regulate different molecular and biochemical targets. They include green tea, grapes, and cocoa seed extracts which are rich in polyphenols, flavonoids, and proanthocyanidins, thus preventing oral diseases. Probiotics are helpful in preventing dental cavity, periodontitis, etc. (Gupta et al. 2015).

7.6.10 *Eye Disorders*

Lutein also known as helenin is one among the carotenoids found in fruits and vegetables and is used for the treatment of visual disorders. Zeaxanthin found in corn, egg yolks, green vegetables, and fruits is employed within the treatment of eye disorders (Rajasekaran et al. 2008). Green tea, carotenoids, flavonoids, vitamin E, and coenzyme Q10 possess antioxidant activity and are effective for presbyopia and cataracts. Omega 3, 6, and 9 fatty acids and folic acid in rice bran also promote eye health (Dutta et al. 2018).

7.6.11 *Migraine*

Migraine is a recurrent throbbing headache that typically affects one side of the top and is often accompanied by nausea and disturbed vision. Nutraceuticals used for the treatment of migraine are feverfew, petasites, coenzyme Q10, etc. Dried leaves of feverfew contain melatonin and chrysanthenyl acetate, which play a role in treating migraine. *Petasites hybridus* commonly known as butterbur is found to be safe and effective for long-term use in management of migraine (Gupta et al. 2015).

7.6.12 *Stress Management*

Stress is a vital part of our psychological makeup and may be a threat to our existence. The natural bioactive compounds called adaptogens help to cope up against stress-related cellular damages. They cause nonspecific increase in resistance of an organism to noxious influences and exert to normalize and provide balancing action for both stress and mental health. Herbal nutraceuticals like ashwagandha, rhodiola, L-theanine, and ginseng are effective adaptogens that activate the production of stress-suppressing heat shock protein 70 (HSP-70) and stabilize physiologic processes, promote homeostasis, increase resistance to environmental stress, reduce moderate to severe anxiety, improve sleep, reduce depression, and improve secondary memory (Kalra 2003).

7.6.13 Prolonging Life Span

Nutraceuticals present in citrus fruits and soybean has effect on epigenetic modifications, autophagy, and necrosis. Caffeic acid and rosmarinic acid present in fruits, vegetables, and herbs are anticarcinogenic, antioxidant, antirheumatic, and antimicrobial in nature and can prolong healthy lifetime (Pietsch et al. 2011).

7.6.14 Oral Health

Oral infection such as dental cavity, teeth loss, periodontal diseases, etc. can greatly affect the human health. Dental cavity could also be caused due to the various infections related to the numerous reasons such as bacterial infections and nutritional deficiencies. There are many growing evidence that the use of green tea and its polyphenol plays a beneficial role in oral health. It is suggested that intake of green tea defends healthy cells from transformation into malignant cells; additionally, it also abolishes halitosis through modification of odorant sulfur components. Nonetheless, consumption of green tea is often useful in prevention and treatment of oral pathologies (Narotzki et al. 2012).

7.6.15 Bone Health

Osteoarthritis is a degenerative disorder with little or less curative treatment. Nutritive approach has proven to be useful in maintenance and joint integrity and/or health. Many researchers found that oral joint supplements work as anti-inflammatory agent acting as building block for the formation and maintenance of normal joint cartilage. It has been established that oral joint supplements contain chondroitin sulfate and/or glucosamine. From the experimental evidence, it was witnessed that nutraceuticals are capable of promoting or improving cartilage health, when it reaches to the joint through the blood stream demonstrating effective improvement and management in the osteoarthritis/joint health (Baici et al. 1992) (Fig. 7.3) and (Table 7.3).

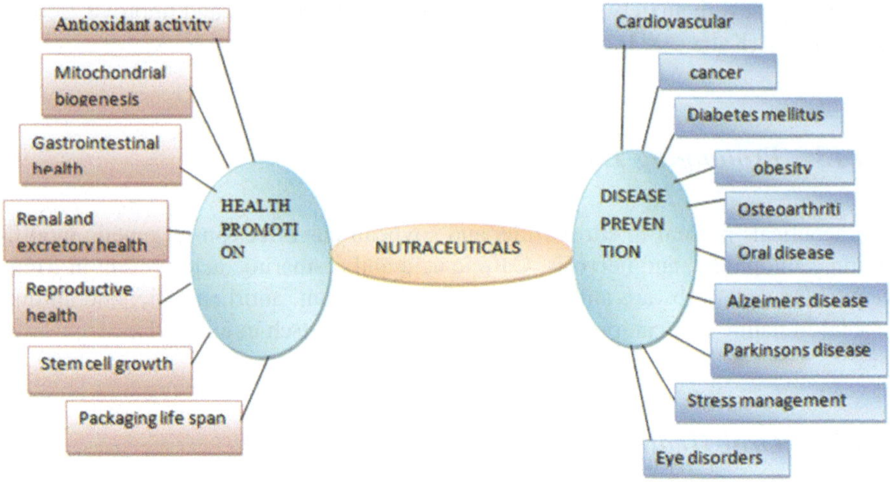

Fig. 7.3 Role of nutraceuticals in health promotion and disease prevention

Table. 7.3 Role of nutraceuticals on disease prevention

Nutraceutics	Disease prevention	Reference
Ginseng, beta carotene, sulfur compounds in garlic.	Cancer	(Cencic and Chingwaru 2010)
Soy Isoflavones, omega 3 fatty acid, lipoic acid, Catechins, spices like fenugreek and cinnamon, bitter melon, pomegranate	Diabetes mellitus	(Stephen 2012)
Diacerein, banana, ginger, green tea, pomegranate, *Boswellia*, oxaceprol, tipi, willow bark, curcumin, avocado, soybean, collagen hydrolysate, chondroitin sulfate, and glucosamine	Osteoarthritis	(Akhtar and Haqqi 2012)
Plant polyphenols, stilbenes, soybean and other phytoestrogens, vitamin C, vitamin D, vitamin E, coenzyme Q 10, unsaturated fatty acid, Brahmi, and inosine	Parkinson's disease	(Chao et al. 2012)
Curcumin, lutein, lycopene, Lavandula, Beta carotene, folic acid, and vit. B12	Alzheimer's disease	
Odontonutraceuticals, green tea, grapes, cocoa seed extracts rich in polyphenols, flavonoids, and proanthocyanidins	Oral diseases	(Janczarek et al. 2016)
Lutein, DHA, green tea, carotenoids, flavonoids, vitamin E, coenzyme Q10, zeaxanthin, melatonin, spirulina, flavonoids, ascorbic acid, tocopherol, carotenoids, caffeine, pyruvate	Eye disorders	(Varoni and Iriti 2016)
Adaptogens (ashwagandha, rhodiola, L-theanine, ginseng)	Stress management	(Kalra 2003)
Flavonoids, flavones, flavonones, quercetin in onion, cruciferous vegetables, blackberries, cherries, berries, apples, and allicin	Cardiovascular diseases	(Cicero et al. 2017)

7.7 Functional Foods

Functional foods denote to a new and challenging concept which is slightly different from the nutraceutical so these can be regarded as food products or ingredient, consumed as a part of usual diet providing the beneficial effects beyond the basic nutrients that traditionally it contains are called as functional foods (Whitman 2001) such as nuts, garlic, and green tea.

Functional foods provide carbohydrates, proteins, fat, and vitamins to the body in required amount for healthy survival of the body. When functional food contributes in the prevention and treatment of disease, then it is referred to as nutraceuticals. Otherwise, the difference between the functional food and nutraceutical is not exactly clear, but the main difference is the form in which they are consumed, that is, the nutraceutical, is ingested in the form of tablet, capsules, or pills, while functional foods are ingested as "ordinary foods." When phytochemical is added to the food formulation, then they are considered as functional foods, and when phytochemical is included in the capsule or any dosage forms, then they are established as nutraceutical (Espín et al. 2007) (Fig. 7.4) and (Table 7.4).

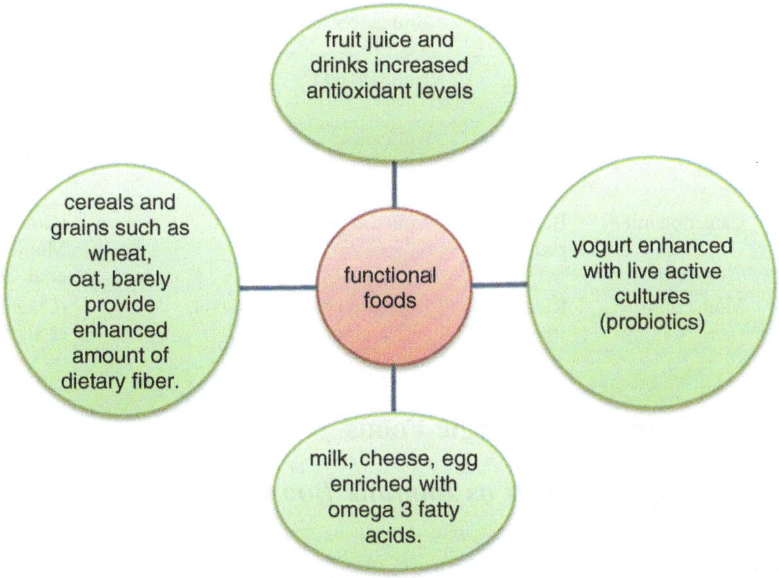

Fig. 7.4 Few examples of functional foods

Table. 7.4 Bioactive compound-based nutraceuticals possessing health benefits

S. No.	Nutraceuticals	Natural sources	Health benefits	Reference
1	Glycerolipids	Seed oils	Skin care	(Mendiola et al. 2007)
2	Amino acids	Sprouts, sports drinks	Effect on nervous system, antioxidant, muscles energy	(Kim et al. 2006)
3	Milk proteins, peptides	Milk derived product	Immunostimulating, antihypertensive, antimicrobial	(Lacomba et al. 2010)
4	Vitamin B1 and vitamin B2	Mushrooms	Antioxidant	
5	Tocopherols (vitamin E)	Vegetable and vegetable oils	Antitumor, antioxidant, treatment of cardiovascular disease	(Bakowska-Barczak et al. 2009)
6	Glycosides (glycyrrhetic acid, glycyrrhizin)	Plants	Diuretic, anticancer, antioxidant properties	(D'Antuono et al. 2008)
7	Saponins	Vegetables	Increase testosterone levels, and stimulate muscle growth and antidiabetic or antiobesity effects	(Brady et al. 2007)
8	Flavonoids	Citrus peel	Anticancer, anti-inflammatory	(Li et al. 2006)
9	Phenolics	Pepper	Antioxidant	(Chatterjee et al. 2007)
10	Lignans	Myristica fragrans (nutmeg)	Anticancer	(Chung et al. 2006)
11	Catecholamines	Banana peel	Antioxidant	(González-Montelongo et al. 2010)
12	Monacolins	Rice	Decrease cholesterol level, anticarcinogenic	(Chairote et al. 2008)

7.8 Significance of Synergic Foods

7.8.1 Synergistic Foods as Immune Boosters

Immunity, infection, and malnutrition are always interlinked. Malnutrition and nutritional alterations, common complications of human immunodeficiency viral infection, include disorders of food intake, nutrient absorption, and intermediary metabolism which play a significant and independent role in morbidity and mortality. The immune system may be a complex network of focused tissues, cells, organs, proteins, and chemicals. Hence, immune competence can be regarded as a measure of adequate nutrition. Interindividual variations in many immune functions existing within the normal healthy population are due to age, genetics, gender, ethnic background, socioeconomic situation, diet, stress, habitual levels of exercise, alcohol consumption, smoking habits, etc. Human cells are continuously exposed to diverse oxidizing agents of which few are necessary for all time. The imbalance within the production of oxidants may lead to oxidative stress, including viral, bacterial, and parasitic infections in the human system (Liu and Hotchkiss 1995). Intake of certain foods may help to keep our immune system strong. A balanced mixture of vitamins and minerals over time reflects a healthy immune system.

Foods like tea and black pepper work during a synergistic way and enhance the bioavailability of epigallocatechin gallate (EGCG) which is present in green tea. Lambert et al. (Lambert et al. 2004) conducted an in vivo study which reveals piperine present in black pepper inhibits the glucuronidation of EGCG, thereby lowering the transit rate of EGCG within the gastrointestinal (GI) tract. The reduced transit rate in turn reduces the resident time of EGCG in GI tract which allows maximum absorption. High amount of catechin present in green tea is related with improved immune tolerance thereby leading to lower incidences of cancer, cardiovascular diseases, high cholesterol, and lots more. Also, lemon along with green tea enhances the absorption of EGCG ten times more when compared to the absorption when green tea is drunk alone (Tewari et al. 2000). A study published in Food Chemistry ensured that vitamin C promotes the absorption and utilization of antioxidant in green tea five times more when compared to green tea consumed alone. The catechins and vitamin C act synergistically for better absorption of antioxidants. Raspberries and chocolates show a better synergistic effect for their antioxidant capacity. Todorovic et al. (2015) discovered significant increase in antioxidant capacity in cocoa, polyphenol, proanthocyanidin, and flavonoids.

7.8.2 Synergistic Foods to Prevent Infections

Malnutrition is the primary cause of immunodeficiency and infections worldwide, with infants, children, adolescents, and, therefore, the elderly most affected. There is a strong relationship between malnutrition and infection and infant death rate, because poor nutrition leaves children underweight, weakened, and susceptible to infections, primarily due to epithelial integrity and inflammation. In our understanding of this interaction between infection and malnutrition, it is important to remember that a decreased immune function is not always a defective one, and many indicators of nutritional status are not reliable during infection. Food synergy plays an important role in preventing infections.

Healing was more rapid in the skin of these rats treated with curcumin and ginger extract which was found to be a completely unique approach in improving the structure of the skin in rats (Bhagavathula et al. 2009). Through multiple signaling of proteins, curcumin inhibits cell proliferation, metastasis, invasion, and angiogenesis in different types of cancers (Saw et al. 2010).

Turmeric has its own potential anti-inflammatory effects, anticancer properties, and tumor-fighting properties. The active component within the turmeric is curcumin. Black pepper has its own hot and active component called piperine. While adding black pepper to turmeric or turmeric-based foods, piperine increases the bioavailability of curcumin to 1000 times. Piperine inhibits the metabolic breakdown of curcumin compounds in the gut and liver. This increases the bioavailability of curcumin compounds in the body (Shoba et al. 1998).

A significant contribution for the antibacterial activity is proved in garlic and honey with phenols and fatty acids which act synergistically in higher growth reduction of bacteria, enhancing the killing activity. This also helps to enhance the shelf life of each other (Saad and Mona 2013). Both are powerful antibiotics that preserve our immune system.

7.8.3 Synergistic Foods to Increase Bioavailability of Nutrients

Plant foods are enriched with micronutrients, but a standard understanding on the bioavailability of nutrients is primarily very essential. When we consume a food or drink, certain nutrients are absorbed into the bloodstream and transported to their respective target tissues. The nutrient supply within the physical body depends mainly on the bioavailability of the actual nutrient instead of the quantity consumed. The bioavailability of the nutrient is predicted on various factors like its reacting medium, promoters and inhibitors, and, therefore, the host environment. Understanding nutrient bioavailability helps in setting appropriate dietary recommendations.

Garlic and honey have a good synergistic effect; when garlic reaches the stomach, it promotes the assembly of gastric juices which are essential for absorption of iron and enrich the bloodstream (Shoba et al. 1998).

Flavonoids with almond skin predominantly benefit health and act synergistically with vitamin C and vitamin E to protect LDL oxidation (Chen et al. 2005).

The carotenoids found within the salads include lycopene, lutein, beta-carotene, alpha-carotene, and zeaxanthin. The egg yolk also contains zeaxanthin and lutein. Eating boiled eggs with the salad of tomatoes, carrots, and green leafy vegetables increased absorption of carotenoids three–ninefold.

Lemon and green leafy vegetables address iron deficiency anemia by increasing hemoglobin. It helps in improving the bioavailability of iron within the blood volume. Ascorbic acid or vitamin C enhances the dietary absorption of non-heme iron. First, it forms non-absorbable iron, and later, ferric iron is converted to ferrous iron which helps in better absorption of iron into the mucosal cells (Hallberg et al. 1989).

Yoghurt and banana play a big role in mutual benefits of probiotic and prebiotic. Probiotics introduce good bacteria into the gut, whereas prebiotics act as a fertilizer for the good bacteria. So, consumption of probiotics along with prebiotics is good for the gut bacteria which enhance gastrointestinal digestion (Hallberg et al. 1989). Inulin present in bananas energizes the expansion of good bacteria in yoghurt, which helps to enhance immunity and regulate digestion. Inulin with probiotics (good bacteria) found in yoghurt helps to increase the level of calcium in our body (Fernandez and Marette 2017).

7.8.4 Synergistic Foods to Reduce Chronic Diseases

Prospective epidemiological studies, some randomized prevention trials, and many short-term studies of intermediate endpoints like blood pressure and lipids have revealed an honest deal about the precise dietary and lifestyle determinants of major chronic diseases. Most of those studies were in part with the historical importance of those diseases. A general opinion is that reducing identified, modifiable dietary, and lifestyle risk factors could prevent most cases of chronic diseases like coronary artery disease, stroke, diabetes, and lots of cancers among high-income populations. The essential bioactive compounds present in the foods help in preventing diseases and betterment of health. Studies have shown that synergistic combination of foods plays a crucial role within the prevention of chronic diseases.

7.8.5 Cardiovascular Diseases

Combination of honey and garlic potentially reduces the blood cholesterol and triglyceride levels and helps in improving cardiovascular problems (Saad and Mona 2013). Studies prove the better synergistic roles of honey with garlic to regulate LDL levels (Chen et al. 2005).

Garlic and fish synergistically boost immunity and improve cardiovascular health. Effects of fish oil and omega-3 fatty acids are modulated by the inhibition of hepatic very-low-density lipoprotein (VLDL)-triglyceride synthesis and a rise within the fractional catabolic rate of VLDL. Various co-factors like selenium, copper, zinc, iron, and iodine work well with EPA and DHA for their anti-inflammatory and cholesterol-reducing properties (Morcos 1997).

For tomatoes and olive oil, tomatoes contain carotenoids which are fat soluble, and hence, absorption is improved with a fat medium like olive oil. Lycopenes, an antioxidant in the carotenoids group, reduce the risk of cardiovascular diseases by improving the serum lipid profile in high-fat diet in comparison to low-fat lycopene-rich diet. Lycopene content increases by 5–6 times on cooking instead of eating them raw (Ahuja et al. 2006; Story et al. 2010; Fielding et al. 2005).

Onion and grape combination resulted during a synergistic antiproliferative effect (APE) instead of consuming onion or grape alone. Black grapes are rich in polyphenol antioxidant catechin, which helps to prevent cardiovascular disease, cancer, and neurological disorders and in weight management. Together, onion and grapes inhibit blood clots and boost cardiovascular health. Studies have shown that this combination helps to relieve allergy symptoms and increases cardiovascular protection by improving circulation (Wang et al. 2013).

7.8.6 *Diabetes Mellitus*

An in vitro study by Agustinah et al. (2016) reveals that a mixture of 80% apple cider and 20% whole blueberry juice has potential antihypertensive and antihyperglycemic properties. Increased blueberry juice proportion increases the entire phenol content in apples with the inhibitory activity of angiotensin-1-converting enzyme (ACE), α-amylase, and α-glucosidase; the ACE inhibitory activity is decreased.

Onion and garlic contain active components like methiin and S-allyl cysteine sulfoxide (SACS), which stimulate the insulin production from the pancreas and decrease the blood glucose level. This interference with dietary glucose absorption by the insulin helps to control diabetes mellitus (Corzo-Martínez et al. 2007; Chung et al. 2007).

Vitamin D and vitamin K supplements help to upregulate the insulin receptor genes and promote secretion of insulin from pancreatic cell, thus enhancing blood glucose metabolism [28, 29]. Interaction of vitamin D and K supplements can also upregulate vascular smooth muscle cells (Van Ballegooijen et al. 2017).

Anthocyanin-rich black currant combined with rowanberry enriched in chlorogenic acids may be a synergic combination to enhance diabetes mellitus. As a synergic combination, both black currant and rowanberry extracts could replace the inhibition lost by reducing the acarbose dose and help to take care or maintain glycemic level for type 2 diabetes (Boath et al. 2012).

7.8.7 *Cancer*

Whole foods, like broccoli and tomatoes, have antitumorogenesis property by themselves which can lower the growth of cancer cells. Prostate tumors grew much less in rats that were fed with tomatoes and broccoli (Canene-Adams et al. 2005).

Apples contain variety of phytochemicals, like phloridzin, quercetin, chlorogenic acid, and catechin; the peel of apple has potential phytochemical with anticancer properties. Eating apple with skin inhibits neoplastic cell proliferation, reduces lipid oxidation, and lowers cholesterol in cancer patients (Wolfe and Liu 2003). Almonds with skin rich in antioxidants also benefits in the reduction of cancer and cardiovascular risks.

Apple and berry juices potentially benefit colon cancer within the presence of dietary compounds like vitamins, minerals, phytochemicals, and fiber. They are consumed among various ethnic groups widely based on their dual nutritional value. Adequate intake of these phytochemicals may hinder the expansion of cancer cells by enhancing DNA repair, thereby reducing the DNA damage by oxidative stress (Jaganathan et al. 2014).

7.8.8 Synergic Foods to Improve Mental and Reproductive Health

Diverse symptoms, including changes in mood, behavior, fluid retention, and certain aspects of mental or physical functioning, are common among women within the luteal phase of the menstrual cycle. Various reports, many of a preliminary nature or based on clinical experience, suggest that these symptoms are associated to diet, in particular, to high intakes of sugar or to deficiency of certain vitamins and minerals or to both. Several studies are reported since the1970s on the effect of high-dose vitamin B6 supplementation on the relief of premenstrual symptoms. Several studies were carried out to find the effect of synergic foods on mental and reproductive health.

Blueberries and strawberries have potential protective effects of antioxidant on neuronal functioning. Antioxidant-rich blueberries and strawberries reduce oxidative stress and inflammation and thus improve neuronal signal processing. Apples are rich in flavonoids, and green leafy vegetables are rich in dietary nitrate. Studies reveal that combination of flavonoids and nitrate could improve nitrous oxide production. The increase in nitrous oxide following consumption of flavonoids and dietary nitrate could improve cognitive function and mood (Bondonno et al. 2014).

Ethnic post-partum care nutritional practices act synergistically on rejuvenating post-partum needs and addressing the nutritional support for delivered mother like improving lactation, smooth bowel movement, immunity, better nourishment, preventing excess bleeding, preventing infections, wound healing, and strengthening of bones and muscles. Synergic combinations of foods would better help in relieving post-partum symptoms. Chick pea and beetroot relieve anxiety associated with menstrual symptoms. Vitamin B6 is required for maintaining normal intracellular magnesium concentrations as vitamin B6 helps in transport of magnesium across the cell membranes (De Souza et al. 2000).

7.9 Current and Future Development of Nutraceuticals

Nutraceutical presents one of the most exciting areas for health innovation, offering inexpensive, safe, and effective results for the today's most challenging health problem. The expanding nutraceutical market indicates that end users are seeking minimally processed food with extra nutritional benefits and organoleptic value. This development, in turn, is propelling expansion in the nutraceutical markets globally. The emerging nutraceuticals industry seems destined to occupy the landscape in the new millennium. Its tremendous growth has implications for the food, pharmaceutical, healthcare, and agricultural industries Many scientists believe that enzymes represent another exciting frontier in nutraceuticals. "Enzymes have been underemployed. They are going to be a hot area in the future." Fermentation technology using microbes to create new food products also represents potential. Global trends to healthy products cannot be reversed. Companies taking the lead by investing strategically in science, product development, marketing, and consumer education will not go unrewarded (Pandey et al. 2010). Nutraceuticals supplied through oral or transdermal delivery system would provide well-targeted health benefits with optimal bioavailability. With the evolution of "Smart Nutraceuticals," a Futuristic "Physician's Desk Reference" would contain information on individual genetic profiles to be matched with specific nutritional interventions as well. This would be a vast improvement over current nutritional recommendations which being too generalized are reported to benefit only 60% of population (Ball 2003).

7.10 Some Marketed Products Available as Nutraceuticals (Chaturvedi et al. 2011; Sarin et al. 2012; Pandey et al. 2010).

Many pharmaceutical companies are attempting to manufacture nutraceuticals in therapeutic areas. Some of the commercially available nutraceuticals are discussed in Table 7.5.

7.11 Conclusion

It is very imperative that nutrients found in many foods, fruits, and vegetables are responsible for the well-documented health benefits. Nutraceuticals are widely accepted by all the age groups due to their safety, efficacy, high quality, and purity. They play a crucial role in protection against the pathologies of numerous age-related or chronic diseases. Future demand of nutraceuticals depends on consumer perception of the relationship between diet and disease. Nutraceutical industry in India is one of the rapid growing markets. Because of their numerous health

Table. 7.5 Marketed products

Sr. no	Product	Category	Source	Ingredients	Benefits	Manufacturer
1	Glucon –D	Energy supplement	Fortified glucose	Glucose	Provide instant energy and rejuvenation	Heinz
2	Glucose – D	Energy supplement	Fortified glucose	Dextrose monohydrate, calcium phosphates, vitamin D	Provide instant energy and rejuvenation	Dabar
3	GRD	Nutritional supplement	–	Vitamins, carbohydrates, proteins	Helps in faster tissue growth and repair	Zydus
4	Proteinex	Protein and nutritional supplements	–	Essential vitamins, minerals, protein hydrolysate, sugar, malt, extract	Supports healthy physical and mental growth	Pfizer ltd
5	Tropicana	Energy drinks	Vitamin	B-vitamin, thiamine, folate	Nutrient rich juice and good source for heart health	Tropicana product Inc.
6	Frooti	Energy drinks	Mango	Water, mango pulp, sugar, anti-oxidant	Refreshing juice	Parle agro. Pvt.ltd
7	Abcor	Heart supplement	–	–	Lowers cholesterol	Nutin pharma
8	Fish oil plus	Brain supplement	Salmon	Omega 3 fatty acids	Reduce risk of heart disease maintain blood pressure and cholesterol	Pacific health Inc.
9	Vectomega	Dietary supplement	Salmon	Omega 3 fatty acids	Enhance the absorption of omega-3	Europharma
10	Proplus	Immune supplement	–	Mushrooms, polysaccharide, and folic acid	Provides powerful immune system	Native remedies

benefits, they must be taken regularly to decrease risk factors like high cholesterol, high blood pressure, and diabetes. With a little bit of careful handling, the future of both plant and animal origin nutraceuticals holds exciting opportunities in the medical field.

References

Agarwal, S., 2017. Leading pharmaceutical consultant, Nutraceuticals and osteoarthritis. Available at: https://www.drsanjayagarwal.com Accessed 14 July 2020

Agustinah W, Sarkar D, Woods F, Shetty K (2016) Apple and blueberry synergies for designing bioactive ingredients for the management of early stages of type 2 diabetes. J Food Qual 39(4): 370–382

Ahuja KD, Pittaway JK, Ball MJ (2006) Effects of olive oil and tomato lycopene combination on serum lycopene, lipid profile, and lipid oxidation. Nutrition 22(3):259–265

Akhtar N, Haqqi TM (2012) Current nutraceuticals in the management of osteoarthritis: a review. Ther Adv Musculoskelet Dis 4(3):181–207

Anderson JW, Randles KM, Kendall CW, Jenkins DJ (2004) Carbohydrate and fiber recommendations for individuals with diabetes: a quantitative assessment and meta-analysis of the evidence. J Am Coll Nutr 23(1):5–17

Bach A, Guasch I, Elcoso G, Chaucheyras-Durand F, Castex M, Fàbregas F, Garcia-Fruitos E, Aris A (2018) Changes in gene expression in the rumen and colon epithelia during the dry period through lactation of dairy cows and effects of live yeast supplementation. J Dairy Sci 101(3): 2631–2640

Bahadoran Z, Mirmiran P, Azizi F (2013) Dietary polyphenols as potential nutraceuticals in management of diabetes: a review. J Diabetes Metab Disord 12(1):1–9

Baici A, Hörler D, Moser B, Hofer HO, Fehr K, Wagenhäuser FJ (1992) Analysis of glycosaminoglycans in human serum after oral administration of chondroitin sulfate. Rheumatol Int 12(3): 81–88

Bakowska-Barczak AM, Schieber A, Kolodziejczyk P (2009) Characterization of Canadian black currant (Ribes nigrum L.) seed oils and residues. J Agric Food Chem 57(24):11528–11536

Ball D (2003) Foods of the future may be tailored to fit. Wall Street J 1:116–121

Baradaran A, Madihi Y, Merrikhi A, Rafieian-Kopaei M, Nasri H (2013) Serum lipoprotein (a) in diabetic patients with various renal function not yet on dialysis. Pak J Med Sci 29(S):354–357

Bhagavathula N, Warner RL, DaSilva M, McClintock SD, Barron A, Aslam MN, Johnson KJ, Varani J (2009) A combination of curcumin and ginger extract improves abrasion wound healing in corticosteroid-impaired hairless rat skin. Wound Repair Regen 17(3):360–366

Boath AS, Stewart D, McDougall GJ (2012) Berry components inhibit α-glucosidase in vitro: synergies between acarbose and polyphenols from black currant and rowanberry. Food Chem 135(3):929–936

Bondonno CP, Downey LA, Croft KD, Scholey A, Stough C, Yang X, Considine MJ, Ward NC, Puddey IB, Swinny E, Mubarak A (2014) The acute effect of flavonoid-rich apples and nitrate-rich spinach on cognitive performance and mood in healthy men and women. Food Funct 5(5): 849–858

Brady K, Ho CT, Rosen RT, Sang S, Karwe MV (2007) Effects of processing on the nutraceutical profile of quinoa. Food Chem 100(3):1209–1216

Brouns F (2002) Soya isoflavones: a new and promising ingredient for the health foods sector. Food Res Int 35(2–3):187–193

Brower V (1998) Nutraceuticals: poised for a healthy slice of the healthcare market? Nat Biotechnol 16(8):728–731

Brown L, Rosner B, Willett WW, Sacks FM (1999) Cholesterol-lowering effects of dietary fiber: a meta-analysis. Am J Clin Nutr 69(1):30–42

Bucher HC, Hengstler P, Schindler C, Meier G (2002) N-3 polyunsaturated fatty acids in coronary heart disease: a meta-analysis of randomized controlled trials. Am J Med 112(4):298–304

Canene-Adams K, Campbell JK, Zaripheh S, Jeffery EH, Erdman JW Jr (2005) The tomato as a functional food. J Nutr 135(5):1226–1230

Caprio M, Infante M, Calanchini M, Mammi C, Fabbri A (2017) Vitamin D: not just the bone. Evidence for beneficial pleiotropic extraskeletal effects. Eat Weight Disord 22(1):27–41

Carlson SE (1999) Long-chain polyunsaturated fatty acids and development of human infants. Acta Paediatr Suppl 88:72–77
Casey CF, Slawson DC, Neal LR (2010) Vitamin D supplementation in infants, children, and adolescents. Am Fam Physician 81(6):745–748
Cencic A, Chingwaru W (2010) The role of functional foods, nutraceuticals, and food supplements in intestinal health. Nutrients 2(6):611–625
Chairote EO, Chairote G, Niamsup H, Lumyong S (2008) The presence and the content of monacolins in red yeast rice prepared from Thai glutinous rice. World J Microbiol Biotechnol 24(12):3039–3047
Chanda S, Tiwari RK, Kumar A, Singh K (2019) Nutraceuticals inspiring the current therapy for lifestyle diseases. Adv Pharmacol Sci 2019:6908716
Chao J, Leung Y, Wang M, Chang RCC (2012) Nutraceuticals and their preventive or potential therapeutic value in Parkinson's disease. Nutr Rev 70(7):373–386
Chatterjee S, Niaz Z, Gautam S, Adhikari S, Variyar PS, Sharma A (2007) Antioxidant activity of some phenolic constituents from green pepper (Piper nigrum L.) and fresh nutmeg mace (Myristica fragrans). Food Chem 101(2):515–523
Chaturvedi S, Sharma PK, Garg VK, Bansal M (2011) Role of nutraceuticals in health promotion. Int J PharmTech Res *3*(1):442À8
Chauhan B, Kumar G, Kalam N, Ansari SH (2013) Current concepts and prospects of herbal nutraceutical: a review. J Adv Pharm Technol Res *4*(1):4
Chelule PK, Mokoena MP, Gqaleni N (2010) Advantages of traditional lactic acid bacteria fermentation of food in Africa. In: Current research, technology and education topics in applied microbiology and microbial biotechnology, vol 2. Formatex Research Center, pp 1160–1167
Chen CY, Milbury PE, Lapsley K, Blumberg JB (2005) Flavonoids from almond skins are bioavailable and act synergistically with vitamins C and E to enhance hamster and human LDL resistance to oxidation. J Nutr 135(6):1366–1373
Cherdshewasart W, Sutjit W, Pulcharoen K, Chulasiri M (2009) The mutagenic and antimutagenic effects of the traditional phytoestrogen-rich herbs, Pueraria mirifica and Pueraria lobata. Braz J Med Biol Res 42(9):816–823
Chung I, Kwon SH, Shim ST, Kyung KH (2007) Synergistic antiyeast activity of garlic oil and allyl alcohol derived from alliin in garlic. J Food Sci 72(9):M437–M440
Chung JY, Choo JH, Lee MH, Hwang JK (2006) Anticariogenic activity of macelignan isolated from Myristica fragrans (nutmeg) against Streptococcus mutans. Phytomedicine 13(4):261–266
Cicero AF, Colletti A, Bajraktari G, Descamps O, Djuric DM, Ezhov M, Fras Z, Katsiki N, Langlois M, Latkovskis G, Panagiotakos DB (2017) Lipid-lowering nutraceuticals in clinical practice: position paper from an international lipid expert panel. Nutr Rev 75(9):731–767
Coleman MD, Eason RC, Bailey CJ (2001) The therapeutic use of lipoic acid in diabetes: a current perspective. Environ Toxicol Pharmacol 10(4):167–172
Corzo-Martínez M, Corzo N, Villamiel M (2007) Biological properties of onions and garlic. Trends Food Sci Technol 18(12):609–625
Cummings JH (2001) The effect of dietary fiber on fecal weight and composition. In: CRC handbook of dietary fiber in human nutrition, vol 3. CRC Press, pp 183–252
D'Antuono LF, Elementi S, Neri R (2008) Glucosinolates in Diplotaxis and Eruca leaves: diversity, taxonomic relations and applied aspects. Phytochemistry 69(1):187–199
De Souza MC, Walker AF, Robinson PA, Bolland K (2000) A synergistic effect of a daily supplement for 1 month of 200 mg magnesium plus 50 mg vitamin B6 for the relief of anxiety-related premenstrual symptoms: a randomized, double-blind, crossover study. J Womens Health Gend Based Med 9(2):131–139
Dutta S, Ali KM, Dash SK, Giri B (2018) Role of nutraceuticals on health promotion and disease prevention: a review. J Drug Deliv Therap 8(4):42–47
Edwards R, Peet M, Shay J, Horrobin D (1998) Omega-3 polyunsaturated fatty acid levels in the diet and in red blood cell membranes of depressed patients. J Affect Disord 48(2–3):149–155

Ehrlich SD (2008) Private practice specializing in complementary and alternative medicine, Review. VeriMed Healthcare Network, Phoenix, AZ

Ehrlich SD (2009) Peppermint (Mentha Piperita), Private Practice Specializing in Complementary and Alternative Medicine, Review. VeriMed Healthcare Network, Phoenix, AZ

Elliott JG (1999) Application of antioxidant vitamins in foods and beverages: developing nutraceuticals for the new millennium. Food Technol (Chicago) 53(2):46–48

Escott S, Mahan LK (2000) Krause's food, nutrition and diet therapy, 10th edn. WB Saunders Company, Philadelphia, pp 553–559

Espín JC, García-Conesa MT, Tomás-Barberán FA (2007) Nutraceuticals: facts and fiction. Phytochemistry 68(22–24):2986–3008

Fabiani A, Morosetti C, Filosa A, Principi E, Lepri L, Maurelli V, Fioretti F, Servi L (2018) Effect on prostatic specific antigen by a short time treatment with a curcuma extract: a real life experience and implications for prostate biopsy. Arch Ital Urol Andro 90(2):107–111

Fabiansen C, Yaméogo CW, Iuel-Brockdorf AS, Cichon B, Rytter MJ, Kurpad A, Wells JC, Ritz C, Ashorn P, Filteau S, Briend A (2017) Effectiveness of food supplements in increasing fat-free tissue accretion in children with moderate acute malnutrition: a randomised 2× 2× 3 factorial trial in Burkina Faso. PLoS Med 14(9):e1002387

Fernandez MA, Marette A (2017) Potential health benefits of combining yogurt and fruits based on their probiotic and prebiotic properties. Adv Nutr 8(1):155S–164S

Fielding JM, Rowley KG, Cooper P, O'Dea K (2005) Increases in plasma lycopene concentration after consumption of tomatoes cooked with olive oil. Asia Pac J Clin Nutr 14(2):131–136

Food and Agricultural Org World Health Org (2001). Health and nutritional properties of probiotics in food including powder milk with live lactic acid bacteria. http://www.who.int/foodsafety/publications/fsmanagement/probiotics/en/index.html Accessed 15 Oct 2020

Formica JV, Regelson W (1995) Review of the biology of quercetin and related bioflavonoids. Food Chem Toxicol 33(12):1061–1080

Genís S, Sánchez-Chardi A, Bach À, Fàbregas F, Arís A (2017) A combination of lactic acid bacteria regulates Escherichia coli infection and inflammation of the bovine endometrium. J Dairy Sci 100(1):479–492

Glore SR, Van Treeck D, Knehans AW, Guild M (1994) Soluble fiber and serum lipids: a literature review. J Am Diet Assoc 94(4):425–436

González-Montelongo R, Lobo MG, González M (2010) Antioxidant activity in banana peel extracts: testing extraction conditions and related bioactive compounds. Food Chem 119(3): 1030–1039

Guarner F (2005) Inulin and oligofructose: impact on intestinal diseases and disorders. Br J Nutr 93 (S1):S61–S65

Gupta S, Parvez N, Sharma PK (2015) Nutraceuticals as functional foods. J Nutr Therap 4(2):64–72

Hallberg LEIF, Brune M, Rossander LENA (1989) The role of vitamin C in iron absorption. Int J Vitam Nutr Res Suppl 30:103–108

Hodge L, Salome CM, Peat JK, Haby MM, Xuan W, Woolcock AJ (1996) Consumption of oily fish and childhood asthma risk. Med J Aust 164(3):137–140

Holzapfel WH, Haberer P, Geisen R, Björkroth J, Schillinger U (2001) Taxonomy and important features of probiotic microorganisms in food and nutrition. Am J Clin Nutr 73(2):365s–373s

Hord NG (2008) Eukaryotic-microbiota crosstalk: potential mechanisms for health benefits of prebiotics and probiotics. Annu Rev Nutr 28:215–231

Houston MC (2005) Nutraceuticals, vitamins, antioxidants, and minerals in the prevention and treatment of hypertension. Prog Cardiovasc Dis 47(6):396–449

Hu FB, Willett WC (2002) Optimal diets for prevention of coronary heart disease. JAMA 288(20): 2569–2578

Iriti M, Faoro F (2006) Grape phytochemicals: a bouquet of old and new nutraceuticals for human health. Med Hypotheses 67(4):833–838

Isolauri E, Rautanen T, Juntunen M, Sillanaukee P, Koivula T (1991) A human lactobacillus strain (lactobacillus casei sp strain GG) promotes recovery from acute diarrhea in children. Pediatrics 88(1):90–97

Jaganathan SK, Vellayappan MV, Narasimhan G, Supriyanto E, Dewi DEO, Narayanan ALT, Balaji A, Subramanian AP, Yusof M (2014) Chemopreventive effect of apple and berry fruits against colon cancer. World J Gastroenterol 20(45):17029

Janczarek M, Bachanek T, Mazur E, Chałas R (2016) The role of probiotics in prevention of oral diseases. Postepy Hig Med Dosw 70:850–857

Kalia AN (2005) Textbook of industrial Pharmacognocy. CBS Publisher and Distributor, New Delhi, pp 204–208

Kalra EK (2003) Nutraceutical-definition and introduction. AAPS PharmSci 5(3):27–28

Kiewiet MBG, van Esch BC, Garssen J, Faas MM, de Vos P (2017) Partially hydrolyzed whey proteins prevent clinical symptoms in a cow's milk allergy mouse model and enhance regulatory T and B cell frequencies. Mol Nutr Food Res 61(11):1700340

Kim SJ, Kawaharada C, Suzuki T, Saito K, Hashimoto N, Takigawa S, Noda T, Matsuura-Endo C, Yamauchi H (2006) Effect of natural light periods on rutin, free amino acid and vitamin C contents in the sprouts of common (Fagopyrum esculentum Moench) and tartary (F. tataricum Gaertn.) buckwheats. Food Sci Technol Res 12(3):199–205

Klatte ET, Scharre DW, Nagaraja HN, Davis RA, Beversdorf DQ (2003) Combination therapy of donepezil and vitamin E in Alzheimer disease. Alzheimer Dis Assoc Disord 17(2):113–116

Kochhar KP (2008) Dietary spices in health and diseases: I. Indian J Physiol Pharmacol 52(2): 106–122

Kokate CK, Purohit AP, Gokhale SB (2002) Nutraceutical and Cosmeceutical. In: Pharmacognosy, 21st edn. Nirali Prakashan, Pune, pp 542–549

Lacomba R, Salcedo J, Alegría A, Lagarda MJ, Barberá R, Matencio E (2010) Determination of sialic acid and gangliosides in biological samples and dairy products: a review. J Pharm Biomed Anal 51(2):346–357

Lairon D, Arnault N, Bertrais S, Planells R, Clero E, Hercberg S, Boutron-Ruault MC (2005) Dietary fiber intake and risk factors for cardiovascular disease in French adults. Am J Clin Nutr 82(6):1185–1194

Lakshmana PS, Suriyaprakash TNK, Dinesh Kumar C, SureshKumar S, Ragavendran T (2012) Nutraceuticals: a review. Elixir Pharmacy 46:8372–8377

Lambert JD, Hong J, Kim DH, Mishin VM, Yang CS (2004) Piperine enhances the bioavailability of the tea polyphenol (−)-epigallocatechin-3-gallate in mice. J Nutr 134(8):1948–1952

Lampe JW (2003) Spicing up a vegetarian diet: chemopreventive effects of phytochemicals. Am J Clin Nutr 78(3):579S–583S

Laparra JM, Sanz Y (2010) Interactions of gut microbiota with functional food components and nutraceuticals. Pharmacol Res 61(3):219–225

Lee J, Koo N, Min DB (2004) Reactive oxygen species, aging, and antioxidative nutraceuticals. Compr Rev Food Sci Food Saf 3(1):21–33

Lenoir-Wijnkoop I, Sanders ME, Cabana MD, Caglar E, Corthier G, Rayes N, Sherman PM, Timmerman HM, Vaneechoutte M, Van Loo J, Wolvers DA (2007) Probiotic and prebiotic influence beyond the intestinal tract. Nutr Rev 65(11):469–489

Li S, Lo CY, Ho CT (2006) Hydroxylated polymethoxyflavones and methylated flavonoids in sweet orange (Citrus sinensis) peel. J Agric Food Chem 54(12):4176–4185

Liu RH, Hotchkiss JH (1995) Potential genotoxicity of chronically elevated nitric oxide: a review. Mutat Res Rev Genet Toxicol 339(2):73–89

Macfarlane S, Macfarlane GT, Cummings JH (2006) Review article: prebiotics in the gastrointestinal tract. Aliment Pharmacol Ther 24:701–714

Mandala F, Guida R, Guida GF (2010) Nutraceuticals: what they are and how they work. Eur Soc Cardiol Pract 9(11)

Manfredi S, Di Ianni F, Di Girolamo N, Canello S, Gnudi G, Guidetti G, Miduri F, Fabbi M, Daga E, Parmigiani E, Centenaro S (2018) Effect of a commercially available fish-based dog

food enriched with nutraceuticals on hip and elbow dysplasia in growing Labrador retrievers. Can J Vet Res 82(2):154–158
Matsungo TM, Kruger HS, Smuts CM, Faber M (2017) Lipid-based nutrient supplements and linear growth in children under 2 years: a review. Proc Nutr Soc 76(4):580–588
Mendiola JA, Torres CF, Toré A, Martín-Álvarez PJ, Santoyo S, Arredondo BO, Señoráns FJ, Cifuentes A, Ibáñez E (2007) Use of supercritical CO 2 to obtain extracts with antimicrobial activity from Chaetoceros muelleri microalga. A correlation with their lipidic content. Eur Food Res Technol 224(4):505–510
Metchinkoff E (1907) The prolongation of life. Putmans Sons, New York, pp 151–183
Meydani M (2000) Effect of functional food ingredients: vitamin E modulation of cardiovascular diseases and immune status in the elderly. Am J Clin Nutr 71(6):1665S–1668S
Montonen J, Knekt P, Järvinen R, Aromaa A, Reunanen A (2003) Whole-grain and fiber intake and the incidence of type 2 diabetes. Am J Clin Nutr 77(3):622–629
Morcos NC (1997) Modulation of lipid profile by fish oil and garlic combination. J Natl Med Assoc *89*(10):673
Narotzki B, Reznick AZ, Aizenbud D, Levy Y (2012) Green tea: a promising natural product in oral health. Arch Oral Biol 57(5):429–435
Nasri H, Baradaran A, Shirzad H, Rafieian-Kopaei M (2014) New concepts in nutraceuticals as alternative for pharmaceuticals. Int J Prevent Med 5(12):1487
Padmavathi D (2018) A general review on "nutraceuticals": its golden health impact over human community. Int J Food Sci Nut 3(2):214–217
Pandey M, Verma RK, Saraf SA (2010) Nutraceuticals: new era of medicine and health. Asian J Pharm Clin Res 3(1):11–15
Pandey N, Meena RP, Rai SK, Rai SP (2011) Medicinal plants derived nutraceuticals: a re-emerging health aid. Intern J Pharma and Bio Sci 2(4):419–441
Petruzziello L, Iacopini F, Bulajic M, Shah S, Costamagna G (2006) Uncomplicated diverticular disease of the colon. Aliment Pharmacol Ther 23(10):1379–1391
Pietsch K, Saul N, Chakrabarti S, Stürzenbaum SR, Menzel R, Steinberg CE (2011) Hormetins, antioxidants and prooxidants: defining quercetin-, caffeic acid-and rosmarinic acid-mediated life extension in C. elegans. Biogerontology 12(4):329–347
Pineiro M, Stanton C (2007) Probiotic bacteria: legislative framework—requirements to evidence basis. J Nutr 137(3):850S–853S
Prakash D, Gupta C (2011) Role of phytoestrogens as nutraceuticals in human health: a review. Biotechnol Indian J 5(3):1–8
Prakash D, Upadhyay G, Pushpangadan P, Gupta C (2011) Antioxidant and free radical scavenging activities of some fruits. J Complement Integrat Med 8(1):1–16
Profumo E, Buttari B, D'Arcangelo D, Tinaburri L, Dettori MA, Fabbri D, Delogu G, Riganò R (2016) The nutraceutical dehydrozingerone and its dimer counteract inflammation-and oxidative stress-induced dysfunction of in vitro cultured human endothelial cells: a novel perspective for the prevention and therapy of atherosclerosis. Oxidat Med Cell Longev 2016:1246485
Radke M, Picaud JC, Loui A, Cambonie G, Faas D, Lafeber HN, de Groot N, Pecquet SS, Steenhout PG, Hascoet JM (2017) Starter formula enriched in prebiotics and probiotics ensures normal growth of infants and promotes gut health: a randomized clinical trial. Pediatr Res 81(4): 622–631
Rajasekaran A, Sivagnanam G, Xavier R (2008) Nutraceuticals as therapeutic agents: a review. Res J Pharm Technol 1(4):328–340
Ross S (2000) Functional foods: the Food and Drug Administration perspective. Am J Clin Nutr 71(6):1735S–1738S
Russo M, Fabersani E, Abeijón-Mukdsi MC, Ross R, Fontana C, Benítez-Páez A, Gauffin-Cano P, Medina RB (2016) Lactobacillus fermentum CRL1446 ameliorates oxidative and metabolic parameters by increasing intestinal feruloyl esterase activity and modulating microbiota in caloric-restricted mice. Nutrients 8(7):415

Saad B, Mona O (2013) Antimicrobial activity of garlic juice (Allium sativum), honey, and garlic-honey mixture on some sensetive and multiresistant microorganisms. Life Sci J 10(4): 2429–2435

Sabita NS, Trygve OT (2012) The role of nutraceuticals in chemoprevention and chemotherapy and their clinical outcomes. J Oncol 64:1–23

Sacco SM, Horcajada MN, Offord E (2013) Phytonutrients for bone health during ageing. Br J Clin Pharmacol 75(3):697–707

Saibil F (1989) Diarrhea due to fiber overload. N Engl J Med 320(9):599

Sangiovanni E, Piazza S, Vrhovsek U, Fumagalli M, Khalilpour S, Masuero D, Di Lorenzo C, Colombo L, Mattivi F, De Fabiani E, Dell'Agli M (2018) A bio-guided approach for the development of a chestnut-based proanthocyanidin-enriched nutraceutical with potential anti-gastritis properties. Pharmacol Res 134:145–155

Santamaria A, Di Benedetto A, Petrella E, Pintaudi B, Corrado F, D'Anna R, Neri I, Facchinetti F (2016) Myo-inositol may prevent gestational diabetes onset in overweight women: a randomized, controlled trial. J Matern Fetal Neonatal Med 29(19):3234–3237

Sarin R, Sharma M, Singh R (2012) Nutraceuticals: a review. Int Res J Pharm 3(4):9599

Saw CLL, Huang Y, Kong AN (2010) Synergistic anti-inflammatory effects of low doses of curcumin in combination with polyunsaturated fatty acids: docosahexaenoic acid or eicosapentaenoic acid. Biochem Pharmacol 79(3):421–430

Sayeed S., 2015. Nutraceuticals – let food be your medicine. Available at: https://www.slideshare.net/nimatanamikaze/nutraceuticals-45465067 Accessed 14 July 2020

Scalbert A, Johnson IT, Saltmarsh M (2005) Polyphenols: antioxidants and beyond. Am J Clin Nutr 81(1):215S–217S

Schrezenmeir J, de Vrese M (2001) Probiotics, prebiotics, and synbiotics—approaching a definition. Am J Clin Nutr *73*(2):361s–364s

Shinde N, Bangar B, Deshmukh S, Kumbhar P (2014) Nutraceuticals: a review on current status. Res J Pharm Technol 7(1):110–113

Shoba G, Joy D, Joseph T, Majeed M, Rajendran R, Srinivas PSSR (1998) Influence of piperine on the pharmacokinetics of curcumin in animals and human volunteers. Planta Med 64(04): 353–356

Shukla Y, Singh M (2007) Cancer preventive properties of ginger: a brief review. Food Chem Toxicol 45(5):683–690

Simopoulos AP (1991) Omega-3 fatty acids in health and disease and in growth and development. Am J Clin Nutr 54(3):438–463

Singh J, Sinha S (2012) Classification, regulatory acts and applications of nutraceuticals for health. Int J Pharm Bio Sci 2(1):177–187

Sirtori CR, Galli C (2002) N-3 fatty acids and diabetes. Biomed Pharmacother 56(8):397–406

Stahl W, Sies H (2005) Bioactivity and protective effects of natural carotenoids. Biochim Biophys Acta 1740(2):101–107

Steffen LM, Jacobs DR Jr, Stevens J, Shahar E, Carithers T, Folsom AR (2003) Associations of whole-grain, refined-grain, and fruit and vegetable consumption with risks of all-cause mortality and incident coronary artery disease and ischemic stroke: the atherosclerosis risk in communities (ARIC) study. Am J Clin Nutr 78(3):383–390

Stephen D (2012) A report of National Nutraceutical Centre. Nutraceuticals India 2012, Webinar, pp 1–22

Story EN, Kopec RE, Schwartz SJ, Harris GK (2010) An update on the health effects of tomato lycopene. Annu Rev Food Sci Technol 1:189–210

Suvarna VC, Boby VU (2005) Probiotics in human health: a current assessment. Curr Sci 88(11): 1744–1748

Tewari S, Gupta V, Bhattacharya S (2000) Comparative study of antioxidant potential of tea with and without additives. Indian J Physiol Pharmacol 44(2):215–219

Thomas B, Ghebremeskel K, Lowy C, Crawford M, Offley-Shore B (2006) Nutrient intake of women with and without gestational diabetes with a specific focus on fatty acids. Nutrition 22(3):230–236

Todorovic V, Redovnikovic IR, Todorovic Z, Jankovic G, Dodevska M, Sobajic S (2015) Polyphenols, methylxanthines, and antioxidant capacity of chocolates produced in Serbia. J Food Compos Anal 41:137–143

Van Ballegooijen AJ, Cepelis A, Visser M, Brouwer IA, Van Schoor NM, Beulens JW (2017) Joint association of low vitamin D and vitamin K status with blood pressure and hypertension. Hypertension 69(6):1165–1172

Van Vliet D, Bruinenberg VM, Mazzola PN, Van Faassen MH, De Blaauw P, Pascucci T, Puglisi-Allegra S, Kema IP, Heiner-Fokkema MR, Van Der Zee EA, Van Spronsen FJ (2016) Therapeutic brain modulation with targeted large neutral amino acid supplements in the Pah-enu2 phenylketonuria mouse model. Am J Clin Nutr 104(5):1292–1300

Varoni EM, Iriti M (2016) Odontonutraceuticals: pleiotropic phytotherapeutic agents for oral health. Pharmaceuticals *9*(1):10

Vogt LM, Elderman ME, Borghuis T, de Haan BJ, Faas MM, de Vos P (2017) Chain length-dependent effects of inulin-type fructan dietary fiber on human systemic immune responses against hepatitis-B. Mol Nutr Food Res *61*(10):1700171

Wang S, Zhu F, Meckling KA, Marcone MF (2013) Antioxidant capacity of food mixtures is not correlated with their antiproliferative activity against MCF-7 breast cancer cells. J Med Food 16(12):1138–1145

Watzl B, Girrbach S, Roller M (2005) Inulin, oligofructose and immunomodulation. Br J Nutr 93 (S1):S49–S55

Whitman MM (2001) Understanding the perceived need for complementary and alternative nutraceuticals: lifestyle issues. Clin J Oncol Nurs 5(5):190–194

Wolfe KL, Liu RH (2003) Apple peels as a value-added food ingredient. J Agric Food Chem 51(6): 1676–1683

Zhao J (2007) Nutraceuticals, nutritional therapy, phytonutrients, and phytotherapy for improvement of human health: a perspective on plant biotechnology application. Recent Pat Biotechnol 1(1):75–97

Part II
Micronutrients and Macronutrients

Chapter 8
Current and Future Prospects of Flavonoids for Human Immune System

Sippy Singh and Durgesh Singh

Abstract Flavonoids are large group of plant-derived aromatic compounds which have diverse functional capabilities. These polyphenolic compounds promote human immune system and protect body against several health ailments. A lot is needed to explore in terms of their bioavailability and their metabolism so as to utilize their broad range potential. Flavonoids are potent substances which are widely used at present because of their antioxidative, anti-inflammatory, antiviral/bacterial, cardioprotective, antidiabetic, antiaging, and anticarcinogenic properties and still have more to be explored and used in the future. This chapter provides view of current and futuristic application of flavonoids.

Keywords Flavonoids · Immune system · Phytonutrient · Immunomodulator

8.1 Introduction

Flavonoids include diverse group of phytonutrients present in fruits and vegetables which are utilized for their growth and protection against plaques (Havsteen 2002). These are low-molecular-weight phenolic compounds having a benzo-γ-pyrone structure distributed in all parts of plants (Dewick 2001). Because of their presence in foods and beverages of plant origin, they are termed as dietary flavonoids. These hydroxylated phenolic compounds are synthesized by plants to combat microbial infection (Dixon et al. 1983) which exhibits antioxidant properties (Kumar et al. 2013; Kumar and Pandey 2013). Flavonoids consist of 15 carbon skeleton comprising of two benzene rings: A and B which are linked through a heterocyclic pyrane ring-C (Kumar and Pandey 2013). These can be further divided into subgroups like flavones, flavonols, flavanones, isoflavones, anthocyanins, and chalcones which are based on the carbon of C ring to which the B ring is attached and oxidation of the C ring (Middleton 1998). Since these subgroups share a common precursor, chalcone,

S. Singh (✉) · D. Singh
Department of Zoology, S.S. Khanna Girls' Degree College, Prayagraj (A Constituent College of University of Allahabad, Prayagraj), Prayagraj, UP, India

R. K. Kesharwani et al. (eds.), *Nutraceuticals and Functional Foods as Immunomodulators*, https://doi.org/10.1007/978-981-19-2507-8_8

they all are biogenetically and structurally related (Grotewold 2006; Samanta et al. 2011; Pandey and Sohng 2013).

Flavonoid term has been derived from Latin word "flavous" which means yellow. Flavonoids perform various activities in different organisms. In plants, they are impart color and aroma to flowers and help in attracting pollinators towards colorful fruits thereby helping in seed dispersal and their growth (Griesbach 2005). Apart from these, flavonoids also protect plants from different forms of environmental stresses (Takahashi and Ohnishi 2004), act as signaling molecules, help in detoxification, and possess antimicrobial activity. Gene silencing in plants has also been correlated with synthesis of flavonoid (Jorgensen 1995).

The hydroxyl and oxo group present in flavonoids have chelating properties which possess pharmacokinetic and pharmacological properties. More interestingly, flavonoids have been reported to have positive impact on human and animal health and are widely used in treating diseases (Kumar and Pandey 2013; Panche et al. 2015). Flavonoids have several health-promoting effects because they exhibit hepatoprotective (Kumar and Pandey 2013), antioxidative, anti-inflammatory, antiviral/bacterial, cardioprotective, antidiabetic, antiaging, and anticarcinogenic properties and can inhibit enzymes like cyclooxygenase, phosphoinositide 3-kinase, etc. (Hayashi et al. 1988; Walker et al. 2000). Flavonoids show several antitumorous activities by one of the following mechanisms of action: antiproliferation activity, preventing carcinogen metabolic activation, arresting cell cycle, inducing apoptosis, promoting differentiation, inhibiting angiogenesis, and modulating multidrug resistance (Patil and Masand 2019). Because of their broad spectrum of utility and importance in health maintenance, they form an important component of pharmaceutical, nutraceutical, and medicine factories (Keservani et al. 2010; Keservani et al. 2010; Rane and Kesarwani 2020). They can easily interact with other nutrients (Scholz and Williamson 2007; Jakobek 2015) and lead to various effects, for example, they lower down the absorption of glucose by suppressing carbohydrate-hydrolyzing enzymes and glucose transporters (Gonzales et al. 2015); intake of fat has been reported to improve bioavailability of flavonoids and increase in their absorption within the body (Gonzales et al. 2015). On the contrary, intake of protein leads to decrease in bioavailability of flavonoids thereby affecting antioxidant property of flavonoids and protein digestion as well (Swieca et al. 2013).

The presence of flavonoids can be detected by various tests such as Fast blue reagent that is widely used to detect phenolic compounds, ferric chloride test, and neutral lead acetate test. Shinoda tests can be also performed wherein alcoholic or aqueous solution of extract is mixed with piece of magnesium ribbon and conc. HCl. The presence of flavonoids can be confirmed by development of pink/magenta color. Use of NaOH and H_2SO_4 can also confirm the presence of flavonoids in food (Mondal and Rahaman 2020).

Flavonoids readily interact with many nutrients (Scholz and Williamson 2007) and can reduce transportation of glucose and glucose absorption. Bioavailability of flavonoids is of major concern, and it has been observed that intake of fat helps in improving bioavailability of flavonoids and also promotes their absorption through secretion of bile salts (Gonzales et al. 2015). On the contrary, protein intake decreases the bioavailability of flavonoids (Swieca et al. 2013).

Immune system helps in protecting human body from various infections and diseases. It helps in combating against pathogens on two basis; it can be either innate or acquired. The innate immunity helps in fighting against broad range of infectious agents as it lacks specific recognition and does not depend on nature of pathogen (Singh et al. 2014; Baranowski et al. 2012), whereas acquired immunity shows pathogen-specific response. Proper diet and nutrition can enhance immune system as it has been reported that improper diet/malnutrition can have negative and adverse effect on immune system (Gerriets and Rathmell 2012). Important dietary components which have been of prime interest are polyphenols; these flavonoids (phytochemicals) have been suggested to act as immunomodulators and affect immune responses (Akiyama et al. 2005; Abril-Gil et al. 2012; Hosseinzade et al. 2019). Because of their potent activity and immunomodulatory effects, the flavonoids have been currently promoted in pharmaceutical industry.

8.2 Use of Flavonoids and Their Immunomodulatory Effects

Flavonoids exist as free aglycones and glycosidic conjugates (Middleton et al. 2000) with varied structural ranges (Ravishankar et al. 2013). These diverse structures enable them to harbor anticancerous, antibacterial, enzyme modulation, antioxidant, and anti-inflammatory properties (Middleton et al. 2000; Ravishankar et al. 2013; Cardenas et al. 2011). Generally, flavonoids exert immunomodulatory effect on the components of both innate and acquired immune system which can be summarized further.

8.2.1 Effect of Flavonoids on Cells of Innate Immune System

Flavonoid poses immunosuppressive and immunomodulatory effects on various cells of innate immune system. Many flavonoids have been reported to stop these allergic responses by inhibiting degranulation in mast cells and basophils (Matsuda et al. 2016; Tanaka 2014). Quercetin (an aglycone) found in apples, berries, grapes, tea, tomatoes, and various nuts (Rogerio et al. 2010) has been studied to inhibit histamine production in rat basophilic leukemia-2H3 cell line (Matsuda et al. 2016; Tanaka 2014). Some of the flavanols such as fisetin, quercetin, rutin, etc. inhibits IgE-mediated effect, histamine release, and intracellular calcium level (Castell et al. 2014; Hagenlocher and Lorentz 2015). Many studies propose that flavonoids such as fisetin block transcription and secretion of Th2-type cytokines leads to inhibition of various transcriptional factors (Matsuda et al. 2016; Tanaka 2014; Castell et al.

2014; Hagenlocher and Lorentz 2015). These flavonoids act as anti-allergic substances which inhibited activation of kinase and transcription (Che et al. 2018) and, therefore, can be used in treating allergic reactions.

Neutrophils mount immune response against bacterial and fungal infection in humans. These cells on stimulation lead to production of reactive oxygen species, reactive nitrogen species, and many toxic mediators causing inflammation (Hellebrekers et al. 2018). Polverino et al. (2017) reported that in many respiratory diseases there is release of enzymes like neutrophil elastase (NE) and proteases which prevents repair process. Chalcones, a potent flavonoid, have the ability to inhibit the production of NE in humans and therefore act as promising anti-inflammatory substance (Reddy et al. 2011). Most of the flavonoids having anti-inflammatory activity binds to reactive oxygen species formed by neutrophils and decreases their production (Ciz et al. 2012; Ribeiro et al. 2015).

Eosinophils are involved in Th2-type allergic reactions observed in case of parasitic infection, asthma, and cancer (Ramirez et al. 2018). Eosinophils being important inflammatory cells lead to pathogenesis of allergic inflammation, and many flavonoids have shown inhibitory action on these inflammatory responses brought about by inhibition of release of chemoattractant factors from eosinophils (Sakai-Kashiwabara et al. 2014).

Monocytes shield the body against pathogens by phagocytosis, and in doing so, they also release inflammatory mediators (Collin and Bigley 2016). These cells help in tissue repair, inflammation, and antigen presentation (Murray 2018). Flavonoids exert anti-inflammatory responses by inhibiting inflammatory mediators and TLR4 expression (Rahimifard et al. 2017). These flavonoids prevent activation of inflammasome (Kim et al. 2017) which promotes autophagy and in turn improves inflammation condition and helps in treating autoimmune diseases (Takahama et al. 2018).

Natural killer cells form an important component of immune system as they act against virus-infected and cancer cells (Zhang and Huang 2017). They have cytolytic activity, and their activity can be stimulated on tumorous cells (Lindqvist et al. 2014; Burkard et al. 2017). Dendritic cell performs important function of antigen presentation to T lymphocytes (Dress et al. 2018). Flavonoids modulate immune system by bringing about the apoptosis of dendritic cells (Galleggiante et al. 2017; Nickel et al. 2011).

8.2.2 Effect of Flavonoids on Cells of Adaptive Immune System

Adaptive immune system comprises of two types of cell: B lymphocytes and T lymphocytes. Flavones have been proposed to be important for activation and functioning of T cells in several disease conditions (Kim and Lee 2013). These flavones inhibit transcription factors (Kim and Lee 2013; Yan et al. 2017).

Supplement of oral genistein in Hashimoto's thyroiditis patients leads to reduction of symptoms of this autoimmune disease (Zhang et al. 2017).

Flavonoids such as quercetin and astilbin have been reported to modulate proliferation of B cells and production of antibody (Gong et al. 2011; Guo et al. 2015). These increase B-cell activation and antibody production. Similarly, flavonoids have their impact on T cells of immune system. In vivo and in vitro studies have shown the direct effect of flavonoids on T CD8+ cells (Liu et al. 2010; Leischner et al. 2016). Compounds like gallotannin and apigenin stimulate CD8 cells and decrease growth of tumor cells either by increasing apoptosis or promoting antigen presentation (Mantena et al. 2005; Zhang et al. 2018). Hence, these flavones and flavonoids and their derivatives are promising its direct impact and positive application to regulate immune system in humans.

8.2.3 Flavonoids as Radical Scavengers

Flavonoids show promising efforts in reducing the oxidative stress by scavenging reactive oxygen species (ROS)/free radicals from body. Theses flavonoids on interacting with free radicals get oxidized and deactivate the free radicals by making them less reactive. Korkina and Afanaśev (1996) explained the reaction occurring between flavonoids and free radicals as the hydroxyl group of flavonoids is highly reactive which inactivates the radicals. The flavonoids can also scavenge superoxides (Hanasaki et al. 1994) and inhibit LDL oxidation and hence prevent atherosclerosis (Kerry and Abbey 1997).

Nitric oxide is produced in the body by various cells such as macrophages, endothelial cells, etc. Normal level of Nitric oxide released in body by the action of enzyme nitric oxide synthase helps in dilation of blood vessel, however, higher amount of nitric oxide creates oxidative damage. Nitric oxide reacts with free radicals and forms peroxynitrite which results in irreversible damage to cell membrane (Assreuy et al. 1994). Flavonoids can scavenge free radicals in the body and therefore limit the availability of free radicals to nitric oxide, thus helps in avoiding the harmful effects of higher level of nitric oxide within the body.

8.2.4 Flavonoids and Their Anticancer Activity

Flavonoids present in fruits and vegetables have been reported to have agents that can prevent cancer (Mishra et al. 2013). Flavonoids act on initiation and promotion stages of carcinogenesis and help in downregulating mutant p53, arresting cell cycle, inhibition of tyrosine kinase, heat shock protein, and expression of Ras protein and also affect estrogen receptor binding capacity thereby lowering the cases of cancer to occur. Quercetin has been reported to inhibit tyrosine kinase (Ferry et al. 1996), and genistein has been widely studied to possess anticancer activity both in vivo and in vitro studies (Barnes 1995). Moreover, the anticancer activity of flavonoids can be correlated with radical scavenging activity of these compounds as the reduction in free radicals lowers the cancer-promoting risk within the body.

8.2.5 Effect of Flavonoids on Various Neurodegenerative Diseases

Studies conducted on different plant metabolites have suggested key role of flavonoids in enzyme as well as receptor system of brain or central nervous system. Hence, these competent compounds can be effectively used in treating/managing neurodegenerative diseases. Flavonoids can inhibit various enzyme activity such as calcium ATPase, aldose reductase, lipoxygenase, etc. and, therefore, prevent neurodegenerative diseases (Jäger and Saaby 2011).

Alzheimer's disease, a neurodegenerative disorder leading to loss of memory, decreased cognition, and dementia has been suggested to occur due to many mechanism involving deficiency of cholinesterases, deposition of β-amyloid plaques, and oxidative stress (Ullah et al. 2016). Flavonoids such as quercetin and gossypetin have been studied to block β-amyloid plaques and remove free radicals (Ansari et al. 2009) and, therefore, can be used as promising drug for treating Alzheimer's disease (AD). Taniguchi et al. (2005) reported that flavonoids such as myricetin and epicatechin-5-gallate can stop heparin-mediated tau protein formation thereby helping in averting AD.

There have been reports that suggest the neuroprotective activity of flavonoids such as reduction in risk of Parkinson's diseases and ischemic hippocampal injury due to consumption of green tea which can inhibit the formation of endogenous neurotoxin toxins 5-S-cysteinyldopamine (Vauzour et al. 2007b). Flavonoids also play neuroprotective role against Huntington's disease by mediating cell signaling through ERK pathway (Maher et al. 2011).

Flavonoids also exhibit anti-Parkinson activity. Parkinson's disease includes progressive neurodegeneration in substantia nigra pars compacta and nigrostriatal tract (Tripathi 2013a, b). Neurodegeneration in Parkinson's occurs due to processes like oxidative damage, lipid peroxidation, reduced level of glutathione, increased level of superoxide dismutase, etc. (Alam et al. 1997; Sofic et al. 1992; Dexter et al. 1989). Neuronal cell death is accompanied with all the abovementioned activities which finally lead to one or the other mechanism of cell death. The production of free radicals/reactive oxygen species causes peroxidation of lipids which in turn causes leakage of cytochrome c from mitochondria, and finally the neuronal cell reaches to death. Some of the flavonoids like naringenin has been observed to elevate expression of NF-κ and COX-2 (Park et al. 2012), whereas emodin and genistein suppress TNF-α (Jia et al. 2013; Chen et al. 2012).

8.2.6 Flavonoid Participation in Cognition and Signaling Pathway

Consumption of flavonoid-rich diet such as grape juice, coca, blueberry, etc. has been reported to enhance cognitive abilities (Letenneur et al. 2007; Shukitt-Hale 2012). Fruits rich in quercetin and anthocyanins help in improving cognition, short- and long-term memory, and memory retrieval (Hartman et al. 2006). File et al. (2001) suggested that isoflavones are capable of enhancing learning and memorizing abilities similarly in the way as the estrogen does in the brain. Apart from it, isoflavones also modulate brain-derived neurotropic factors and nerve growth factors in brain regions (Pan et al. 1999a, b).

Flavonoids exhibit neuroprotective activity by binding to various neuronal receptors and mediating signaling pathways (Lee et al. 2010; Incani et al. 2010). They achieve this by stimulating or inhibiting receptors and modulating expression of gene and leading to neuronal dynamics, synaptic protein synthesis, and other morphological changes. Flavonoids can mediate MAPK cascade leading to downstream activation of CREB protein which in turn causes variation in synaptic dynamism and memory formation (Impey et al. 1998). Flavonoids can also modulate PI3 kinase activity such as quercetin which can inhibit Akt/PKB signaling by inhibiting PU3-kinase activity (Spencer et al. 2003) and hesperetin which can activate Akt/PKB signaling pathway by enabling cortical neurons to survive (Vauzour et al. 2007a).

8.2.7 Role of Flavonoids in Apoptosis

Apoptosis also referred to as programmed cell death is an important phenomenon which controls the development in body. Two main signaling pathways of apoptosis are extrinsic and intrinsic where the former is related with tumor necrosis factor, while the latter includes mitochondrial pathway (Abotaleb et al. 2018). The cancerous cells do not undergo apoptosis that is why they undergo repeated cell division by overexpression of oncogenic genes and suppression of tumor suppressor gene (Jan and Chaudhry 2019). Flavonoids have the ability to suppress proliferation of cancerous cells by various means as they can inhibit mitogen-activated protein kinase, epidermal growth factor receptor, PI3K, protein kinase B, and NF-κβ (Neagu et al. 2019; Abotaleb et al. 2018). Flavonoids target apoptotic signal pathway and guide the infected cells towards cell death (Chirumbolo et al. 2018a, b).

8.2.8 Multidimensional Activity of Flavonoids

Apart from the above discussed activities, flavonoids have other interesting activities also such as they exhibit anti-ulcer, spasmolytic, antidepressant, antibacterial, anti-hypertensive, anti-inflammatory, and antidiabetic activity.

Whenever there is imbalance in acid, pepsin, and bile and gastric mucus, nitric oxide, and bicarbonate secretion in the stomach, it leads to peptic ulcers in gastro-intestinal tract. Hesperidin (flavonoids) has been reported to have antioxidant and mucoprotective activity; it slows down the injury caused by oxidation by maintaining the level of enzymes. Hesperidin also allows the ulcerated tissues within gastric layers (Bigoniya and Singh 2014). Similarly, quercetin is also a potent anti-ulcer flavonoids by reducing the formation of histamine in gastric mucosa thereby inhibiting production of HCl (Izzo et al. 1994).

Some of the flavonoids show spasmolytic effect by inhibiting or blocking M_3 receptor; therefore, the visceral smooth muscles fail to contract (Tripathi 2013a, b). Catechin has been reported to influence vasodilation of endothelium layer and stimulate endothelium-dependent synthesis of nitric oxide (Ghayur et al. 2007). In longitudinal muscle of guinea pig, flavonoids have been studied to result in influx of calcium ions from extracellular region which leads to suppression of smooth muscle contraction (Vasconcelos et al. 2015).

During state of depression, the transmission of monoaminergic within the brain is affected, and there is depletion of 5-HT (Tripathi 2013a, b). Monoamine oxidase-A is known for oxidative deamination of 5-HT, and the inhibition of the former can help in amelioration of depressive state (Yamada and Yasuhara 2004). Quercetin can inhibit monoamine oxidase-A and, therefore, can be used as an antidepressant (Saaby et al. 2009). Repeated intake of reserpine can cause cognition deficit and increase the level of oxidative stress, and as a remedy to it, quercetin has been suggested to show protective effect which can avert the effects of reserpine-induced dysfunction (Naidu et al. 2004). Apart from quercetin, fisetin, naringenin, luteolin, and kaempferitrin have also been found to possess antidepressant potency (Zhen et al. 2012; Yi et al. 2010; De la Peña et al. 2014; Cassani et al. 2014).

Flavonoids also show antibacterial activity as apigenin, vitexin, etc. show inhibitory effect on various bacteria like *Enterobacter cloacae* and *Pseudomonas aeruginosa* (Basile et al. 1999). Extract of golnar has been reported to act against gram-positive and gram-negative food poisoning bacteria (Mahboubi et al. 2015).

The antihypertensive, anti-inflammatory, and antidiabetic activity of flavonoids have also been well documented. Quercetin has been found to result in fall in BP in experimental rats (Perez-Vizcaino et al. 2009); naringin, hesperidin, and luteolin also result in better in vitro acetylcholine-induced vasodilation (Gómez-Guzmán et al. 2012; Serafini et al. 2010). Flavonoids mediate anti-inflammatory activity by various mechanisms such as stimulation of protein kinase C and inhibition of pro-inflammatory enzymes and NF-κβ (Noreen et al. 1998; Cho et al. 2000). Chronic inflammation many times continues with tumor development; hence anti-inflammatory property of flavonoids can help in decreasing inflammation and promote antitumorous activity of immune system. Researchers suggest that flavonoids lower the risks of diabetes mellitus by either avoiding absorption of glucose or by improving the glucose tolerance (Johnston et al. 2005). Plaumann et al. (1996) reported that genistein and capsaicin can improve obesity by controlling AMP-activated kinase signaling pathway.

According to previous studies, flavonoids can affect microbes within the body by converting primary into secondary bile acids, production of endotoxin, participating in nutrient absorption, and sustaining immune homeostasis (Ridlon et al. 2014; Shortt et al. 2018). Intake of flavonoids during pregnancy has been studied to have positive impact on health (Powe et al. 2011; Tomimatsu et al. 2017). However, a study on mouse model suggests excessive intake of high amount of catechin, rutin, and epicatechin during pregnancy and lactation adversely affects the fetal development (Baksu et al. 2005).

Citrus flavonoids can reduce the level of triglyceride and apolipoprotein level in plasma of patients suffering from high triglyceride levels. Even the cholesterol level can also be reduce by flavonoids as they directly bind to estrogen receptors (Seremak-Mrozikiewicz and Drews 2004; Sibai 1996). Flavonoids are widely used in field of therapeutics and agriculture as medicine and pesticide, respectively (Emerenciano et al. 2001). The health-promoting ability of flavonoids widens its application be it nutraceuticals, medical, cosmetic, or biomedical use. Within plants, it acts as antimicrobial, visual attractor or repellant, photoreceptor, antioxidant, and antiviral substance (Dixon and Pasinetti 2010; Andersen and Markham 2006). Incorporation of nanotechnology to bioactive compounds like flavonoids can increase the bioavailability of compounds and potentially benefit in cancer treatment (Bondonno et al. 2015; Keservani et al. 2017a, b).

8.2.9 Flavonoids and COVID-19

Humans across the globe have been facing serious threat of COVID-19 infection which causes respiratory tract infection. This pandemic posed huge damage to society, and to cope with it, many antiviral treatments were proposed such as use of remdesivir, arbidol, nucleoside analogs, etc. Xu et al. (2020) reported nelfinavir as the best potential inhibitor of COVID-19. However, currently there is no approved drug for treatment of this infection. The presently used drugs primarily act on main protease, and many herbal compounds are used to combat this infection. Kaempferol, quercetin, naringenin, catechin, curcumin, and epicatechin-gallate are some of the recommended flavonoids which have potential to act as potential inhibitors of COVID-19 infection (Mondal and Rahaman 2020).

8.3 Flavonoids: Futuristic Approach

Flavonoids are potent compounds which have been analyzed and prove to be promising therapeutic agents which can be successfully used in treating many pathological conditions. Apart from this, they can equally be utilized in agricultural sector as pesticides. These flavonoids have wider scope at present as antihypertensive agent, anticholinesterase agent, antioxidant agent, anticancerous agent,

Table. 8.1 Some of the important flavonoids, their source, and potent action (Ayaz et al. 2019)

Flavonoids	Source	Action
Isoflavones	Soy food	Activates brain derived neurotropic factor
Quercetin	Fruit, nut etc.	Decreases neuroinflammation and nitric oxide production, inhibits Akt/PKB signaling and cell cycle arrest in lymphoid cells, shows hepatoprotective activity
Rutin	Green tea, citrus fruits	Shows hepatoprotective activity
Myricetin	Vegetables	Inhibits heparin-induced tau formation
Epigallocatechin	Green tea	Increases removal of free radicals and metal ion chelation
Anthocyanin	Cocoa, pomegranate, etc.	Increases activity of CREB and ACh and improves cognition
Genistein	Green tea	Reduces oxidative stress

antiproliferative agent, antidiabetic agent, anti-inflammatory agent, and a tool for treatment of neurodegenerative disorders. For future, it can act on wider scale in field of nanomedicine for efficient drug delivery (Mondal and Rahaman 2020). Flavonoids are also promising option for antimalarial drugs, as they are expected to show efficacious results with lesser toxicity (Rudrapal and Chetia 2017). Hence, these plant-derived flavonoids have futuristic application in treating malaria. The benefits of flavonoids have not been completely explored, there is very restricted data available in relation to their presence in food items and their metabolites (Spencer et al. 2008). The future of flavonoid application would rely more on developing knowledge regarding bioavailability of flavonoids after processing of food and impact of metabolic pathways; more focus is to be paid on biological activities of metabolites despite on parent compound and also developing in vitro models of bioactivity on conjugated analogues (Kay 2010). Therefore, a multidisciplinary approach including human intervention, molecular approach, and epidemiology is required to cover all the aspects and applicability of flavonoids. Flavonoids are considered as promising medicine, and its use in diet can be helpful in maintaining better human health. Use of molecular modeling and quantitative SAR can be employed for gathering information related to possible target which will enhance the use of flavonoids in therapeutic field.

This is just a fraction of application of these active compounds; much more is yet to be explored, and these bioactive substances which are very much part of a balanced diet have huge options for the future for treatment of acute and chronic diseases (Table 8.1).

References

Abotaleb M, Samuel SM, Varghese E, Varghese S, Kubatka P, Liskova A, Busselberg D (2018) Flavonoids inCancer and Apoptosis. Cancers 11:28

Abril-Gil M, Massot-Cladera M, Perez-Cano FJ, Castellote C, Franch A, Castell MA (2012) Diet enriched with cocoa prevents IgE synthesis in a rat allergy model. Pharmacol Res 65:603–608. https://doi.org/10.1016/j.phrs.2012.02.001

Akiyama H, Sato Y, Watanabe T, Nagaoka MH, Yoshioka Y, Shoji T et al (2005) Dietary unripe apple polyphenol inhibits the development of food allergies in murine models. FEBS Lett 579: 4485–4491. https://doi.org/10.1016/j.febslet.2005.07.019

Alam ZI, Jenner A, Daniel SE, Lees AJ, Cairns N, Marsden CD et al (1997) Oxidative DNA damage in the parkinsonian brain: an apparent selective increase in 8-hydroxyguanine levels in substantia Nigra. J Neurochem 69(3):1196–1203

Andersen M, Markham KR (2006) Flavonoids, Chemistry, biochemistry and applications. CRC Press, Taylor & Francis Group, Boca Raton, FL

Ansari MA, Abdul HM, Joshi G, Opii WO, Butterfield DA (2009) Protective effect of quercetin in primary neurons against AÎ2 (1–42): relevance to Alzheimer's disease. J Nutr Biochem 20:269–275

Assreuy J, Cunha FQ, Epperlein M, Noronha-Dutra A, O'Donnell CA, Liew FY et al (1994) Production of nitric oxide and superoxide by activated macrophages and killing of Leishmania major. Eur J Immunol 24(3):672–676

Ayaz M, Sadiq A, Junaid M, Ullah F, Ovais M, Ullah I, Ahmed J, Shahid M (2019) Flavonoids as prospective neuroprotectants and Their therapeutic propensity in aging associated neurological disorders. Front Aging Neurosci 11:155

Baksu B, Davas I, Baksu A et al (2005) Plasma nitric oxide, endothelin-1 and urinary nitric oxide and cyclic guanosine monophosphate levels in hypertensive pregnant women. Int J Gynecol Obstet 90(2):112–117

Baranowski M, Enns J, Blewett H, Yakandawala U, Zahradka P, Taylor CG (2012) Dietary flaxseed oil reduces adipocyte size, adipose monocyte chemoattractant protein-1 levels and T-cell infiltration in obese, insulin-resistant rats. Cytokine 59:382–391. https://doi.org/10.1016/j.cyto.2012.04.004

Barnes S (1995) Effect of genistein on in vitro and in vivo models of cancer. J Nutr 125 (3 Suppl):777S–783S. https://doi.org/10.1093/jn/125.3_Suppl.777S

Basile A, Giordano S, López-Sáez JA, Cobianchi RC (1999) Antibacterial activity of pure flavonoids isolated from mosses. Phytochemistry 52(8):1479–1482

Bigoniya P, Singh K (2014) Ulcer protective potential of standardized hesperidin, a citrus flavonoid isolated from Citrus sinensis. Rev Bras 24(3):330–340

Bondonno NP, Bondonno CP, Hodgson JM et al (2015) The Efcacy of quercetin in cardiovascular health. Curr Nutr Rep 4(4):290–303

Burkard M, Leischner C, Lauer UM, Busch C, Venturelli S, Frank J (2017) Dietary flavonoids and modulation of natural killer cells: implications in malignant and viral diseases. J Nutr Biochem 46:1–12

Cardenas C, Quesada AR, Medina MA (2011) Anti-angiogenic and anti-inflammatory properties of kahweol, a coffee diterpene. PLoS One 6:e23407. https://doi.org/10.1371/journal.pone.0023407

Cassani J, Dorantes-Barrón AM, Novales LM, Real GA, Estrada-Reyes R (2014) Anti-depressant-like effect of kaempferitrin isolated from Justicia spicigera Schltdl (Acanthaceae) in two behavior models in mice: evidence for the involvement of the serotonergic system. Molecules 19(12):21442–21461

Castell M, Pérez-Cano FJ, Abril-Gil M, Franch À (2014) Flavonoids on allergy. Curr Pharm Des 20(6):973–987

Che DN, Cho BO, Shin JY, Kang HJ, Kim YS, Jang SI (2018) Fisetin inhibits IL-31 production in stimulated human mast cells: possibilities of fisetin being exploited to treat histamine-independent pruritus. Life Sci 201:121–129

Chen X, Yang X, Liu T, Guan M, Feng X, Dong W et al (2012) Kaempferol regulates MAPKs and NF-κB signaling pathways to attenuate LPS-induced acute lung injury in mice. Int Immunopharmacol 14(2):209–216

Chirumbolo S, Bjørklund G, Andrea S, Antonio V (2018a) The role of basophils as innate immune regulatory cells in allergy and immunotherapy. Hum Vaccin Immunotherapy 14(4):815–831. https://doi.org/10.1080/21645515.2017.1417711

Chirumbolo S, Bjorklund G, Lysiuk R, Vella A, Lenchyk L, Upyr T (2018b) Targeting cancer with phytochemicals via their fine tuning of the cell survival signaling pathways. Int J Mol Sci 19: 3568

Cho JY, Kim PS, Park J, Yoo ES, Baik KU, Kim YK et al (2000) Inhibitor of tumor necrosis factor-α production in lipopolysaccharide-stimulated RAW264. 7 cells from Amorpha fruticosa. J Ethnopharmacol 70(2):127–133

Ciz M, Denev P, Kratchanova M, Vasicek O, Ambrozova G, Lojek A (2012) Flavonoids inhibit the respiratory burst of neutrophils in mammals. Oxid Med Cell Longev 2012:181295

Collin M, Bigley V (2016) Monocyte, macrophage, and dendritic cell development: the human perspective. Microbiol Spectr 154(1):3–20

De la Peña JB, Kim CA, Lee HL, Yoon SY, Kim HJ, Hong EY et al (2014) Luteolin mediates the antidepressant-like effects of *Cirsium japonicum* in mice, possibly through modulation of the GABA a receptor. Arch Pharm Res 37(2):263–269

Dewick PM (2001) The shikimate pathway: aromatic amino acids and phenylpropanoids. In: Dewick PM (ed) Medicinal natural products: a biosynthetic approach, 2nd edn. John Wiley, Chichester, pp 137–186

Dexter DT, Carter CJ, Wells FR, Javoy-Agid F, Agid Y, Lees A et al (1989) Basal lipid peroxidation in substantia nigra is increased in Parkinson's disease. J Neurochem 52(2): 381–389

Dixon RA, Dey PM, Lamb CJ (1983) Phytoalexins: enzymology and molecular biology. Adv Enzymol Relat Areas Mol Biol 55(1):69

Dixon RA, Pasinetti GM (2010) Flavonoids and isoflavonoids: from plant biology to agriculture and neuroscience. Plant Physiol 154(2):453–457

Dress RJ, Wong AY, Ginhoux F (2018) Homeostatic control of dendritic cell numbers and differentiation. Immunol Cell Biol 96(5):463–476

Emerenciano V, Militao J, Campos C et al (2001) Flavonoids as chemotaxonomic markers for Asteraceae. Biochem Syst Ecol 29(9):947–957

Ferry DR, Smith A, Malkhandi J et al (1996) Phase I clinical trial of the flavonoid quercetin: pharmacokinetics and evidence for in vivo tyrosine kinase inhibition. Clin Cancer Res 2(4): 659–668

File SE, Jarrett N, Fluck E, Duffy R, Casey K, Wiseman H (2001) Eating soya improves human memory. Psychopharmacology 157:430–436

Galleggiante V, De Santis S, Cavalcanti E, Scarano A, De Benedictis M, Serino G et al (2017) Dendritic cells modulate iron homeostasis and inflammatory abilities following quercetin exposure. Curr Pharm Des 23(14):2139–2146

Gerriets VA, Rathmell JC (2012) Metabolic pathways in T cell fate and function. Trends Immunol 33:168–173. https://doi.org/10.1016/j.it.2012.01.010

Ghayur MN, Khan H, Gilani AH (2007) Antispasmodic, bronchodilator and vasodilator activities of (+)-catechin, a naturally occurring flavonoid. Arch Pharm Res 30(8):970–975

Gómez-Guzmán M, Jiménez R, Sánchez M, Zarzuelo MJ, Galindo P, Quintela AM et al (2012) Epicatechin lowers blood pressure, restores endothelial function and decreases oxidative stress and endothelin-1 and NADPH oxidase activity in DOCA-salt hypertension. Free Radic Biol Med 52(1):70–79

Gong SQ, Sun W, Wang M, Fu YY (2011) Role of TLR4 and TCR or BCR against baicalin-induced responses in T and B cells. Int Immunopharmacol 11(12):2176–2180

Gonzales GB, Smagghe G, Grootaert C, Zotti M, Raes K, Van Camp J (2015) Flavonoid interactions during digestion, absorption, distribution and metabolism: a sequential structure-activity/property relationship-based approach in the study of bioavailability and bioactivity. Drug Metab Rev 47:175–190

Griesbach R (2005) Biochemistry and genetics of flower color. Plant Breed Rev 25:89–114

Grotewold E (2006) In: Grotewold E (ed) The science of flavonoids. Springer-Verlag, New York. https://doi.org/10.1007/978-0-387-28822-2

Guo L, Liu W, Lu T, Guo W, Gao J, Luo Q et al (2015) Decrease of functional activated T and B cells and treatment of glomerulonephritis in lupus-prone mice using a natural flavonoid astilbin. PLoS One 10(4):e0124002

Hagenlocher Y, Lorentz A (2015) Immunomodulation of mast cells by nutrients. Mol Immunol 63: 25–31

Hanasaki Y, Ogawa S, Fukui S (1994) The correlation between active oxygens scavenging and antioxidative effects of flavonoids. Free Radic Biol Med 16(6):845–850

Hartman RE, Shah A, Fagan AM, Schwetye KE, Parsadanian M, Schulman RN et al (2006) Pomegranate juice decreases amyloid load and improves behavior in a mouse model of Alzheimer's disease. Neurobiol Dis 24:506–515

Havsteen B (2002) The biochemistry and medical significance of the flavonoids. Pharmacol Ther 96:67–202

Hayashi T, Sawa K, Kawasaki M et al (1988) Inhibition of cow's milk xanthine oxidase by flavonoids. J Nat Prod 5(1):345–348

Hellebrekers P, Vrisekoop N, Koenderman L (2018) Neutrophil phenotypes in health and disease. Eur J Clin Investig 2018:e12943

Hosseinzade A, Sadeghi O, Biregani AN, Soukhtehzari S, Brandt GS, Esmaillzadeh A (2019) Immunomodulatory effects of flavonoids: possible induction of T CD4+ regulatory cells through suppression of mTOR pathway signaling activity. Front Immunol 10:51. https://doi.org/10.3389/fimmu.2019.00051

Impey S, Smith DM, Obrietan K, Donahue R, Wade C, Storm DR (1998) Stimulation of cAMP response element (CRE)-mediated transcription during contextual learning. Nat Neurosci 1: 595–601. https://doi.org/10.1038/2830

Incani A, Deiana M, Corona G, Vafeiadou K, Vauzour D, Dessì MA et al (2010) Involvement of ERK, Akt and JNK signalling in H2O2-induced cell injury and protection by hydroxytyrosol and its metabolite homovanillic alcohol. Mol Nutr Food Res 54:788–796. https://doi.org/10.1002/mnfr.200900098

Izzo AA, Carlo GD, Mascolo N, Capasso F, Autore G (1994) Antiulcer effect of flavonoids. Role of endogenous PAF. Phytother Res 8(3):179–181

Jäger AK, Saaby L (2011) Flavonoids and the CNS. Molecules 16(2):1471–1485

Jakobek L (2015) Interactions of polyphenols with carbohydrates, lipids and proteins. Food Chem 175:556–567

Jan R, Chaudhry GE (2019) Understanding apoptosis and apoptotic pathways targeted cancer therapeutics. Adv Pharm Bull 9:205–218

Jia Z, Babu PV, Si H, Nallasamy P, Zhu H, Zhen W et al (2013) Genistein inhibits TNF-α-induced endothelial inflammation through the protein kinase pathway a and improves vascular inflammation in C57BL/6 mice. Int J Cardiol 168(3):2637–2645

Johnston K, Sharp P, Clifford M, Morgan L (2005) Dietary polyphenols decrease glucose uptake by human intestinal Caco-2 cells. FEBS Lett 579(7):1653–1657

Jorgensen R (1995) Co-suppression, flower color patterns, and metastable gene expression states. Science 268:686–691

Kay CD (2010) The future of flavonoid research. Br J Nutr 104:S91–S95

Kerry NL, Abbey M (1997) Red wine and fractionated phenolic compounds prepared from red wine inhibit low density lipoprotein oxidation in vitro1. Atherosclerosis 135(1):93–102

Keservani RK, Kesharwani RK, Vyas N, Jain S, Raghuvanshi R, Sharma AK (2010) Nutraceutical and functional food as future food: a review. Pharm Lett 2(1):106–116

Keservani RK, Sharma AK, Kesharwani RK (2017a) An overview and therapeutic applications of nutraceutical and functional foods. Recent Adv Drug Deliv Technol:160–201. ISBN13: 9781522532675

Keservani RK, Kesharwani RK, Sharma AK (2017b) Introduction to nanotechnology in drug delivery. In: Drug delivery approaches and Nanosystems. Apple Academic Press, pp 1–19. ISBN: 9781771885836

Kim JK, Jin HS, Suh HW, Jo EK (2017) Negative regulators and their mechanisms in NLRP3 inflammasome activation and signaling. Immunol Cell Biol 95:584–592

Kim W, Lee H (2013) Advances in nutritional research on regulatory T cells. Nutrients 5(11): 4305–4315

Korkina LG, Afanaśev I (1996) Antioxidant and chelating properties of flavonoids. In: Sies H (ed) Advances in Pharmacology. Academic Press, New York, pp 151–63. 47

Kumar S, Mishra A, Pandey AK (2013) Antioxidant mediated protective effect of *Parthenium hysterophorus* against oxidative damage using in vitro models. BMC Complement Altern Med 13. Article 120

Kumar S, Pandey AK (2013) Phenolic content, reducing power and membrane protective activities of *Solanum xanthocarpum* root extracts. Vegetos 26:301–307

Lee B-H, Choi S-H, Shin T-J, Pyo MK, Hwang S-H, Kim B-R et al (2010) Quercetin enhances human a7 nicotinic acetylcholine receptor-mediated ion current through interactions with Ca2C binding sites. Mol Cells 30:245–253

Leischner C, Burkard M, Pfeiffer MM, Lauer UM, Busch C, Venturelli S (2016) Nutritional immunology: function of natural killer cells and their modulation by resveratrol for cancer prevention and treatment. Nutr J 15(1):47

Letenneur L, Proust-Lima C, Le Gouge A, Dartigues J-F, Barberger- Gateau P (2007) Flavonoid intake and cognitive decline over a 10-year period. Am J Epidemiol 165:1364–1371

Lindqvist C, Bobrowska-Hägerstrand M, Mrówczyńska L, Engblom C, Hägerstrand H (2014) Potentiation of natural killer cell activity with myricetin. Anticancer Res 34(8):3975–3979

Liu YZ, Cao YG, Ye JQ, Wang WG, Song KJ, Wang XL et al (2010) Immunomodulatory effects of proanthocyanidin A-1 derived in vitro from *Rhododendron spiciferum*. Fitoterapia 81(2): 108–114

Mahboubi A, Asgarpanah J, Sadaghiyani PN, Faizi M (2015) Total phenolic and flavonoid content and antibacterial activity of *Punica granatum* L. var. pleniflora flowers (Golnar) against bacterial strains causing food borne diseases. BMC Complement Altern Med 15(1):1–7

Maher P, Dargusch R, Bodai L, Gerard P, Purcell J, Marsh JL (2011) ERK activation by the polyphenols fisetin and resveratrol provides neuroprotection in multiple models of Huntington's disease. Hum Mol Genet 20:261–270

Mantena SK, Meeran SM, Elmets CA, Katiyar SK (2005) Orally administered green tea polyphenols prevent ultraviolet radiation-induced skin cancer in mice through activation of cytotoxic T cells and inhibition of angiogenesis in tumors. J Nutr 135:2871–2877

Matsuda H, Nakamura S, Yoshikawa M (2016) Degranulation inhibitors from medicinal plants in antigen-stimulated rat basophilic leukemia (RBL-2H3) cells. Chem Pharm Bull 64(2):96–103

Middleton E, Kandaswami C, Theoharides TC (2000) The effects of plant flavonoids on mammalian cells: implications for inflammation, heart disease, and cancer. Pharmacol Rev 52:673–751

Middleton EJ (1998) Effect of plant flavonoids on immune and inflammatory cell function. Adv Exp Med Biol 439:175–182

Mishra A, Sharma AK, Kumar S, Saxena AK, Pandey AK (2013) Bauhinia variegata leaf extracts exhibit considerable antibacterial, antioxidant and anticancer activities. BioMed Res Int 2013., Article ID 915436:10

Mondal S, Rahaman ST (2020) Flavonoids: a vital resource in healthcare and medicine. Pharma Pharmacol Int J 8(2):91–104

Murray PJ (2018) Immune regulation by monocytes. Semin Immunol 35:12–18

Naidu PS, Singh A, Kulkarni SK (2004) Reversal of reserpine-induced orofacial dyskinesia and cognitive dysfunction by quercetin. Pharmacology 70(2):59–67

Neagu M, Constantin C, Popescu ID, Zipeto D, Tzanakakis G, Nikitovic D, Fenga C, Stratakis CA, Spandidos DA, Tsatsakis AM (2019) Inflammation and metabolism in cancer cell-mitochondria key player. Front Oncol 9:348

Nickel T, Hanssen H, Sisic Z, Pfeiler S, Summo C, Schmauss D (2011) Het Immuno regulatory effects of the flavonolquercetin in vitro and in vivo. Eur J Nutr 50(3):163–172

Noreen Y, Serrano G, Perera P, Bohlin L (1998) Flavan-3-ols isolated from some medicinal plants inhibiting COX-1 and COX-2 catalysed prostaglandin biosynthesis. Planta Med 64(06): 520–524

Pan Y, Anthony M, Clarkson TB (1999a) Effect of estradiol and soy phytoestrogens on choline acetyltransferase and nerve growth factor mRNAs in the frontal cortex and hippocampus of female rats. Proc Soc Exp Biol Med 221:118–125. https://doi.org/10.3181/00379727-221-44393

Pan Y, Anthony M, Clarkson TB (1999b) Evidence for up-regulation of brain-derived neurotrophic factor mRNA by soy phytoestrogens in the frontal cortex of retired breeder female rats. Neurosci Lett 261:17–20. https://doi.org/10.1016/s0304-3940(98)00994-x

Panche A, Chandra S, Diwan A et al (2015) Alzheimer's and current therapeutics: a review. Asian J Pharm Clin Res 8:14–19

Pandey RP, Sohng JK (2013) Genetics of flavonoids. In: Ramawat K, Mérillon JM (eds) Natural products. Springer, Berlin, Heidelberg. https://doi.org/10.1007/978-3-642-22144-6_52

Park HY, Kim GY, Choi YH (2012) Naringenin attenuates the release of pro-inflammatory mediators from lipopolysaccharide-stimulated BV2 microglia by inactivating nuclear factor-κ B and inhibiting mitogen-activated protein kinases. Int J Mol Med 30(1):204–210

Patil VM, Masand N (2019) Anticancer potential of flavonoids: chemistry, biological activities, and future perspectives. Stud Nat Prod Chem 59:401–429

Perez-Vizcaino F, Duarte J, Jimenez R, Santos-Buelga C, Osuna A (2009) Antihypertensive effects of the flavonoid quercetin. Pharmacol Rep 61(1):67–75

Plaumann B, Fritsche M, Rimpler H, Brandner G, Hess RD (1996) Flavonoids activate wild-type p53. Oncogene 13(8):1605–1614

Polverino E, Rosales-Mayor E, Dale GE, Dembowsky K, Torres A (2017) The role of neutrophil elastase inhibitors in lung diseases. Chest 152(2):249–262

Powe CE, Levine RJ, Karumanchi SA (2011) Preeclampsia, a disease of the maternal endothelium: the role of antiangiogenic factors and implications for later cardiovascular disease. Circulation 123(24):2856–2869

Rahimifard M, Maqbool F, Moeini-Nodeh S, Niaz K, Abdollahi M, Braidy N et al (2017) Targeting the TLR4 signalling pathway by polyphenols: a novel therapeutic strategy for neuroinflammation. Ageing Res Rev 36:11–19

Ramirez GA, Yacoub MR, Ripa M, Mannina D, Cariddi A, Saporiti N et al (2018) Eosinophils from physiology to disease: a comprehensive review. Biomed Res Int 2018:9095275

Rane BR, Kesarwani R (2020) Introduction to nutraceuticals and dietary supplements. In: Nutraceuticals and dietary supplements. Apple Academic Press, pp 35–56. ISBN: 9781771888738

Ravishankar D, Rajora AK, Greco F, Osborn HM (2013) Flavonoids as prospective compounds for anti-cancer therapy. Int J Biochem Cell Biol 45:2821–2831. https://doi.org/10.1016/j.biocel.2013.10.004

Reddy MV, Hwang TL, Leu YL, Chiou WF, Wu TS (2011) Inhibitory effects of Mannich bases of heterocyclic chalcones on NO production by activated RAW 264.7 macrophages and superoxide anion generation and elastase release by activated human neutrophils. Bioorg Med Chem 19(8):2751–2756

Ribeiro D, Freitas M, Tomé SM, Silva AM, Laufer S, Lima JL, Fernandes M (2015) Flavonoids inhibit COX-1 and COX-2 enzymes and cytokine/chemokine production in human whole blood. Inflammation 38(2):858–870

Ridlon JM, Kang DJ, Hylemon PB, Bajaj JS (2014) Bile acids and the gut microbiome. Curr Opin Gastroenterol 30:332–338

Rogerio AP, DoraCL AEL, Chaves JS, Silva LF, Lemos-Senna E et al (2010) Anti-inflammatory effect of quercetin-loaded microemulsion in the airways allergic inflammatory model in mice. Pharmacol Res 61:288–297. https://doi.org/10.1016/j.phrs.2009.10.005

Rudrapal M, Chetia D (2017) Plant flavonoids as potential source of future antimalarial leads. Sys Rev Pharm 8(1):13–18

Saaby L, Rasmussen HB, Jäger AK (2009) MAO-A inhibitory activity of quercetin from *Calluna vulgaris* (L.) Hull. J Ethnopharmacol 121(1):178–181

Sakai-Kashiwabara M, Abe S, Asano K (2014) Suppressive activity of quercetin on the production of eosinophil chemoattractants from eosinophils in vitro. In Vivo 28(4):515–522

Samanta A, Das G, Das S (2011) Roles of flavonoids in plants. Int J Pharm Sci Tech 6:12–35

Scholz S, Williamson G (2007) Interactions affecting the bioavailability of dietary polyphenols in vivo. Int J Vitam Nutr Res 77:224–235

Serafini M, Peluso I, Raguzzini A (2010) Flavonoids as anti-inflammatory agents. Proc Nutr Soc 69(3):273–278

Seremak-Mrozikiewicz A, Drews K (2004) Methyldopa in therapy of hypertension in pregnant women. Ginekol Pol 75(2):160–165

Shortt C, Hasselwander O, Meynier A, Nauta A, Fernandez EN, Putz P, Rowland I, Swann J, Turk J, Vermeiren J et al (2018) Systematic review of the effects of the intestinal microbiota on selected nutrients and non-nutrients. Eur J Nutr 57:25–49

Shukitt-Hale B (2012) Blueberries and neuronal aging. Gerontology 58:518–523

Sibai BM (1996) Treatment of hypertension in pregnant women. N Engl J Med 335(4):257–265

Singh P, Kesharwani K, R., Misra, K., & I Rizvi, S. (2014) In silico validation for the modulatory effect of tea Catechins on erythrocyte Na+/K+-ATPase. Natural Products J 4(3):173–182

Sofic E, Lange KW, Jellinger K, Riederer P (1992) Reduced and oxidized glutathione in the substantia nigra of patients with Parkinson's disease. Neurosci Lett 142(2):128–130

Spencer JP, Abd El Mohsen NM, Minihane AM et al (2008) Biomarkers of the intake of dietary polyphenols: strengths, limitations and application in nutrition research. Br J Nutr 99:12–22

Spencer JP, Rice-Evans C, Williams RJ (2003) Modulation of pro-survival Akt/protein kinase B and ERK1/2 signaling cascades by quercetin and its in vivo metabolites underlie their action on neuronal viability. J Biol Chem 278:34783–34793

Swieca M, Gawlik-Dziki U, Dziki D, Baraniak B, Czyz J (2013) The influence of protein-flavonoid interactions on protein digestibility in vitro and the antioxidant quality of breads enriched with onion skin. Food Chem 141:451–458

Takahama M, Akira S, Saitoh T (2018) Autophagy limits activation of the inflammasomes. Immunol Rev 281(1):62–70

Takahashi A, Ohnishi T (2004) The significance of the study about the biological effects of solar ultraviolet radiation using the exposed facility on the international space station. BiolSci Space 18:255–260

Tanaka T (2014) Flavonoids for allergic diseases: present evidence and future perspective. Curr Pharm Des 20(6):879–885

Taniguchi S, Suzuki N, Masuda M, Hisanaga S-I, Iwatsubo T, Goedert M et al (2005) Inhibition of heparin-induced tau filament formation by phenothiazines, polyphenols, and porphyrins. J Biol Chem 280:7614–7623

Tomimatsu T, Mimura K, Endo M et al (2017) Pathophysiology of preeclampsia: An angiogenic imbalance and long-lasting systemic vascular dysfunction. Hypertens Res 40(4):305–310

Tripathi KD (2013a) Antiparkinsonian drugs. In: Essentials of medical pharmacology, 7th edn. JP Medical Ltd., New Delhi, p 425

Tripathi KD (2013b) Drugs used in mental illness: antidepressant and antianxiety drugs. In: Essentials of medical pharmacology, 7th edn. JP Medical Ltd., New Delhi, p 454

Ullah F, Ayaz M, Sadiq A, Hussain A, Ahmad S, Imran M et al (2016) Phenolic, flavonoid contents, anticholinesterase and antioxidant evaluation of *Iris germanica* var; florentina. Nat Prod Res 30: 1440–1444

Vasconcelos LH, Correia AC, de Souza IL, da Cc SM, Paredes-Gamero EJ, Bárbara VD et al (2015) Flavonoid galetin 3, 6-dimethyl ether attenuates Guinea pig ileum contraction through K+ channel activation and decrease in cytosolic calcium concentration. Eur J Pharmacol 767:52–60

Vauzour D, Vafeiadou K, Rice-Evans C, Williams RJ, Spencer JP (2007a) Activation of pro-survival Akt and ERK1/2 signalling pathways underlie the anti-apoptotic effects of flavanones in cortical neurons. J Neurochem 103:1355–1367. https://doi.org/10.1111/j.1471-4159.2007.04841.x

Vauzour D, Vafeiadou K, Spencer JP (2007b) Inhibition of the formation of the neurotoxin 5-S-cysteinyl-dopamine by polyphenols. Biochem Biophys Res Commun 362:340–346

Walker E, Pacold M, Perisic O et al (2000) Structural determinations of phosphoinositide 3-kinase inhibition by wortmannin, LY294002, quercetin, myricetin, and staurosporine. Mol Cell 6:909–919

Xu, Z., Peng, C., Shi, Y., Zhu, Z., Mu, K., Wang, X., et al. (2020). Nelfinavir was predicted to be a potential inhibitor of 2019-nCov main protease by an integrative approach combining homology modelling, molecular docking and binding free energy calculation. bioRxiv. https://doi.org/10.1101/2020.01.27.921627

Yamada M, Yasuhara H (2004) Clinical pharmacology of MAO inhibitors: safety and future. Neurotoxicology 25(1–2):215–221

Yan F, Yang F, Wang R, Yao XJ, Bai L, Zeng X et al (2017) Isoliquiritigenin suppresses human T lymphocyte activation via covalently binding cysteine 46 of IκB kinase. Oncotarget 8(21): 3422335

Yi LT, Li CF, Zhan X, Cui CC, Xiao F, Zhou LP et al (2010) Involvement of monoaminergic system in the antidepressant-like effect of the flavonoid naringenin in mice. Prog Neuro-Psychopharmacol Biol Psychiatry 34(7):1223–1228

Zhang B, Wang B, Cao S, Wang Y, Wu D (2017) Silybin attenuates LPS induced lung injury in mice by inhibiting NF-κB signaling and NLRP3 activation. Int J Mol Med 39(5):1111–1118

Zhang X, Kang Z, Li Q, Zhang J, Cheng S, Chang H et al (2018) Antigen-adjuvant effects of icariin in enhancing tumor-specific immunity in mastocytoma-bearing DBA/2J mice. Biomed Pharmacother 99:810–816

Zhang Y, Huang B (2017) The development and diversity of ILCs, NK cells and their relevance in health and diseases. Adv Exp Med Biol 1024:225–244

Zhen L, Zhu J, Zhao X, Huang W, An Y, Li S et al (2012) The antidepressant-like effect of fisetin involves the serotonergic and noradrenergic system. Behav Brain Res 228(2):359–366

Chapter 9
Resveratrol and Immunomodulation

Mayela Govea Salas, Leticia A. Menchaca Castro, Rafael G. Aráujo, Sujey A. Niño Herrera, Anna Iliná, Roberto Arredondo Valdés, María del Carmen Rodríguez Salazar, Mónica L. Chávez González, E. Patricia Segura Ceniceros, and Cristóbal N. Aguilar González

Abstract Resveratrol is a stilbenoid polyphenolic molecule widely known for its biological properties, which is available in nature in various varieties of grapes, berries, and some plants. It has been shown to play an important role in the prevention and treatment of chronic diseases such as cancer, heart disease, metabolic, degenerative, autoimmune diseases, and even viral infections. One of the main mechanisms of action that it presents is linked to its antioxidant capacity as it is a strict polyphenol, which gives it the ability to stimulate an immune response in the host by regulating and differentiating immune cells, promoting the synthesis of specific proteins, activate apoptosis, stimulate the secretion of pro-inflammatory cytokines, and even modulate gene expression. These effects have favored the decrease in the progression of various inflammatory and degenerative diseases, thus demonstrating the immunomodulatory capacity of resveratrol on in vitro and in vivo models.

Keywords Resveratrol · Immunomodulation · Immunotherapy · Bioactive compound

9.1 Introduction

The use of products obtained from these natural sources such as food, plants, and beverages has made it possible to use various natural molecules that are present in sources as secondary metabolites mainly. There is a diverse variety of bioactive

M. G. Salas (✉) · A. Iliná · R. A. Valdés · M. L. Chávez González · E. P. S. Ceniceros
Nanobioscience Group, Autonomous University of Coahuila, Saltillo, Coahuila, Mexico

Food Research Department, Autonomous University of Coahuila, Saltillo, Coahuila, Mexico
e-mail: m.govea.salas@uadec.edu.mx

L. A. Menchaca Castro · S. A. Niño Herrera
Nanobioscience Group, Autonomous University of Coahuila, Saltillo, Coahuila, Mexico

R. G. Aráujo · M. del Carmen Rodríguez Salazar · C. N. Aguilar González
Food Research Department, Autonomous University of Coahuila, Saltillo, Coahuila, Mexico

R. K. Kesharwani et al. (eds.), *Nutraceuticals and Functional Foods as Immunomodulators*, https://doi.org/10.1007/978-981-19-2507-8_9

molecules of natural origin and that can be extracted by various physical and chemical processes that are friendly to the environment, which can be purified and characterized for use in various areas of research (Kesharwani and Misra 2010; Keservani et al. 2017). An example of this is resveratrol, a molecule that has been discovered for several years and has been purified and studied in various cell models in vitro and in vivo, the number of investigations that generate contributions of the effects of resveratrol on various pathologies increases. Resveratrol is a polyphenol of natural origin that is found in various varieties of grapes and red berries; therefore, beverages and foods prepared from these fruits are rich in this polyphenol. Its structure is complex and has several structural varieties that are mentioned below.

Several studies with resveratrol have shown that it has various biological properties against various pathologies such as metabolic, cardiac, degenerative, inflammatory, and cancer diseases and even in various viral and microbial infections. The effects generated by resveratrol in these pathologies are attributed mainly by its antioxidant properties, because the mechanism of action of this compound occurs through immunomodulation processes, which allow the activation of various immunological pathways and cells of the immune system so that they can fight disease, or even promote the death of damaged or diseased cells, as well as the activation of metabolic, cellular, and molecular processes linked to cellular oxidative stress.

This immunomodulatory effect has been widely used in the studies and developments of immunotherapies, since modifications of immune responses can be used in various pathologies through the activation, attenuation, or induction of immunological effects for therapeutic purposes. In the development of this chapter, different biological activities related to the immunomodulatory effect of resveratrol will be addressed on different in vitro and in vivo study models, elucidating the signaling pathways, metabolism, and molecular processes related to resveratrol and its effect at the cellular and molecular level, confirming that it is one of the nutraceuticals with the highest number of therapeutic effects attributed to its structural properties.

9.2 Structure and Characteristics of Resveratrol

Resveratrol, also known as trans-resveratrol, is a hydroxylated stilbene derivative with two phenolic rings. IUPAC nomenclature is 5-[(E)-2-(4-hydroxyphenyl) ethenyl]benzene-1,3-diol. Being a stilbenoid compound, it is considered as an aromatic lipidic polycide molecule. It presents a condensed chemical formula of $C_{14}H_{12}0_3$ equivalent to 228.25 g/mol. This compound was first described in the 1940s when it was isolated from the plant *Veratrum grandiflorum* (Gambini et al. 2013). There are two isomers of this compound, *cis* and *trans*-resveratrol. The *trans* isomer is the most stable form and the one commonly found in nature. Resveratrol is considered a phytoalexin, because it is present in a wide variety of plant materials and has been shown to be toxic to a wide variety of microorganisms.

In nature, it can be found as a glycosylated compound in fruits such as berries, grosella, and peanuts (Catalogna et al. 2019). Another food with a higher resveratrol

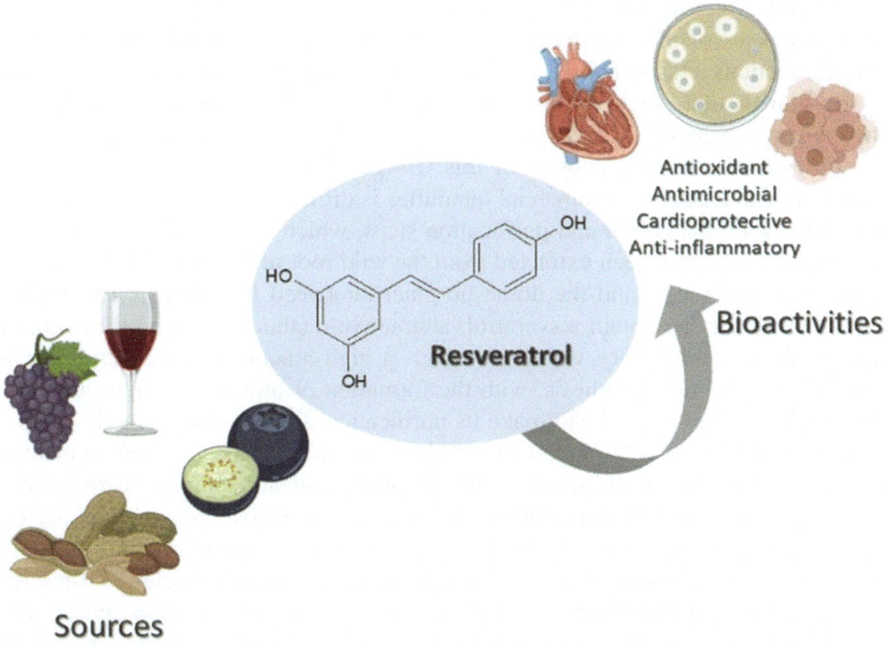

Fig. 9.1 Chemical structure of resveratrol, natural sources, and some bioactivities associated

content is grapes, and therefore, products derived from this fruit have a considerable amount of resveratrol such as juices and wines.

This polyphenol presents interesting biological activities, the most important being that of being a powerful antioxidant; another of the bioactivities shown is that of being an excellent antimicrobial, and tests have been carried out to combat cancer; it presents cardioprotective properties and helps in the reduction of inflammatory processes, anti-aging, and anti-diabetes and promotes neuroprotection (Moshawih et al. 2019; Raskovic et al. 2019) (Fig. 9.1).

9.3 Sources/Production and Applications in Health

Resveratrol RESV (trans-3,4,5-trihydroxystilbene) is a natural polyphenolic phytoalexin composed of two phenyl rings linked by a double bond, produced by plants in response to stress, and belongs to the group of phytoestrogens, it is attributed antioxidant and enzyme inhibition functions (cyclooxygenase, lipoxygenase and xanthine). In nature, it can be found in two isoforms, trans-resveratrol and cis-resveratrol, where the former is the most stable form found mainly in grape skins and red wine (Thapa et al. 2019).

Resveratrol is a powerful antioxidant, a natural polyphenol produced by more than 70 species of plants in response to stressful situations (ultraviolet radiation, fungal infections, etc.) (Thapa et al. 2019); it is present in several fruits that are part of the human diet, such as blueberries (*Vaccinium* spp.), currant, blackberries (*Morus* spp.), peanuts (*Arachis hypogaea*), as well as red wine which is the one that provides the highest content of this (Borge et al. 2013; Salehi et al. 2018). Extracting resveratrol in commercial quantities is difficult due to its low concentration and several extraction and purification steps, which is harmful to the environment. However, it has been extracted from the wild root of *Polygonum cuspidatum*, grape skins and seeds, and the domestic giant knotweed from China, the world's leading producer. To obtain resveratrol, alternative methods are needed as it has a high medicinal and dietary value. However, a high amount of resveratrol can be produced by chemical synthesis, with the formation of unwanted by-products that contaminate the resveratrol and make its purification difficult (Salehi et al. 2018).

Resveratrol has a wide range of biological activities such as anti-inflammatory, antiviral, and antitumor properties (rats). Proposed anticancer mechanisms for resveratrol include (a) inhibition of ribonucleotide reductase, DNA polymerase, protein kinase C, or cyclooxygenase-2 activities, (b) inhibition of cell proliferation and free radical-induced carcinogenesis, and (c) induction of apoptosis (Vélez-Marín et al. 2012). Resveratrol eliminates free radicals, such as reactive oxygen species and reactive nitrogen species, produced by the metabolism of the body, minimizes radical damage to delicate organs, inhibits lipid peroxidation, and increases cholesterol in the blood, which improves neurological and cardiovascular activities (Table 9.1) (Kuršvietienė et al. 2016).

Resveratrol is rapidly absorbed, metabolized by glucuronides or sulfate conjugates, and distributed to various organs. In general, polyphenol metabolites are rapidly eliminated from plasma, which indicates that a daily consumption of plant products is necessary to maintain high concentrations of these metabolites in the blood. The pharmacokinetic parameters of resveratrol are half-life of 9.2 h, absorption of 75%, and low oral bioavailability, less than 1%; sulfation and glucuronidation in the small intestine appear to be the limiting steps in bioavailability. Its most important metabolites are glucuronide and resveratrol sulfate (Vélez-Marín et al. 2012). It is extensively metabolized in the liver mostly by glucuronidation and sulfation, the latter carried out by phenol sulfotransferases.

There are multiple positive effects of resveratrol already demonstrated on in vitro models and clinical trials in humans: antiatherosclerotic and cardioprotective effect, antiplatelet, anticancer, antiviral, lipid-lowering, antioxidant, and neuroprotective effect. Resveratrol is available for sale to the public as a dietary supplement, with the premise of losing weight, as a cardiovascular protector and antioxidant, and even to prolong longevity. Most of the current evidence is in animals, and human clinical studies are needed to demonstrate its safety and tolerability (Gambini et al. 2013).

Flavonoids, including resveratrol, are associated with numerous mechanisms of action, which exert beneficial effects in various conditions including obesity, metabolic syndrome, type 2 diabetes mellitus, cancer, dementia, and Alzheimer's. Resveratrol metabolites are found primarily in the liver, compared to muscle and

Table 9.1 Immunomodulation effects of resveratrol in various pathologies

Bioactivity	Effect	References
Antioxidant activity	The antioxidant capacity of resveratrol is due to the inhibition of nicotinamide adenine dinucleotide phosphate oxidases, reducing ROS and increasing the levels of mRNA of antioxidant enzymes such as superoxide dismutase and catalase	Inglés et al. (2014), Banez et al. (2020)
Anticancer effect	Resveratrol inhibits the expression of FAK, which stops carcinogenesis, and decreases the expression of H-RAS, which controls the cell cycle of tumors. It increases the expression of Bcl-2 and the activation of caspase-3 within cells and increases the expression of Fas, which controls cell death outside them. It also can suppress the protumor activation of tumor-associated macrophages (TAMS) which is associated with malignant and metastatic cancers	Choi et al. (2013), Yoon et al. (2015), Nitulescu et al. (2018)
Anti-inflammatory activity	RSV blocks TyrRS and COX-1, which are chemoattractant molecules, activates SIRT1 which decreases the production of pro-inflammatory molecules such as IL-8 or TNF-α, and also reduces the expression of ICAM-1, decreasing the adhesion of white cells	Latruffe et al. (2015), Berman et al. (2017)
Antihypertensive activity	Resveratrol causes vasodilation of blood vessels by activating AMP/PK, reducing blood pressure by preventing smooth muscle contraction; it also reduces ICAM-1 expression, preventing HA	Cao et al. (2014), Berman et al. (2017)
Antiviral effect	MERS-CoV, RSV can block the NF-kB pathway preventing the production of inflammation molecules. In HSV, it increases ROS, essential to stop viral infection. It induces the death of B cells infected with EBV by inhibiting the expression of LPM1, an essential protein for infection by this virus; in an infection with IAV, it increases the expression of MHC-1 to stimulate the immune cells that will kill the infected cells	Chen et al. (2012), Teijaro (2016), Taniguchi and Karin. (2018), Saha and Robertson. (2019)
Resveratrol effect in metabolic diseases	RSV plays an important role in obesity by activating SIRT1, which controls the metabolism of lipids and	Howitz et al. (2003), Petrovski et al. (2011), Huang et al. (2020)

(continued)

Table 9.1 (continued)

Bioactivity	Effect	References
	carbohydrates, replacing the lack of physical activity, and increases the absorption of glucose into the cells due to the lack of insulin in DM and the production of NO in the endothelial cells of the blood vessels, acting as a vasodilator and preventing platelet aggregation	
Resveratrol effect in autoimmune diseases	Resveratrol acts on DMT1, binding to insulin to improve glucose absorption; in Crohn's disease, it decreases the expression of MDSCs, which inhibits the activation of effector lymphocytes in the intestine. Reduces the production of IL-17 and IL-19 and activates the apoptosis of keratinocytes which cause psoriasis	Lee et al. (2016), Oliveira et al. (2017)
Resveratrol effect in degenerative diseases	In AD, resveratrol reduces MMP9 expression, decreasing the entry of leukocytes to the brain, and restores memory by eliminating the formaldehyde accumulated by age. In AMD, COPD and DMD have an anti-inflammatory role by reducing the expression of IL-6 and IL-8, preventing the accumulation of leucocytes in damaged sites. Furthermore, it is capable of decreasing serum concentrations of VEGF and CPR, preventing the development of atherosclerotic plaques	Gordon et al. (2013), Lançon et al. (2016), Moussa et al. (2017), Wang et al. (2017a, b), Figueira and González (2018), Liu et al. (2019)

adipose tissue. Resveratrol can be consumed in pills or food, especially fresh foods (Nicol et al. 2020). However, the consumption of functional foods, including fruits and vegetables, is lower than the recommendations (at least five servings a day) in adults.

9.4 Immunomodulatory Role of Resveratrol in Various Pathologies

9.4.1 Antioxidant Activity

Antioxidant compounds are substances that can reduce, modulate, or prevent the oxidation of essential molecules in biological systems, neutralizing pro-oxidant molecules. Antioxidants prevent the reactive species as free radicals and reactive oxygen and nitrogen species from acting on DNA, lipids, and proteins through neutralizing its oxidative capacity by being oxidized themselves (Ajith et al. 2017; Olszowy 2019). The role of antioxidant compounds can be done by different forms such as scavenging the free radicals; preventing the reactive species formation; forming chelate complexes with pro-oxidant metals, singlet oxygen, and photosensitizers quenching; and removing and repairing damages caused by the reactive species and enzyme deactivation or activation (Oroian and Escriche 2015; Olszowy 2019; Banez et al. 2020; Song et al. 2021).

Resveratrol is a recognized antioxidant compound, with many health benefits associated with low oxidative stress. Resveratrol allows the inhibition of the formation of oxygen free radicals inhibiting nicotinamide adenine dinucleotide phosphate oxidases and subsequent production of reactive oxygen species and induces expression of antioxidant enzymes and their substrates (Banez et al. 2020).

Antioxidants have been widely studied for the treatment of various diseases since the last decades, mainly in the treatment and prevention of cardiovascular and neurodegenerative diseases. Many studies have shown that some antioxidant compounds may also have potential immunoregulatory activity. The effect of immunomodulation resulting by antioxidant effects is less explored; however, the mechanism and effects of all health benefits and immunomodulations of resveratrol have predominantly been attributed to its antioxidant activity and to some extent direct target interaction (Ajith et al. 2017; Prysyazhna et al. 2019).

Some antioxidant molecular mechanisms of resveratrol have been elucidated, as the positive regulation of the phosphatase and tensin homolog, which decreased Akt phosphorylation, leading to an upregulation of antioxidant enzyme mRNA levels such as superoxide dismutase and catalase (Inglés et al. 2014). Resveratrol activates adenosine monophosphate-activated protein kinase to maintain the structural stability of forkhead box O1, facilitating its translocation, and accomplish its transcriptional function (Yun and Lee 2018), and it can reduce the ischemia-reperfusion injury-induced oxidative stress by inhibiting the activation of the p38 mitogen-activated protein kinase pathway; therefore, the levels of antioxidants like glutathione increase, and free radicals are directly reduced (Fu et al. 2018; Meng et al. 2020). Also, resveratrol exhibited antioxidant bioactivities by regulating antioxidant gene expression via the Kelch-like ECH-associated protein 1 pathway and sirtuin 1 (Li et al. 2016).

Currently, resveratrol was associated to attenuate oxidative injury owing to the induced autophagy via the AMPK-mediated inhibition of mammalian target of rapamycin signaling or via the activation of transcription factor EB, which promoted the formation of lysosomes and autophagosomes as well as their fusion into an autolysosome (Zhou et al. 2019).

9.4.2 Anticancer Effect

In recent years, the use of resveratrol and other polyphenols for cancer treatment has been widely investigated. The research that was held by Jang et al. (1997) was one of the first reports that mentioned the anticancer activity of resveratrol. They reported that resveratrol could act as a chemopreventive agent by inhibiting tumor formation on murine models of skin cancer. Even though the mechanism was not understandable at all, this study was the pioneer to be able to acknowledge different mechanisms of action that resveratrol exerts on distinct types of cancer.

The anticancer activity of resveratrol is mainly due to it can intervene in different cellular pathways related to apoptosis (Lin et al. 2011). Nevertheless, there are other processes than resveratrol carries out to act against cancer cells, for example, by reducing ROS production and the damage induced by oxidative stress or the contrary, by promoting the production of ROS in cancer cells (Heo et al. 2018; Yan et al. 2018) by increasing immunity in cosurveillance, cell cycle arrest (Ko et al. 2017), stopping the transformation of procarcinogen to carcinogen, among others.

Takashina et al. (2017) reported that resveratrol could induce apoptosis in leukemia cells U937 (Human leukemic monocyte lymphoma) and MOLT-4 (Human acute T lymphoblastic leukemia) by reducing the expression of H-Ras and consequently diminish the activation of Akt. H-Ras activates the PI3/Akt pathway, which controls processes such as apoptosis, cell cycle progression, cell size, and transcription; studies have proven that this pathway is hyperactivated in different types of cancer, resulting in cell survival and apoptosis inhibition. In the same way as the PI3K/Akt pathway, overactivation of Ras is also found in many cancers (Asati et al. 2016; Nitulescu et al. 2018). Other authors have also mentioned the anticancer activity of resveratrol in leukemia. For example, Fan et al. (2018) described that resveratrol could induce cell death in HL-60 cells through autophagy and apoptosis. Resveratrol could activate the intrinsic and extrinsic apoptotic pathways. The first one was triggered by increasing the production of Bcl-2 and following the activation of caspase-3. Contrarily, the extrinsic pathway was stimulated by increasing the expression of Fas and FasL; then caspase-8 was cleavaged to activate other caspases and perform cell death. Also, resveratrol could induce autophagy by inhibiting the negative regulator of autophagy mTOR through the blockage activation of the PI3/Akt pathway. mTOR and PI3K/Akt may activate some oncogenes resulting in the suppression of autophagy and improvement of cancer formation (Choi et al. 2013).

Autophagy plays an essential role in cancer, is related to tumor promotion and suppression, and contributes to cancer cell development and proliferation. Miki et al. (2012) also reported that resveratrol could induce autophagy and apoptosis in colon cancer cells HT-29 and COLO 201. In that study, the activity of Caspase-8/Caspase-3 was increased leading to apoptosis through the death-receptor pathway. Autophagy was also induced by resveratrol and acted as a cell death mechanism; protein LC3-II which is part of autophagosomal membranes was also found in the analysis. Both apoptosis and autophagy were triggered by ROS production; even though resveratrol is a natural antioxidant, there is evidence that supports that resveratrol and other polyphenols can induce the production of ROS to lead apoptosis in some types of cancer (Juan et al. 2008).

Resveratrol has affected other types of cancer. Buhrmann et al. (2017) investigated the effect of this stilbene in colorectal cancer cells (CRC) HCT116 and SW480. They found that resveratrol reduced CRC invasion and proliferation by enhancing the expression of Sirt1, inhibiting of NF-κB-mediated inflammatory pathway, and suppressing focal adhesion kinase (FAK) activity. That results in a loss of focal adhesion molecules, a planar surface of the CRC cells, and an increase in apoptosis. FAK is a protein that regulates cell adhesion, migration, metastasis, motility, proliferation, and survival (Yoon et al. 2015). In many types of cancers, FAK is overexpressed, including CRC (Kong et al. 2015). The inhibition of FAK results in the suppression of the early stages of carcinogenesis. In addition to this, Buhrmann et al. (2017) reported that resveratrol showed to suppress the activation of NF-κβ, suppressing the transcription of products involved in invasion (MMPs), metastasis (CXCR4) and activating those involved in apoptosis such as cleavage of caspase-3. The blockage of NF-κβ in this study was related to the upregulation of Sirt1. Sirt1 plays a fundamental role in cell death, survival, and immune tolerance. One of the principal substrates of Sirt1 is the p65/RelA, an NF-κB subunit which is the primary regulator of leukocyte activation and inflammatory cytokines signaling. The activation of Sirt1 by resveratrol generates the deacetylation of Re1A. That action attenuates NF-κB-mediated gene transcription of factors that intervene in inflammation pathways, including TNF, IL-1, IL-6, Cox-2, metalloproteases (MMP)-1, and MMP3 (Yamamoto and Gaynor 2001; Lee et al. 2009a, b).

Chai et al. (2017) also mentioned the enhanced expression of Sirt1 by using resveratrol on human hepatocellular carcinoma (HCC) cell lines. They reported that the activation of Sirt1 is related to the deacetylation of FoxO1, which leads to tumor suppression in HCC cells. Also, Sirt1 acts as a negative regulator of the PI3K/Akt pathway, which is involved in cell proliferation, tumor growth, and metastasis (Jiang et al. 2020). Resveratrol also showed to diminish the production of antiapoptotic proteins Bcl-2/Bax and enhanced the activation of caspase-3 and caspase-7. In another study, resveratrol also had activity against liver cancer. Bishayee et al. (2010) demonstrated that resveratrol combats oxidative-nitrosative stress and suppresses the inflammatory cascade during liver carcinogenesis through the overexpression of Nrf2. Nrf2 is a protein that protects cells against oxidative-nitrosative stress. The generation of ROS and reactive species of nitrogen (RON) is a crucial factor in the initiation and progression of hepatocarcinoma. Oxidative-nitrosative stress also is involved in invasion, migration, and metastasis of HCC (Fu and Chung 2018).

Furthermore, Sun et al. (2017) suggested that resveratrol could effectively inhibit lung cancer progression by suppressing the protumor activation of tumor-associated macrophages (TAMs). TAMs play a crucial role in cancer progression, evasion of immunity, and dissemination of cancer cells. They are associated with tumor malignancy and relapse. Sun et al. (2017) reported that resveratrol inhibited the alternatively activated macrophage (M2) polarization of TAMs through the inhibition of STAT 3 activation, a transcription factor that is involved in M2 polarization cell cycle progression and promotion of antiapoptosis and proliferation. The inhibition of STATs signaling pathways can suppress tumor growth and metastasis by inhibiting M2-like polarization of macrophages, further suggesting that TAMs are a possible target in cancer therapy. Additional evidence suggests the role of resveratrol in tumor growth, Kimura and Sumiyoshi (2016) reported that resveratrol prevented tumor growth and metastasis in mice models with osteosarcoma. According to this report, resveratrol prevented lung and liver metastasis by different mechanisms of action. One of them was by stopping the activation and differentiation of M2 in TAMs through the suppression of STAT3. The other was by diminishing the expression of lymphatic vessel endothelial hyaluronan receptor-1 (LYVE-1). This protein is widely expressed in the endothelium of lymphatic vessels of tumors and promotes lymphangiogenesis, metastasis, and tumor growth (Hara et al. 2018).

Even though the potential of resveratrol as a chemopreventive and chemotherapeutic agent is promising, there are still crucial points that must be studied thoroughly. Among these points are resveratrol's bioavailability, route administration, formulation, and possible interaction with other compounds (Ko et al. 2017).

9.4.3 *Anti-inflammatory Activity*

Inflammation is an immune response to an injury or the presence of an antigen in the body; it is characterized by pain, redness, heat, swelling, and even the loss of tissue function (De Sá-Coutinho et al. 2018); it is normally linked to a large number of diseases (Latruffe et al. 2015). Inflammation is the result of the signaling of chemokines and interleukins that move leukocytes to the damaged site; this is possible due to the release of pro-inflammatory molecules such as prostaglandins and leukotrienes (Latruffe et al. 2014). Resveratrol is a molecule capable of modulating different enzymes that intervene in this immune response like kinases, lipoxygenases, cyclooxygenases, and sirtuins, in addition to eliminating free radicals present in cells (De Sá-Coutinho et al. 2018).

Resveratrol can suppress inflammation by different mechanisms, which stand out: aminoacyl-tRNA synthetases enzymes play an important role in inflammation; tyrosine-tRNA-ligase (TyrRS) is normally inactive, but through natural proteolysis, it fragments and forms mini-TyrRS which is a chemoattractant of leukocytes acting like IL-8; and resveratrol has the ability to bind to the catalytic site of mini-TyrRS and inhibit it causing an anti-inflammatory effect (Latruffe et al. 2015). Enzymes such as prostaglandin-H synthetase, cyclooxygenase (COX), and lipoxygenase are involved in the main route of synthesis of pro-inflammatory mediators from the arachidonic acid (AA). Resveratrol can bind to the COX-1 catalytic site preventing AA from catalysis. In addition, it decreases the expression of COX-2 through two transcription factors: NF-κB and the AP-1 complex, which are controlled by an IKKB/MAPK signaling cascade. Resveratrol prevents the phosphorylation of these kinases causing the decrease in the secretion of pro-inflammatory cytokines such as IL-6, IL-8, and TNF-α and increasing the anti-inflammatory cytokines, also reducing the expression of adhesion proteins (ICAM-1) that help to the mobilization of leukocytes (Latruffe et al. 2015). Resveratrol also activates the silent information regulator T1 (SIRT1), affecting the nuclear factor kB (NF-kB) signaling pathway, which deacetylates the RelA/p65 subunit of nuclear factor kB. NF-kB is present in the cytoplasm of macrophages linked to an inhibitory protein Ikβ; when it is phosphorylated, NF-kB is released and goes to the nucleus to activate the transcription of genes involved in the inflammatory response, such as tumor necrosis factor-α (TNF-α), interleukin 1β (IL-1β), nitric oxide synthase (iNOS), and increasing vascular permeability (Berman et al. 2017). Resveratrol is capable of attenuating the activity of the microsomal enzyme-prostaglandinE-synthetase-1 (mPGES-1) which is responsible for the synthesis of PGE. (Moller et al. 2015). Another mechanism involves the toll-like receptor 4; it is an important initiator in the inflammatory response that, once activated, myeloid differentiation factor 88 (MyD88) binds to it and activates NF-kB and therefore begins the production of cytokines pro-inflammatory; resveratrol significantly reduces the levels of TLR4 and MyD88, having an anti-inflammatory effect (Zhang et al. 2016). Resveratrol is a powerful activator of poly (ADP-Ribose) polymerase 1 (PARP1), which is capable of inhibiting the action of COX-2 and increases the production of B-cell lymphoma-6 protein (BCL6) which has the ability to recruit monocytes in an inflammatory environment (Yanez et al. 2019).

Latruffe et al. (2015) observed the action of resveratrol in THP-1 monocyte cells; this increased the expression of miRNA-663, which is directed to suppress genes involved in the immune response, such as JUNB and JUND transcription factors, and also decreases the transcription factor AP-1 activity, which is normally activated by inflammatory cytokine stimuli. Wang et al. (2013) experimented with obese mice giving them a resveratrol supplement to study its anti-inflammatory potential; the results showed a decrease in TNF-α and IL-6 and an increase in IL-10 in a dose-dependent manner.

Szkudelsk et al. (2020) studied the anti-inflammatory properties of RSV in diabetic rats, he noted that the content of inflammatory mediators such as IL-6, IL-1β, TNF-α, and NF-κB was low in skeletal muscle; however, in the adipose tissue, release of these mediators was increased. Szkudelsk et al. (2020) reported that the efficacy of resveratrol is tissue-specific. Wang et al. (2017a, b) did studies with acute inflammation in rats, comparing the effectiveness of dexamethasone (DXM) and resveratrol at different concentrations; in a xylene-induced edema, the degree of inflammation was dose dependent, reaching the same effectiveness as the DMX. He also caused an allergic inflammation, where mast cells release inflammatory mediators and attract leukocytes; however, when treating them with resveratrol, the number of leukocytes found in the area was low, which confirms the anti-inflammatory capacity of resveratrol.

Senturk et al. (2018) made a wound in the spine of mice, treated it with resveratrol, and had a notable reduction in the levels of pro-inflammatory cytokines; this was confirmed by the increase in healthy glial cells and motor neurons.

There is an inflammatory response in the pancreas due to insulin resistance as one ages, Calvo et al. (2017) administered resveratrol in elderly mice, which had an overexpression of pro-inflammatory cytokines; after treatment, a lower expression of TNF-α was observed and IL-1β, as well as NF-κB p52; however, there was no increase in anti-inflammatory interleukins such as IL-10.

Zhou et al. (2018) induced acute pharyngitis in rabbits, treatment with RSV inhibited the activity of macrophage inflammatory protein 2 (MIP-2) and COX-2, and it also stopped the increase in caspase-3/9 activity, which acts on the cell death. In addition, low serum levels of TNF-α and IL-6 were observed. Likewise, Zhou et al. (2018) demonstrated that resveratrol decreases the expression of TLR4 and MyD88, which represent a pro-inflammatory signaling pathway.

Asthma is an inflammatory disease characterized by excessive infiltration of leukocytes, mainly eosinophils, in the airways, causing bronchoconstriction. Lee et al. (2009a, b) experimented in mice, administering resveratrol as a treatment, and observed a decrease in the migration of white cells; the amount of immunoglobulin E and the concentration of IL-4 and IL-5 in the lungs were also reduced. IL-5 is known to stimulate eosinophils's degranulation, so resveratrol stops or decreases the inflammatory response.

Owing to its great anti-inflammatory capacity, resveratrol could be used as a drug for different diseases closely linked to inflammation; however, its bioavailability within the body is very low, which limits its application (Malhotra et al. 2015).

9.4.4 Antihypertensive Activity

Hypertension is a disease that affects approximately 25% of adults worldwide, and some hypertensive patients are resistant to blood pressure reduction with actual antihypertensive drugs. Different complications are associated with hypertension, including end organ damage, stroke, left ventricular hypertrophy, and arteriosclerosis. The number of people with metabolic syndrome is growing and contributes to the increase in hypertension cases, which is creating an important alarm for the search for new compounds or treatments to avoid cardiovascular complications (Dolinsky et al. 2013).

A recent study showed the ability of resveratrol to induce vasodilation by oxidative activation of protein kinase 1α and reduced blood pressure in hypertensive wild-type mice. This activity was explained by the oxidation of phenolic rings of resveratrol that paradoxically leads to oxidative modification of proteins, elucidated by formation of a reactive quinone that oxidizes the thiolate side chain of cysteine residues, which were improved in cells under oxidative stress (Prysyazhna et al. 2019).

Resveratrol showed to affect the activity of vascular smooth muscle cells and to inhibit the expression of vascular cell adhesion molecules, which are responsible for hypertension and the development of atherosclerosis, respectively (Berman et al. 2017).

Grujic-Milanovic et al. (2017) demonstrated that resveratrol from red wine improves the hemodynamics oxidative defense and aortal structure in essential and malignant hypertension. Resveratrol exerts vasorelaxant properties, provides protection of endothelial milieu, acts as pleiotropic cellular effector, and developed hypertrophy in female rat aorta. The results suggest that resveratrol may act in vivo by reduction of free radicals, activation of antioxidant enzymes, removing of superoxide anions, and promotion of nitric oxide availability, hydroxyl radicals, and lipid peroxyl radicals. An important finding was the activation of AMP-activated protein kinase induced by resveratrol and reduces blood pressure inhibiting human vascular smooth muscle cells contractility through the suppression of myosin phosphatase-targeting subunit 1 and myosin light chain phosphorylation (Cao et al. 2014).

The treatment of hypertensive rats with resveratrol attenuates development of hypertension and prevents endothelial dysfunction. The mechanisms proposed by resveratrol effect was the attenuation of vascular oxidative stress resulting in increased nitric oxide bioavailability, increased expression of important proteins in nitric oxide pathway, and prevention of endothelial nitric oxide synthase uncoupling possibly via inhibition of cofactor tetrahydrobiopterin oxidation by free radicals (Bhatt et al. 2011). A meta-analysis of six randomized controlled human trials (including 247 subjects) showed that high doses of resveratrol (>150 mg/day) significantly decreased blood pressure, while lower doses had no effect (Liu et al. 2015).

9.4.5 Antiviral Effect

Resveratrol is one of the most well-known stilbenes and, as mentioned in previous lines, is found in several natural sources. According to different studies, this compound has shown multiple biological activities (Malaguarnera 2019); most of them, including antiviral activity, are related to the immunomodulatory role that resveratrol can exhibit. The antiviral activity of resveratrol has widely investigated, in different studies, this effect is associated with cell signaling pathways, its interaction with inflammatory signaling transduction, and the upregulating and downregulating of genes involved in apoptosis (Abba et al. 2015; Malaguarnera 2019). In the following paragraphs, a few examples will be discussed.

Lin et al. (2017) reported that resveratrol was able to inhibit the infection of Middle East respiratory syndrome coronavirus (MERS-CoV) by different mechanisms of action. One of these mechanisms was by blocking the NF-κβ pathway, which involves the downregulating inflammatory signaling. The activation of NF-κβ is necessary to produce inflammatory cytokines. It is worth mentioning that MERS-CoV infection induces the production of pro-inflammatory cytokines such as TNF-α, IL-1, IL-6, and IL-8, among others causing several effects or even death in patients (Lau et al. 2013). By blocking the NF-κβ pathway, cells are not able to carry out the transcriptional activity to produce these kinds of molecules (Taniguchi and Karin

2018). Xie et al. (2012) also mentioned the anti-inflammatory effect of resveratrol. It is reported in their study that the stilbene inhibited the production of IL-6 in 9HTEo cell lines infected with the respiratory syncytial virus (RSV) by controlling the expression of Toll-like receptor 3 (TLR-3) and, thus, the suppression of Toll/IL-1R domain-containing adaptor-inducing IFN-β (TRIF). TRIF is an adaptor of TL3 and is involved in the activation of NF-κβ. As mentioned before, NF-κβ is necessary to produce different cytokines; in consequence, the changes in the expression TL-3 and TRIF may decrease the production of IL-6. Also, Xie et al. (2012) reported that resveratrol could suppress the expression of TANK-binding kinase 1 (TBK1). This molecule also participates in NF-κβ activation by inducing Iκβ degradation. Therefore, the suppression of TBK1 also is involved in the inhibition of IL-6 production. Other authors also explain the blockade of NF-κβ by resveratrol. For example, Faith et al. (2006) and Annunziata et al. (2018) reported that resveratrol blocks NF-κβ inhibiting the replication of herpes simplex virus (HSV). However, some reports suggest other antiviral mechanisms against HSV by resveratrol and its derivatives. For example, Chen et al. (2012) reported that a derivate from resveratrol was able to inhibit HSV infection by increasing the production of reactive oxygen species (ROS). Docherty et al. also proposed this mechanism against HSV by using resveratrol in cutaneous lesions. ROS play an essential role during the activation of the innate immune response during viral infection, including autophagy, gene expression, and signal transduction. It is well-known that polyphenols are antioxidants, but they can also act as pro-oxidants in a dose-dependent manner (Soares et al. 2018). That could explain the mechanism of action of resveratrol during the infection of HSV according to different authors. Another member that belongs to *Herpesviridae* family and has been inhibited by resveratrol is Epstein-Barr virus (EBV). According to Espinoza et al. (2012), resveratrol was able to interrupt the immortalization process associated with EBV infection in human B cells by inducing apoptosis in infected cells. This mechanism starts when resveratrol inhibits the expression of the latent membrane protein 1 (LPM1), which is essential for EBV-transformation of B cells since it acts as a homolog of CD40 and activates multiple cellular pathways such as NF-κβ, AKT, JNK, STAT-3, and p38. They are involved in the regulation of cellular apoptosis and proliferation, for example, by elevating the expression of Bcl2, an antiapoptotic protein (Saha and Robertson 2019). Furthermore, Espinoza et al. (2012) described that resveratrol blocks the activation of NF-κβ and suppresses the production of IL-6 and IL-10; both of these interleukins also contribute to the proliferation and survival of infected B cells.

Resveratrol has also counteracted other viral infections. Lin et al. (2015) reported that resveratrol was able to inhibit influenza A virus (IAV) replication by increasing the production of IFN-β in infected cells. They suggest that resveratrol enhances the expression of TLR-9, which is involved in the activation of interferon-alpha and beta (IFN-α/β) genes through IFN regulatory factor 7 (IRF7) phosphorylation (Sun and Reddy 2013). The activation of IRF7 mainly leads to the production of IFN-β. IFN-β can inhibit virus propagation and stimulate the adaptative immune response by promoting the MHC-I expression on various cell populations. MHC-I is essential for T-cell stimulation, differentiation, and expansion and killing of cells that have infected with the virus (Teijaro 2016). Summarizing this, resveratrol and IFN-β work synergistically to inhibit the replication of influenza A virus in lung cells. Even though IFN-α/β response is one of the first lines of defense against viral infections, it is reported that in some cases of infection by influenza viruses, this response is diminished due to these viruses employ mechanisms to evade and antagonize the effect through the nonstructural protein 1 (NS1) (Killip et al. 2015), PB1-F2 protein (Dudek et al. 2011), and viral polymerase (Iwai et al. 2010). Resveratrol has tested against other respiratory viruses. Mastromarino et al. (2015) studied the effect of this polyphenol facing a strain of human rhinovirus (HRV). They reported reduced levels of inflammatory cytokines such as IL-6, IL-8, and RANTES by using resveratrol. This reduction of interleukins is related to the blockage of NF-κβ. Also, it is worth to mention that RANTES and IL-8 are produced in high quantities during HRV infection comparing to other viruses such as adenovirus and resveratrol (Chun et al. 2013). The elevated production of these cytokines produces severe inflammation in the host. In particular, IL-8 acts as a mediator to activate neutrophils, and they contribute to airway obstruction following severe asthma (Krawiec et al. 2001; Wark et al. 2002). According to Mastromarino et al. (2015), resveratrol was also able to inhibit HRV replication by diminishing the expression of intercellular adhesion molecule 1 (ICAM-1), also known as CD54. During HRV infection, ICAM-1 is overexpressed and is used by HRV to enter into the respiratory epithelium and start the replication (Xing et al. 2003; Shukla et al. 2017).

The several studies that prove the immunomodulatory-antiviral role of resveratrol could lead to the development of alternative therapies for the treatment of viral diseases.

9.4.6 Resveratrol Effect in Metabolic Diseases

Obesity is a global health problem. The World Health Organization (WHO) indicates that obesity has almost tripled since 1975. In 2016, more than 1.9 billion adults over 18 years of age were reported overweight of these. The world health organization estimates that 650 million are obese. Obesity is an abnormal or excessive accumulation of fat that is detrimental to health (WHO 2020a).

Obesity damages people's health because it is a significant risk factor for developing cardiovascular diseases, insulin resistance, and type 2 diabetes mellitus (Fruh 2017). Obesity is characterized by the accumulation of fat in adipose tissue. However, there is a risk that fat accumulates in non-adipose tissues such as the liver and skeletal muscle. Causing other diseases such as the fatty liver and decreased sensitivity to insulin, it is related to reduced insulin sensitivity (de Ligt et al. 2015).

Resveratrol is a polyphenolic phytoalexin widely studied for multiple properties. One of these approaches is the use of resveratrol for the treatment of metabolic diseases. Therefore, the mechanisms of action of resveratrol in cells as a regulator of metabolism have been elucidated; however, it is not yet fully understood, and more human clinical studies are lacking (Szkudelska and Szkudelski 2010).

Resveratrol is an activator of the deacetylase sirtuin 1 (SIRT1) protein gene that is believed to regulate lipid and carbohydrate metabolism; its primary function is to maintain normal cell energy levels (Howitz et al. 2003). Resveratrol can activate the expression of SIRT1 directly or indirectly through the activation of AMP-activated protein kinase (AMPK) (Howitz et al. 2003; Baur et al. 2006). In addition to resveratrol, SIRT1 activation can be induced by exercise and calorie restriction in the diet. The resveratrol resembles the mechanism that physical activity and hypocaloric diet can perform in cellular metabolism (Ruderman et al. 2010).

The mitochondria of muscle cells show a reduction in the oxidative capacity of fat in obesity. Peroxisome proliferator-activated receptor γ co-activator 1 α (PGC-1 α) is a regulatory protein of mitochondrial biogenesis and metabolism. In addition, it binds to specific transcription factors that participate in the expression of mitochondrial proteins that increase their activity. The participation of PGC-1 α as an effector protein in the activation of the AMPK-SIRT1 signaling cascade of lipid and carbohydrate metabolism has been reported (Mihaylova and Shaw 2011).

A reduction in the expression of the PGC-1 α gene has been found in people with type 2 diabetes mellitus, suggesting that the decrease in the expression of PGC-1 α leads to a reduction in mitochondrial function and, consequently, a decrease in sensitivity to insulin in the muscle (Mootha et al. 2003). The physical activity and reduction of the calorie intake improve the function of the oxidative capacity of the mitochondria of the muscle cells and increase the sensitivity to insulin. Calorie restriction for 6 months increases the expression of AMPK, PGC-1 α, and SIRT1 and also reduces insulin levels (Civitarese et al. 2007).

Animal studies have been conducted to understand the effect of resveratrol on the oxidative activity of the mitochondria of muscle cells. Resveratrol influences mitochondrial biogenesis by activating the AMPK-SIRT1-PGC-1 α signaling pathway. Resveratrol induces a reduction in adipogenesis and decreases the size of adipocytes. This was tested in mice on a high-calorie diet (Lagouge et al. 2006). Resveratrol, combined with a hypercaloric diet, increases the survival and motor function of mice, in addition to changing gene expression towards normal expression in mice with a normal diet (Baur et al. 2006; Lagouge et al. 2006). Resveratrol has also been found to inhibit adipogenesis (Rayalam et al. 2008).

Diabetes mellitus is a chronic disease that occurs because the pancreas does not produce insulin or defects in the use of insulin. It is classified as type 1 and 2. Type 1 diabetes is caused by the destruction of the beta cells of the pancreas by autoantibodies, there is no insulin production. Type 2 diabetes is due to abnormal secretion and action of insulin. Both types cause hyperglycemia, which, over time and poor control, can damage organs (WHO 2020b).

Resveratrol regulates the metabolism of carbohydrates in different ways according to what has been documented, mainly by reducing the concentration of glucose in the blood; it has also been reported to improve the action of insulin and protect the beta cells of the pancreas (Szkudelski and Szkudelska 2011).

Animal models with diabetes have been used to assess the effect of resveratrol. Finding a decrease in the levels of glycated hemoglobin (HbA1c) reflects the reduction in blood glucose levels (Palsamy and Subramanian 2010). The reduction in blood glucose is because resveratrol increases glucose uptake by cells in the absence of insulin (Huang et al. 2020). Also, there are reports that it stimulates the expression of insulin-dependent glucose transporter (GLUT4), which was tested in a study in rats with induced diabetes and treated with resveratrol compared to the control group of rats with induced diabetes without resveratrol (Penumathsa et al. 2008). Insulin resistance occurs when cells do not respond to insulin. It mainly affects people who are overweight or obese. Exercise and a low-calorie diet decrease adipose tissue thus improving insulin sensitivity. A study in mice fed a high-fat diet in conjunction with resveratrol improved insulin sensitivity compared to the group that only received the high-fat diet without resveratrol (Baur et al. 2006).

On the other hand, a clinical study carried out in male patients with obesity found the administration of resveratrol, 75 mg twice a day for 30 days, had effects on carbohydrate metabolism, reducing the blood glucose and improving insulin sensitivity. In muscle cells, improvement in mitochondrial function by activating the AMPK-SIRT1- PGC-1a pathway decreased levels of triglycerides, inflammation markers, and intrahepatic lipids (Timmers et al. 2011).

Cardiovascular disease is a health problem. This group of diseases includes high blood pressure, coronary artery disease, and atherosclerosis. Genetic factors and environmental factors such as poor diet and sedentary lifestyle influence the development of these pathologies. In atherosclerosis, the use of resveratrol at physiological concentrations in the consumption of wine increases the expression of the enzyme NO synthase that participates in the production of nitric oxide (NO) that functions as a vasodilator in the endothelial cells that form the blood vessels and decreases the expression of endothelin (ET) that functions as a vasoconstrictor. Resveratrol works as a cardioprotective agent by inhibiting platelet aggregation. In atherosclerosis, resveratrol decreases the production of cytokines IL-6 and IL-8 and granulocyte-macrophage colony-stimulating factor (GM-CSF). Also, resveratrol prevents the oxidation of low-density lipoprotein (LDL) from preventing its adherence to the blood vessel. Hypertension is characterized by increased heart weight, elevated blood endothelin (ET) and angiotensin II, and decreased NO (nitric oxide). Resveratrol reduces heart weight, ET, and angiotensin II expression, while increasing NO concentration (Petrovski et al. 2011).

9.4.7 Resveratrol Effect in Autoimmune Diseases

The immune system is responsible for maintaining homeostasis in the body by protecting it from pathogenic microorganisms and developing abnormal cells through different mechanisms that lead to the activation of specialized cells. However, when these mechanisms are abnormal, the immune system attacks cells, organs, and tissues. Currently, more than 80 autoimmune diseases are known.

Autoimmune diseases happen by losing tolerance towards self-antigens, pro-inflammatory cells, and auto-antibodies (Ramos et al. 2015). Resveratrol as a possible therapeutic alternative for autoimmune diseases has had favorable results according to various studies.

Type I diabetes mellitus is a disease in which autoantibodies are produced that destroy the pancreas beta cells, which are cells specialized in the production of insulin, which is an essential hormone for the transport of glucose to the cells and its deficiency causes hyperglycemia. Hyperglycemia is related to a decrease in antioxidant agents and an increase in oxidative stress due to the activation of the NF-kappaB pathway, promoting pro-inflammatory cytokines and adhesion molecules (Chang et al. 2012).

The immunological process that takes place in this disease is pro-inflammatory activity. The activation of pro-inflammatory cytokines promotes apoptosis of beta cells of the pancreas islets in an environment of oxidative and inflammatory stress (Palsamy and Subramanian 2010). Resveratrol has antioxidant properties and can act by scavenging free radicals. It also has anti-inflammatory properties, since it suppresses the activation of the NF-kappaB pathway and expression of cytokines (Zhang Hanrui et al. 2009; Lee et al. 2011).

Resveratrol acts in different ways in DM1: it inhibits Th1 cells by binding to the chemokine receptor 6 (CCR6 receptor). Resveratrol binds to insulin to promote glucose uptake. Resveratrol activates the deacetylase sirtuin 1 (SIRT1) pathway, increasing oxidation during cellular stress to decrease the concentrations of reactive oxygen species (ROS) and prevent damage to the pancreas beta cells (de Oliveira et al. 2017).

Crohn's disease and ulcerative colitis are chronic and inflammatory diseases that affect the intestine, originate from a combination of genetic, environmental, immunological, and microbiological factors. These diseases are caused by changes in the intestinal mucosa, causing the release of bacteria that induce the immune system activation. Crohn's disease is the result of the combination of genetic factors that increase the susceptibility to the disease, environmental factors, and the intestinal microbiota, resulting in an abnormal immune response in the intestinal mucosa that affects the cells that line the intestine due to the inflammation produced by the immune response (Torres et al. 2017).

Activation of the NF-kappaB pathway occurs to produce cytokines in response to bacterial fragments recognized by NOD-like receptors, which causes the activation of Th1 and Th17 cells. In ulcerative colitis, the inflammation affects the colon, rectum, and distal colon. There is the activation of Th2 and Th17 lymphocytes, releasing pro-inflammatory mediators such as ROS and TNF-a (de Oliveira et al. 2017; Shi et al. 2017). Oxidative stress occurs in both diseases with the release of ROS, NO (nitric oxide), MDA (Malondialdehyde), and MPO (myeloperoxidase) and inhibition of antioxidant enzymes such as sod and GP (glutathione peroxidase). It was producing an inflammatory response. Resveratrol works by inhibiting inflammatory cytokines and neutralizing ROS. Preventive treatment with resveratrol works by reducing tissue damage and oxidative stress and was determined by measuring the reduction in malondialdehyde and the increase in glutathione peroxidase (Nunes et al. 2018).

There is also a decrease in pro-inflammatory cytokines. The microbiota increase improves the intestinal health bifidobacteria and lactobacilli (Larrosa et al. 2010). Studies in humans have found that an administration of 500 mg/day of resveratrol for 6 weeks interrupts the activation of NF-kappaB in mononuclear cells of peripheral blood. In plasma, there is a decrease in TNF-a and C-reactive protein (Samsamikor et al. 2016). Block the activation of TH1 lymphocytes by inhibiting the secretion of IL-1, IL-6, and TNF-a. Also, resveratrol induces immunosuppression of myeloid-derived suppressor cells (MDSCs) in the colon, suppressing the activity of effector lymphocytes (Singh et al. 2012).

Psoriasis is a skin disease associated with dysregulations in the interaction of the immune and adaptive system with skin cells. The mechanism of this disease is controlled by the immune system, mainly dendritic cells and T lymphocytes. Cathelicidin peptide LL-37 initiates it, and the host DNA forms complexes recognized by dendritic cells, which produce interferon-a and other inflammatory mediators such as TNF-a and interleukin-23. TNF-a induces secondary mediators and binding molecules that participate in psoriasis, activating the keratinocytes that release antimicrobial peptides, cytokines, and chemokines that act as attractants of immune cells. The most reported symptoms are pain, itching, and bleeding (Boehncke 2015).

Resveratrol has long been used for its anti-inflammatory properties. In vitro studies have shown that resveratrol induces cell death of the keratinocyte cell line HaCaT through the activation of SIRT1 that inhibits protein kinase B (Akt). This protein participates in the regulation of cell proliferation (Lee et al. 2016). In vivo studies in mice with induced psoriasis were treated with resveratrol, concluding that resveratrol reduce symptoms such as itching and decreases the production of cytokines that participate in the development of psoriasis such as IL-17 and IL-19 (Kjær et al. 2015).

Rheumatoid arthritis is a disease in which the immune system is involved. Immune complexes are created to mediate complement activation and cellular response against self-antigens. In its development, genetic factors that play an important role in the disease risk, severity, and progression are involved in this development, such as the genetic susceptibility of the HLA (human leukocyte antigen), HLA-DRB1, the gene encoding the non-receptor protein tyrosine phosphatase type 22 (PTPN22) and cytotoxic T-lymphocyte associated protein 4 (CTLA)-4) (de Oliveira et al. 2017). In addition to environmental factors such as stress, smoking and obesity (Firestein and McInnes 2017).

The disease is characterized by chronic inflammation of the joints that can destroy cartilage and bone erosion. In the immunological response, there is a loss of tolerance towards self-antigens. Therefore, there is an activation of T lymphocytes by macrophages. Adhesion molecules (CAMs) participate. Cytokines such as IFN-g, IL-12, IL-21, and IL-23. IL-6 are related to synovial inflammation, cartilage, and bone destruction. TNF-a acts on chondrocytes, which release matrix metalloproteinase (MMP) that degrades cartilage and bone destruction (de Oliveira et al. 2017).

Resveratrol has anti-inflammatory activity by modulating cytokine secretion, TNF-a, and IL-1b (Ma et al. 2015). In vitro studies with fibroblast-like synoviocytes (FLS) treated with resveratrol decrease the generation of ROS and COX/PGE2. It was concluding that resveratrol modulates the inflammatory response (Tsai et al. 2017). In vivo studies concluded that resveratrol administration intra-articularly in an animal model significantly reduces cartilage destruction and reduces inflammation (Elmali et al. 2007).

9.4.8 Resveratrol Effect in Degenerative Diseases

Resveratrol can cross the blood-brain barrier (BBB), which means it has a neuroprotective effect against neurodegenerative diseases (Zhang et al. 2018). The mechanisms of inflammation in the brain are closely linked to age; the main characteristic is the presence of reactive glial cells, which release pro-inflammatory mediators such as prostaglandins, cytokines, and reactive oxygen species; however, it is the response to maintain homeostasis of the brain (Liu et al. 2019).

Alzheimer's (AD) is a disease that causes dementia and decreases the cognitive capacity of patients (Ahmed et al. 2016); it is developed by the presence of dense β-amyloid plaques (Aβ) and neurofibrillary tangles in the CNS (Bastianetto et al. 2015). This disease causes the degeneration of neurons and, therefore, of the neuronal connections, causing an inflammatory environment throughout the brain area, and the activation of glial cells (phagocytic brain cells). The resveratrol had a protective effect by reducing the matrix metalloproteinase 9 (MMP9) of the cerebrospinal fluid, which decrease the permeability of the BBB and therefore limits the entry of leukocytes and inflammatory mediators to the brain (Moussa et al. 2017). Resveratrol will mitigate the action of NF-kB, inhibiting the secretion of PGE, TNF-α, and IL-6 and the expression of MCP-1 mRNA (Lu et al. 2010). Due to the accumulation of Aβ, the expression of nitric oxide synthetase (iNOS) is increased, which leads to cell apoptosis; this pro-inflammatory molecule is decreased with resveratrol in a dose-dependent manner. Resveratrol improves long-term memory by increasing brain-derived neurotrophic factor (BDNF); in addition, with advancing age, formaldehyde accumulates in the brain which is eliminated by resveratrol; this can restore memory (Liu et al. 2019). In addition to this, resveratrol binds to β-amyloid peptides, inhibiting their aggregation, and binds to β-amyloid plaques to break them, which reduces the progress of AD (Ahmed et al. 2016).

Parkinson's disease (PD) is a neurodegenerative disease characterized by the presence of Lewis bodies, formed by α-synuclein, causing the loss of dopaminergic neurons (Feng et al. 2018). The accumulation of Lewis bodies will cause an inflammation response, activating microglial cells which increase the production of pro-inflammatory mediators such as TNF-α and IL-1β, as well as the activation of astrocytes that release large amounts of species reactive oxygen which causes the death of neurons. Different effects of resveratrol are known, one of them is the decrease in miRNA-214 which is involved in the synthesis of α-synuclein (Wang et al. 2015) and another is reducing the levels of activated astrocytes and microglial, in a dose-dependent manner (Zhang et al. 2018). Liu (2019) indicates that resveratrol reduces the levels of IL-1β and TNF-α and inhibits cellular apoptosis by increasing the expression of Bcl-2 and decreasing the expression of Bax, an antiapoptotic molecule.

Age-related macular degeneration (AMD) is one of the main causes of vision loss in adults; the retinal pigment epithelium (RPE) is responsible for phagocytizing or degrading the photoreceptor segments in the eyes, which creates an environment of oxidative stress in addition to inflammation. Lançon (2016) reported that the treatment of AMD with resveratrol has an anti-inflammatory effect; it decreases the expression of IL-6, IL-8 and inhibits the production of cell adhesion molecules (ICAM-1) and protein monocyte chemoattractant (MCP-1), which reduces the accumulation of leukocytes, mainly neutrophils, in the affected area.

Chronic obstructive pulmonary disease (COPD) is a disease characterized by persistent respiratory symptoms and a very noticeable shortness of breath, which can be caused by a respiratory disease, alveolar abnormalities, or the parenchymal destruction of the lungs. Therefore, a severe inflammatory environment is generated in all the patient's airways, which aggravates the disease. Treatments carried out with resveratrol showed an increase in the expression of SIRT1 and PCG-1 mRNA, which are transcription factors that regulate the secretion of inflammatory mediators such as IL-6 or IL-8, which are markedly decreased; in addition, the resveratrol causes a decrease in the infiltration of white cells, freeing space in the pulmonary alveoli, thus improving air flow (Wang et al. 2017a, b).

Atherosclerosis is a thickening of the intimal layer of blood vessels caused by the recruitment of lymphocytes and monocytes when there is some injury to the vessel wall, generating a large amount of cytokines (Chen et al. 2018). This will decrease the lumen of the vessel and, therefore, the supplementation of oxygen, which causes the release of vascular endothelial growth factor (VEGF); therefore, inflammation and angiogenesis play an important role in this disease. Also, the C-reactive protein, which increased in atherosclerosis, promotes the expression of cell adhesion molecules (ICAM), stimulating the migration of leukocytes to the site and inducing the secretion of pro-inflammatory factors such as IL-6, IL-8, and NF-kB. Studies show that treatment with resveratrol was able to reduce serum concentrations of VEGF and CRP, decreasing endothelial permeability and stopping the formation of atherosclerotic plaques (Figueira and González 2018). In addition, Seo (2019) indicates a decrease in the expression of adhesion molecules in the vascular endothelium, which reduces the accumulation of leukocytes.

Osteoporosis is an age-related disease, caused by the resorption of bone tissue, usually by an imbalance in the activity of osteoblasts and osteoclasts present in the bones. Osteoblasts are the cells responsible for the formation of bones; therefore, their inactivation is the main cause. It has been reported that the overexpression or deletion of silent information regulator transcription 1 (SIRT1) increases or decreases bone mass; resveratrol is an activator of SIRT1, so it helps to have a balance between osteoblasts and osteoclasts, which has repercussions in the formation of bones. In addition, it has been registered that resveratrol helps maintain mineral density in bones (Yang et al. 2019). On the other hand, active osteoclasts are responsible for bone resorption, eliminating bone tissue and releasing minerals into the blood; the resveratrol will intervene in the activation of osteoclasts, stimulating the release of the inhibitory factor of osteoclastogenesis (OPG). The Fox01 protein functions for the inhibition of osteoclastogenesis and is required for the proliferation of osteoblasts; resveratrol improves Fox01 transcription by inhibiting the PI3K/AKT signaling cascade which keeps Fox01 phosphorylated, preventing its expression (Feng et al. 2018).

Duchenne muscular dystrophy (DMD) is a genetic disease linked to the X chromosome, which results in the translation of the dysfunctional protein dystrophin; this has an important role to join the intracellular cytoskeleton with the extracellular matrix of the muscles. Without this connection, the muscles are easily damaged causing a persistent state of inflammation. There will be regeneration/degeneration cycles until the muscle fibers lose that capacity and will be replaced by fibrous or fatty tissue. Treatment with resveratrol aims to increase the expression of the peroxisome proliferator-activated receptor-gamma coactivator (PGC-1α)-1 alpha, a transcriptional coactivator that targets the utrophin gene, a protein capable of taking the place of dystrophin. Gordon (2013) confirms the increase in utrophin mRNA and also indicated that there was a reduction in total immune cell infiltration, decreasing inflammation.

Amyotrophic lateral sclerosis (ALS) is a fatal neurodegenerative disease characterized by the selective loss of motor neurons in the brain and spinal cord causing muscle weakness and atrophy, spasms, paralysis, and respiratory failure until death (Mancuso et al. 2014). The causes of this pathology are mutations in SOD1, which causes the accumulation of free radicals; there is also an excessive accumulation of glutamate in the synapse that can kill neurons; in addition, there is an increase in the amount of free metals such as Cu, Zn, and Ca which generates oxidative stress; neurons also have mitochondrial dysfunction by the presence of vacuoles in them. Tellone et al. (2014) report that treating ALS with resveratrol, a favorable effect is obtained, resveratrol will activate STIR1 transcription factor that will deacetylate heat shock factor (HSF1) which reduces the death of neurons; in addition to this activation, cell death factor p53 is inhibited by increasing the expression of pro-apoptotic factors such as Bax. In addition to this, resveratrol activates PGC-1α by activating STIR1, which favors the formation of mitochondria and improves cellular respiration due to the induction of nuclear respiration factors (NRF) 1 and 2.

9.5 Bioavailability and Metabolism of Resveratrol

The recent attention for this compound is expanded by several epidemiological studies that demonstrate an inverse relationship among balanced consumption of different food as wines, peanuts, and select teas, especially for several disease applications (Chimento et al. 2019). Numerous studies have investigated the absorption, transport, and metabolism of resveratrol in vitro and ex vivo. Upon ingestion, resveratrol or its precursors go through the gastrointestinal tract, and it is estimated that around 70–75% of the intake of resveratrol is absorbed (Chaplin et al. 2018). Almost several resveratrol and derived metabolites have been described in animal or human plasma, urine, and various tissues, mostly generated by three major pathways: hepatic and intestinal glucuronidation, hepatic and intestinal sulfation, and gut microbial transformation (Luca et al. 2020). Initially, absorption of resveratrol was evaluated using isolated rat small intestine perfusion models, but other studies were employed about glucuronidation (G), and sulfation (S) in the duodenum was used to establish as comparative models as hepatic cells (Wenzel and Somoza 2005).

The biochemical structure of resveratrol conducts low water solubility (<0.05 mg/mL), which influences its absorption. On the other hand, to increase its solubility, require ethanol or organic solvents (Gambini et al. 2015). Resveratrol also demonstrated its ability to generate organic molecular complexes; these bioactives can be taken up by the intestine as xenobiotics and consequently cross the intestinal epithelium to the blood via a transcellular pathway; this route takes place through the enterocytes in the small intestine. Enterocytes are the first site of reported resveratrol metabolism after being internalized by either passive diffusion (Springer and Moco 2019). After absorption, resveratrol suffers serious and quick metabolization. It is essential to reference that no phase I reactions of resveratrol, such as oxidation, reductions, or hydrolyzes, were observed in the human systems (Wenzel and Somoza 2005). Once resveratrol is absorbed into the enterocyte, it undergoes phase II reactions of drug metabolism, producing polar metabolites, with simpler excretion in the body.

Resveratrol conjugation with sulfate was mediated by sulfotransferases (SULTs) and with glucuronate by uridine 5′-diphospho-glucuronosyltransferases (UGTs) (Springer and Moco 2019); on the other hand, intestinal bacteria also contribute to resveratrol metabolism, having the ability to convert it via hydrogenation to dihydroresveratrol (DHR), which is partly absorbed and further metabolized to conjugated forms (monosulfate DHR, monoglucuronide DHR) that can easily be eliminated in urine (Luca et al. 2020). The family of SULTs demonstrates a broad spectrum of diverse endogenous and exogenous substrate activity (1A1, 1A2, 1A3, and 1E1) and generates resveratrol-3′-*O*-sulfate (R3S), resveratrol-4′-*O*-sulfate (R4S), and resveratrol disulfates (RdS) as metabolite result by the sulfate group to a hydroxyl group in phenolic compounds. On the other hand, the UGTs are a large family of related enzymes involved on detoxification, which activity (1A1,1A9, 1A3, 1A6, 1A7, 1A8, and 1A10) catalyzes the conjugation of resveratrol with glucuronic acid, yielding principally resveratrol-3′-*O*-glucuronide (R3G) and

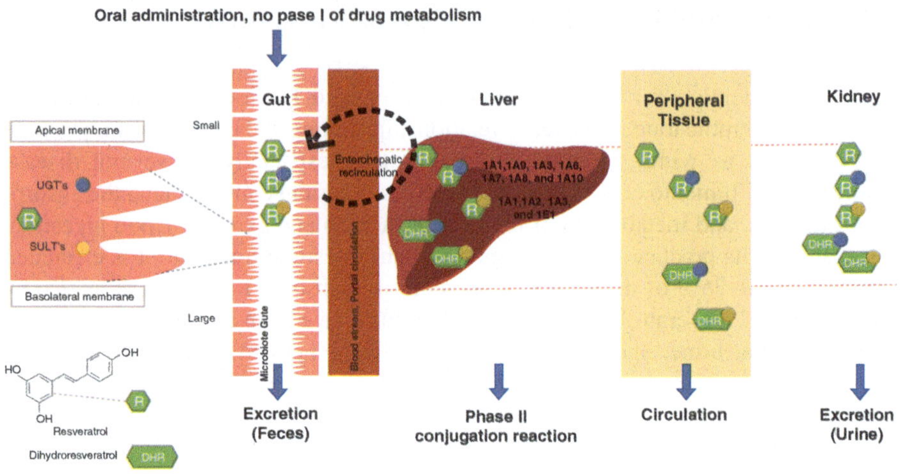

Fig. 9.2 Bioavailability and metabolism of resveratrol

resveratrol-4′-*O*-glucuronide (R4G) (Wang and Sang 2018). Figure 9.2 represented the metabolic fate and biotransformed resveratrol in the human gastrointestinal tract and metabolism in different organs. After conjugation, resveratrol sulfates and glucuronides have two possible fates; they can be transported through the apical membrane and extent the intestinal lumen, or they can pass through the basolateral membrane and enter the bloodstream. On both membranes, the enterocyte presents ABC (ATP-binding cassette) transporters, which are part of a large family of transport proteins that are considered instrumental in drug absorption and response diffusion (Springer and Moco 2019). On the apical and basolateral membrane, transporters are not limited to represent a role in the absorption and distribution of resveratrol metabolites in the small intestine (when resveratrol or his metabolites extent the bloodstream, they can be transferred by binding to blood proteins such as hemoglobin, albumin, and lipoproteins) as they are also expressed in several tissues, such as the liver or kidneys. Resveratrol demonstrate a high degree of lipophilicity; cant circulates single in plasma, required to binding proteins, such as albumins or low-density lipoproteins (LDL), occurs in a great extent (up to 90% of free resveratrol). However, despite being mostly bound to serum proteins, hepatic uptake is still efficient, the liver being considered the primary accumulation site of resveratrol and its metabolites (Gambini et al. 2015). Resveratrol is principally excreted in urine and feces. Total elimination after oral administration of ^{14}C-labeled resveratrol in humans was demonstrated around 70 to 98%, which is 53–85% in urine and 3.3–35% in feces. In urine, metabolites are considered 22–44% of the ingested dose, with 11–31% sulfate conjugates and 9–16% glucuronides (Walle et al. 2004; Tome-Carneiro et al. 2013).

The resveratrol oral administration is the ideal method for treatments in humans; it is known that plasma absorption of the unchanged resveratrol depends on the dosages consumed. Preclinical investigations elucidate the appropriate resveratrol oral dosage and bioavailability by human patients. Several studies demonstrated an oral dose of 25 mg of resveratrol, presented in plasma concentration around 1–5 ng/mL of resveratrol (Chimento et al. 2019). Administration of higher dosages (up to 5 g) indicated the increase in the intensity of unchanged resveratrol up to 530 ng/mL, demonstrating how, after a high resveratrol dosage, only a low quantity of the unchanged resveratrol is present in the plasma. Even if resveratrol seems like to be well-accepted, tolerated, and safe, administration of higher oral doses does not consent to improve therapeutic effects, but, as an alternative, perhaps the cause of the side effects seen at the dose of 1 g/kg (body weight) including diarrhea, abdominal pain, and nausea.

Consequently, based on the findings from clinical studies, it appears that the main obstacle that must be overcome to consider resveratrol as a therapeutic agent is its low bioavailability (Boocock et al. 2007; Almeida et al. 2009; Patel et al. 2011). Wenzel and Somoza in 2005 conclude that resveratrol, in human patients, whether as aglycone or in its glycosidic form, is also absorbed quite fast after oral consumption and resveratrol levels were readily measurable in plasma and urine with the highest level plasma concentrations around 30 to 60 min after ingestion.

In the present day, scientific resveratrol investigation in the world consists of around 140 clinical trials in humans by disease treatments (clinicaltrials.gov), and more than 10,000 scientific papers consist of realizing an effort to elucidate different employs, uses, and effects of resveratrol. Due to differing doses, disparate experimental setups, low statistical power, and a myriad of biological or other types of confounders, the fate and effect of resveratrol in humans remain indefinable. Mechanisms of action are diverse and of potential use for obesity, cardiovascular health, inflammation, metabolic health, antioxidant activity, and cancer management. Therefore, resveratrol is of wide pleiotropy. Systems conducted approaches that help map the impacts of multitargeted compounds such as resveratrol and emphasize links among phase II metabolism, redox pathways, and bioenergetics, linked by protein targets regulated by resveratrol (Springer and Moco 2019).

However, despite these factual scientific data, the resveratrol paradox (defined as low bioavailability and high bioactivity) still raises serious questions, as the final responsible mechanisms incriminated for the observed effects have not yet been elucidated. Since oral bioavailability is mainly dependent on aqueous solubility, membrane permeability, and metabolic stability, these factors are further addressed below to assess its problematic bioavailability. Despite the advances, the scientific community still has yet to continue with efforts to conduct more extensive and more deeply characterized studies which could be an aim of the research community in order to bridge the current gaps in knowledge on resveratrol metabolism and beneficial effects.

9.6 Therapy Resveratrol Carriers

Resveratrol, as described, is a well-studied natural compound with great therapeutic potential (Wiciński et al. 2020). Despite promising results, the widespread use of resveratrol has had limited success, largely due to its instability, low bioavailability and solubility, as well as insufficient systemic administration (Soo et al. 2016). Two geometric isomers of resveratrol are known: *trans* and *cis*. Both isomers often exist in combination. The *trans* form is more biologically active, and this isomer can become *cis* under ultraviolet light. Resveratrol is characterized by low half-life (less than 0.25 h). So, for its proper action, it is difficult to ensure the therapeutic concentration. Previous in vitro studies have shown that 5μmol/L (Sessa et al. 2014) resveratrol is the minimum concentration required to achieve its chemopreventive effects. It is known that 75% of resveratrol is possibly absorbed by transepithelial diffusion under oral administration in humans. However, due to the rapid and extensive metabolism in the intestine and liver, oral bioavailability is low (<1%) (Christakis Sergides et al. 2016). The drug suffers enterohepatic recirculation and extensive first-pass metabolism by CYP3A4 in the liver, resulting in very low bioavailability. Oral administration of repeated or increased doses does not improve compound levels. The pharmacokinetics of resveratrol have shown a circadian variation, with greater bioavailability after morning administration (Christakis Sergides et al. 2016).

Resveratrol glucuronides and sulfates are the main metabolites detected in urine and plasma, in addition to reduced dihydro-resveratrol conjugates and other unknown highly polar products (Pecyna et al. 2020). In addition to metabolites generated in the liver and intestine, the impact on resveratrol transformation can be assigned to colonic bacterial activity (dos Santos et al. 2019). By controlling the activity of β-glucuronidase and sulfatase enzymes, the efficacy of resveratrol at the target site can be improved by tissue-specific accumulation of resveratrol (Lu et al. 2013; Pecyna et al. 2020).

Lu et al. (2009) propose the administration of resveratrol in a stable form conjugated with sulfate, achieving the generation of parent compound gradually for the beneficial effects in vivo. The activities of resveratrol metabolites were also demonstrated, these being pharmacologically active in vivo (Kiskova et al. 2020).

So, pharmacokinetic properties of resveratrol such as low solubility, rapid degradation, and extensive metabolism are known. This translates into insufficient bioavailability in its oral application. Furthermore, it is known that under exposure to light, trans-resveratrol, the biologically active isomer, is rapidly converted to cis-resveratrol (Singh et al. 2014). This presents several problems for the application of resveratrol in therapy that can be solved by such strategies as the co-administration of enzymatic inhibitors of metabolism, the synthesis and use of its analogues, as well as the design of new administration systems (Silva et al. 2014). Moreover, to overcome the described physicochemical and pharmacokinetic restrictions, nanotechnology presents a powerful strategy considering the nanoencapsulation of resveratrol (Wang et al. 2011; Soo et al. 2016). Nanotechnology makes it possible to obtain nanoparticles loaded with biologically active compounds. The use of nanoparticles was approved by the United States Food and Drug Administration (FDA) in 1983. The particles are normally classified according to their size: normally the diameter of nanoparticles loaded with resveratrol is less than 1000 nm, and only in some cases is the diameter of ultrafine particles less than 100 nm (Ansari et al. 2011; Keservani et al. 2017).

The nanoencapsulation of this bioactive natural compound as one of the chemotherapeutic agents provides many advantages in control of degradation and in the interaction with biological systems, as well as improvement in bioavailability, retention time, absorption, and intratissue and intracellular penetration (Penalva et al. 2015; Sharma et al. 2018). Several nanoformulations focused on enhancing therapeutic potential of resveratrol have been developed Keservani et al. 2021) (Table 9.2).

Initially, the nanoencapsulation of resveratrol was performed using chitosan nanoparticles. However, the release of resveratrol was limited in the presence of solidifying agents (Kiskova et al. 2020).

Nanosystems with biocompatible and biodegradable polymers have been applied for resveratrol entrapment (Sanna et al. 2013). These systems made it possible to improve the solubilization of the drug and protect it from degradation. Moreover, different systems of nanoparticles loaded with resveratrol cause a greater effect on the viability of cancer cells even in lower doses than free resveratrol (Table 9.2), mediated by the discrepancy of intracellular reactive oxygen species. Often times, nanocarriers coated with poly (ethylene glycol) (PEG) on their surface as well as mPEG-poly(caprolactone)-based nanoparticles allowed compound accumulation in tumors through the enhanced effect of permeability and retention (Wang et al. 2011; Sanna et al. 2013) (Table 9.2).

Resveratrol-loaded nanoparticles protected cells from damage induced by beta-amyloid peptide (Abeta) in a dose-dependent manner by attenuating intracellular oxidative stress and caspase-3 activity (Lu et al. 2009, 2012) (Table 9.2). Solid lipid nanoparticles (SLN) with a size smaller than 180 nm showed the ability to quickly cross the cell membrane, were distributed throughout the cytosol and moved successively between different cell levels, and were located in the perinuclear region without inducing cytotoxicity. Moreover, in these nanocarriers, solubility, stability, and intracellular delivery were all increased by loading into SLN (Shidhaye et al. 2008; Pandita et al. 2014).

In vitro assays revealed that liposome-encapsulated curcumin and resveratrol successfully inhibited cell growth and induced apoptosis by inhibition of phosphatase and tensin homolog (PTEN), including p-Akt, cyclin D1, mammalian rapamycin, and RA (Narayanan et al. 2009).

All studied nanocarriers are suitable to be used for the delivery of bioactive resveratrol (Table 9.2). The encapsulation of resveratrol in polymeric nanoparticles allows the optimization of its charge, an effective controlled release, and protection against transformation by exposure to light, which opens new perspectives for its use in (nano) chemoprevention/chemotherapy. Thus, solid lipid nanoparticles and nanostructured lipid carriers enhance the oral bioavailability of resveratrol (Gokce et al. 2012; Jose et al. 2014).

The anticancer activity and the molecular mechanism of subcutaneously implanted human ovarian primary carcinoma cells in nude mice were demonstrated for resveratrol-bovine serum albumin nanoparticles. The growth of carcinomas was inhibited to a greater degree when compared to free resveratrol due to the mitochondrial apoptotic pathway activated by applied treatment (Neves et al. 2013). An

Table 9.2 Nanoformulations for resveratrol delivery to enhance its therapeutic effect

Delivery system	Characteristics	References
Poly(lactide) (PLA) nanoparticles	Resveratrol-loaded polysorbate 80-coated poly(lactide) nanoparticles (but not free resveratrol) demonstrated neuroprotective effect in mice with induced Parkinson's disease	Da Rocha et al. (2015)
Poly-(lactic-co-glycolic) acid (PLGA) nanoparticles	Transferrin-conjugated polyethylene glycol-polylactic acid nanoparticles conjugated with resveratrol showed a therapeutic effect for glioma	Guo et al. (2013)
Poly(epsilon-caprolactone) (PCL) nanoparticles	Nanosystems, composed of a biocompatible blend of poly(epsilon-caprolactone) and poly (D,L-lactic-*co*-glycolic acid)-poly(ethylene glycol) conjugate, had antiproliferative efficacy on prostate cancer cells	Sanna et al. (2013)
Polymeric micelles	Resveratrol-loaded polymeric micelles basing on amphiphilic block copolymer protected PC12 cells from Abeta-induced damage in a dose-dependent manner by attenuating intracellular oxidative stress and caspase-3 activity	Lu et al. (2009)
Nanoparticles based on zein, an amphiphilic protein	Zein-based nanoparticles provided high and prolonged plasma levels of the polyphenol, increase in oral bioavailability of resveratrol, decrease of the endotoxic symptoms in mice treated with LPS, decrease in serum tumor necrosis factor-alpha (TNF-α) in comparison with control	Penalva et al. (2015)
Lecithin-based nanoemulsions	Lecithin-based nanoemulsions were able to transport resveratrol through cell monolayers in characteristically shorter times (1–6 h) than those required for their metabolization (3–12 h), allowing for better preservation of the integrity of the emulsion droplets, protecting the resveratrol against chemical degradation, providing the capability of nanoemulsions in sustained release of resveratrol	Sessa et al. (2014)
Cyclodextrins	Betta cyclodextrin-based nanosponges were obtained for buccal delivery and topical application by hyper-cross-linked cyclodextrin polymers with carbonyldiimidazole forming three-dimensional systems. Formulation increased resveratrol solubility, cytotoxicity against HCPC-I cells, permeation in pigskin, and accumulation in rabbit mucosa. Resveratrol solubility was improved by the formation of inclusion complexes with hydropropyl-cyclodextrins and micellar systems. Stable resveratrol system was achieved to incorporate it in the food and pharmaceutical products.	Ansari et al. (2011) Silva et al. (2014) Venuti et al. (2014)

(continued)

Table 9.2 (continued)

Delivery system	Characteristics	References
	Resveratrol/sulfobutylether-β-cyclodextrin inclusion complex was obtained with high affinity for resveratrol with molar ratio 1:1 and high stability	
Liposomes	Resveratrol was stablished in formulations obtained by reverse-phase evaporation method with or without poly (ethylene glycol-2000)—grafted distearoyl phosphatidylethanolamine. Resveratrol-loaded liposomes improve anticancer effect A conjugate of dequalinium polyethylene glycol-distearoylphosphatidylethanolamine with resveratrol was obtained as a complex targeting mitochondria. The complex was considered a potential treatment for lung cancers that induces apoptosis through the mitochondrial signaling pathway	Lu et al. (2012) Wang et al. (2011)
Dual nanoencapsulation in cyclodextrin inside liposomes	Resveratrol was encapsulated in cyclodextrin inclusion complexes in the lipophilic and hydrophilic compartments of the liposomes as an example of a dual carrier system	Soo et al. (2016)
Solid lipid nanoparticles (SLNs) and nanostructured lipid carriers (NLCs)	Solid lipid nanoparticles (SLN) and resveratrol-loaded nanostructured lipid carriers (NLC) were obtained by a modified hot homogenization technique. In vitro resveratrol remained associated with lipid nanoparticles after incubation in digestive fluids for several hours Resveratrol-coated stearic acid-based SLNs coated with poloxamer 188 were obtained by the solvent diffusion and solvent evaporation method. The lipid formulation improved at 8.035-fold the oral bioavailability of resveratrol compared to suspension of the drug	Neves et al. (2013) Pandita et al. (2014)

antiglioma effect was observed in the presence of trans-resveratrol-loaded lipid core nanocapsules in vitro and in vivo. A greater decrease in the viability of C6 glioma cells was observed with encapsulated than free resveratrol, without toxicity to healthy neural cells. The treatment caused apoptotic cell death through early arrest in the S and G1 phases of the cell cycle. This treatment led to a marked decrease in tumor size and reduced the incidence of some features associated with malignant tumors compared to resveratrol in solution in vivo (Guo et al. 2013).

Resveratrol-loaded nanoparticles based on poly(e-caprolactone) and poly(D, L-lactic-co-glycolic acid)-poly(ethylene glycol) were used for prostate cancer treatment. Resveratrol released under simulation from gastrointestinal fluids was 55% resveratrol in the first 2 h in an acid medium and reached 100% in the next 5 h at pH 7.4. Confocal microscopy revealed that CaP cell lines efficiently absorbed these nanoparticles. Treatment of androgen-independent and hormone-sensitive prostate cancer cells was more efficient in nanosystem than free resveratrol (Sanna et al. 2013).

Resveratrol chemopreventive effect was observed as its ability to counteract CSC-induced DNA fragmentation and to activate protective apoptosis in bronchial epithelial cells in both in vitro and ex vivo experiments. Nanoparticles reduced the toxicity of resveratrol and increased its ability to counteract the cytotoxicity of CSC (Da Rocha et al. 2015; Beloqui et al. 2016).

Resveratrol is a suitable treatment for chemoprevention and is a viable option for the control of cancer in the early stages of the carcinogenesis process and also in other degenerative diseases. To take advantage of the beneficial therapeutic potential, pharmacokinetic problems are solved using resveratrol prodrugs and nanostructured delivery systems. The use of the latter leads to improvement in bioavailability, intracellular penetration, and control of administration, as well as protection against degradation and reduction of potential toxicity. Currently, studies are being developed to approve the use of resveratrol-based nanosystems in clinical practice, which will drive the evolution of innovative nanodevices capable of consolidating the chemopreventive potential of resveratrol.

9.7 Conclusions

The biological effects reported from the studies of resveratrol have shown that it is a molecule with an important capacity to generate effects in diverse mechanisms of action of a great variety of cellular models in vitro and in experimental models in vivo. All these studies reveal the importance of this molecule as a target for treatment and even prevention of various diseases that currently exist, such as metabolic, degenerative, cardiovascular, inflammatory, carcinogenic, and neurological diseases and even infections by etiological agents such as viruses and bacteria.

Scientific evidence indicates that the main effects they generate in tumor cells, diseased or infected cells, are related to the chemical and immunological sensitization that it causes, activating different signaling pathways and molecular mechanisms that favor the therapeutic response, even activating the cells themselves of the host to generate the immunological effect. Some of these mechanisms occur through inflammatory processes, activating cells of innate and adaptive immunity to promote their maturation and act as effector cells. In addition, it can stimulate the secretion of pro-inflammatory cytokines, promote apoptosis, activate lipid and carbohydrate metabolism, and even trigger specific responses through the expression of specific proteins and the activation of genes.

All these promising effects of resveratrol place this molecule of natural origin as one of the most promising nutraceuticals for the treatment of various pathologies, having the advantage that being a molecule of natural origin, its bioavailability in nature allows it to be easy access for consumption. However, more studies will always be necessary to further understand the mechanisms on in vivo trials and even clinical trials of resveratrol, since there are also reports of antagonistic effects of resveratrol, and this could influence the desired effects in conjunction with physiological conditions of the organisms in which it is evaluated.

Acknowledgments A special thanks to the Research and Postgraduate Department of the Autonomous University of Coahuila (UAdeC). Also a special thanks to the Laboratorio Nacional BIOBANCO CONACYT (LANBIOBAN) of UAdeC/UANL.

Conflicts of Interest The authors declare no conflict of interest.

References

Abba Y, Hassim H, Hamzah H, Noordin MM (2015) Antiviral activity of resveratrol against human and animal viruses. Adv Virol 2015:1–7

Ahmed T, Javed S, Javed S et al (2016) Resveratrol and Alzheimer's disease: mechanistic insights. Mol Neurobiol 54:2622–2635

Ajith Y, Dimri U, Dixit SK et al (2017) Immunomodulatory basis of antioxidant therapy and its future prospects: an appraisal. Inflammopharmacology 25:487–498

Almeida L, Vaz-da-Silva M, Falcão A et al (2009) Pharmacokinetic and safety profile of trans-resveratrol in a rising multiple-dose study in healthy volunteers. Mol Nutr Food Res 53:7–15

Annunziata G, Maisto M, Schisano C et al (2018) Resveratrol as a novel anti-herpes simplex virus nutraceutical agent: an overview. Viruses 10:473–483

Ansari KA, Vavia PR, Trotta F et al (2011) Cyclodextrin-based nanosponges for delivery of resveratrol: *in vitro* characterization, stability, cytotoxicity and permeation study. AAPS J Am Assoc Pharmaceut Scientists 12:279–286

Asati V, Mahapatra DK, Bharti SK (2016) PI3K/Akt/mTOR and Ras/Raf/MEK/ERK signaling pathways inhibitors as anticancer agents: structural and pharmacological perspectives. Eur J Med Chem 109:314–341

Banez MJ, Geluz MI, Chandra A et al (2020) A systemic review on the antioxidant and anti-inflammatory effects of resveratrol, curcumin, and dietary nitric oxide supplementation on human cardiovascular health. Nutr Res 78:11–26

Bastianetto S, Ménard C, Quirion R (2015) Neuroprotective action of resveratrol. Biochim Biophys Acta (BBA) - Mol Basis Dis 1852:1195–1201

Baur JA, Pearson KJ, Price NL et al (2006) Resveratrol improves health and survival of mice on a high-calorie diet. Nature 444:337–342

Beloqui A, Solinis MA, Rodriguez-Gascon A et al (2016) Nanostructured lipid carriers: promising drug delivery systems for future clinics. Nanomedicine 12:143–161

Berman A, Motechin R, Wiesenfeld M et al (2017) The therapeutic potential of resveratrol: a review of clinical trials. NPJ Precision Oncol 1:1–9

Bhatt SR, Lokhandwala MF, Banday AA (2011) Resveratrol prevents endothelial nitric oxide synthase uncoupling and attenuates development of hypertension in spontaneously hypertensive rats. Eur J Pharmacol 667:258–264

Bishayee A, Barnes KF, Bhatia D et al (2010) Resveratrol suppresses oxidative stress and inflammatory response in diethylnitrosamine-initiated rat hepatocarcinogenesis. Cancer Prev Res 3:753–763

Boehncke W-H (2015) Etiology and pathogenesis of psoriasis. Rheum Dis Clin 41:665–675

Boocock DJ, Faust GES, Patel KR et al (2007) Phase I dose escalation pharmacokinetic study in healthy volunteers of resveratrol, a potential cancer chemopreventive agent. Cancer Epidemiol Biomark Prev 16:1246–1252

Borge AM, Solano MV, Valverde JPS (2013) El resveratrol y sus posibles usos como nueva terapia farmacológica. Rev Méd Costa Rica Centro América LXX 608:679–684

Buhrmann C, Shayan P, Goel A, Shakibaei M (2017) Resveratrol regulates colorectal cancer cell invasion by modulation of focal adhesion molecules. Nutrients 9:1073–1095

Calvo M, Corral P, Lopez G et al (2017) El resveratrol protege contra la respuesta inflamatoria en páncreas secundaria al envejecimiento. Rev Complutense Ciencias Veterinarias 11:216–221

Cao X, Luo T, Luo X, Tang Z (2014) Resveratrol prevents AngII-induced hypertension via AMPK activation and RhoA/ROCK suppression in mice. Hypertens Res 37:803–810

Catalogna G, Moraca F, D'Antona L et al (2019) Review about the multi-target profile of resveratrol and its implication in the SGK1 inhibition. Eur J Med Chem 183:111675

Chai R, Fu H, Zheng Z et al (2017) Resveratrol inhibits proliferation and migration through SIRT1 mediated post-translational modification of PI3K/AKT signaling in hepatocellular carcinoma cells. Mol Med Rep 16:8037–8044

Chang C, Chang C, Huang J, Hung L (2012) Effect of resveratrol on oxidative and inflammatory stress in liver and spleen of streptozotocin-induced type 1 diabetic rats. Chin J Physiol 55:192–201

Chaplin A, Carpéné C, Mercader J (2018) Resveratrol, metabolic syndrome, and gut microbiota. Nutrients 10:1–29

Chen X, Qiao H, Liu T et al (2012) Inhibition of herpes simplex virus infection by oligomeric stilbenoids through ROS generation. Antivir Res 95:30–36

Chen J, Cao X, Cui Y et al (2018) Resveratrol alleviates lysophosphatidylcholine-induced damage and inflammation in vascular endothelial cells. Mol Med Rep 17:4011–4018

Chimento A, De Amicis F, Sirianni R et al (2019) Progress to improve oral bioavailability and beneficial effects of resveratrol. Int J Mol Sci 20:1381

Choi AMK, Ryter SW, Levine B (2013) Mechanisms of disease: autophagy in human health and disease. N Engl J Med 368:651–662

Christakis Sergides C, Chirilă M, Silvestro L et al (2016) Bioavailability and safety study of resveratrol 500 mg tablets in healthy male and female volunteers. Exp Ther Med 11(1):164–170

Chun YH, Park JY, Lee H et al (2013) Rhinovirus-infected epithelial cells produce more IL-8 and rantes compared with other respiratory viruses. Allergy, Asthma Immunol Res 5:216–223

Civitarese AE, Carling S, Heilbronn LK et al (2007) Calorie restriction increases muscle mitochondrial biogenesis in healthy humans. PLoS Med 4:e76

Da Rocha LG, Bonfanti Santos D, Colle D et al (2015) Improved neuroprotective effects of resveratrol-loaded polysorbate 80-coated poly(lactide) nanoparticles in MPTP-induced parkinsonism. Nanomedicine (Lond) 10:1127–1138

De Ligt M, Timmers S, Schrauwen P (2015) Resveratrol and obesity: can resveratrol relieve metabolic disturbances? Biochim Biophys Acta (BBA) - Mol Basis Dis 1852:1137–1144

de Oliveira ALB, Monteiro VVS, Navegantes-Lima KC et al (2017) Resveratrol role in autoimmune disease—a mini-review. Nutrients 9:1306

De Sá-Coutinho D, Pacheco M, Frozza R et al (2018) Anti-inflammatory effects of resveratrol: mechanistic insights. Int J Mol Sci 19:1812–1837

Dolinsky VW, Chakrabarti S, Pereira TJ et al (2013) Resveratrol prevents hypertension and cardiac hypertrophy in hypertensive rats and mice. Biochim Biophys Acta Mol basis Dis 1832:1723–1733

Dos Santos AS, Rodrigues de Albuquerque TM, de Brito Alves JL et al (2019) Effects of quercetin and resveratrol on *in vitro* properties related to the functionality of potentially probiotic *Lactobacillus* strains. Front Microbiol 10:2229–2231

Dudek SE, Wixler L, Nordhoff C et al (2011) The influenza virus PB1-F2 protein has interferon-antagonistic activity. Biol Chem 392:1135–1144

Elmali N, Baysal O, Harma A et al (2007) Effects of resveratrol in inflammatory arthritis. Inflammation 30:1–6

Espinoza JL, Takami A, Trung LQ et al (2012) Resveratrol prevents EBV transformation and inhibits the outgrowth of EBV-immortalized human B cells. PLoS One 7:e513061–e513067

Faith SA, Sweet TJ, Bailey E et al (2006) Resveratrol suppresses nuclear factor-κB in herpes simplex virus infected cells. Antivir Res 72:242–251

Fan Y, Chiu JF, Liu J et al (2018) Resveratrol induces autophagy-dependent apoptosis in HL-60 cells. BMC Cancer 18:1–10

Feng Y, Jiang X, Ma F et al (2018) Resveratrol prevents osteoporosis by upregulating FoxO1 transcriptional activity. Int J Mol Med 41:202–212

Figueira L, González J (2018) Efecto del resveratrol sobre las concentraciones séricas del factor de crecimiento endotelial vascular durante la aterosclerosis. Clín Investig Arteriosclerosis 30(5): 209–216

Firestein GS, McInnes IB (2017) Immunopathogenesis of rheumatoid arthritis. Immunity 46:183–196

Fruh SM (2017) Obesity: risk factors, complications, and strategies for sustainable long-term weight management. J Am Assoc Nurse Pract 29:S3–S14

Fu Y, Chung F-L (2018) Oxidative stress and hepatocarcinogenesis. Hepatoma Res 4:39–50

Fu S, Lv R, Wang L et al (2018) Resveratrol, an antioxidant, protects spinal cord injury in rats by suppressing MAPK pathway. Saudi J Biol Sci 25:259–266

Gambini J, López-Grueso R, Olaso-González G et al (2013) Resveratrol: distribución, propiedades y perspectivas. Rev Esp Geriatr Gerontol 48(2):79–88

Gambini J, Inglés M, Olaso G et al (2015) Properties of resveratrol: in vitro and in vivo studies about metabolism, bioavailability, and biological effects in animal models and humans. Oxidative Med Cell Longev 2015:1–13

Gokce EH, Korkmaz E, Dellera E et al (2012) Resveratrol-loaded solid lipid nanoparticles *versus* nanostructured lipid carriers: evaluation of antioxidant potential for dermal applications. Int J Nanomedicine 7:1841–1850

Gordon B, Delgado D, Kostek M (2013) Resveratrol decreases inflammation and increases utrophin gene expression in the mdx mouse model of duchenne muscular dystrophy. Clin Nutr 32(1): 104–111

Grujic-Milanovic J, Miloradovic Z, Jovovic D et al (2017) The red wine polyphenol, resveratrol improves hemodynamics, oxidative defence and aortal structure in essential and malignant hypertension. J Funct Foods 34:266–276

Guo W, Li A, Jia Z et al (2013) Transferrin modified PEG-PLA-resveratrol conjugates: *in vitro* and *in vivo* studies for glioma. Eur J Pharmacol 718:41–47

Hanrui Z, Jing Z, Zoltan U, Cuihua Z (2009) Resveratrol improves endothelial function. Arterioscler Thromb Vasc Biol 29:1164–1171

Hara Y, Torii R, Ueda S et al (2018) Inhibition of tumor formation and metastasis by a monoclonal antibody against lymphatic vessel endothelial hyaluronan receptor 1. Cancer Sci 109:3171–3182

Heo JR, Kim SM, Hwang KA et al (2018) Resveratrol induced reactive oxygen species and endoplasmic reticulum stress-mediated apoptosis, and cell cycle arrest in the A375SM malignant melanoma cell line. Int J Mol Med 42:1427–1435

Howitz KT, Bitterman KJ, Cohen HY et al (2003) Small molecule activators of sirtuins extend *Saccharomyces cerevisiae* lifespan. Nature 425:191–196

Huang D-D, Shi G, Jiang Y et al (2020) A review on the potential of resveratrol in prevention and therapy of diabetes and diabetic complications. Biomed Pharmacother 125:109767

Inglés M, Gambini J, Miguel MG et al (2014) PTEN mediates the antioxidant effect of resveratrol at nutritionally relevant concentrations. Biomed Research International

Iwai A, Shiozaki T, Kawai T et al (2010) Influenza A virus polymerase inhibits type I interferon induction by binding to interferon β promoter stimulator 1. J Biol Chem 285:32064–32074

Jang M, Cai L, Udeani GO et al (1997) Cancer chemopreventive activity of resveratrol, a natural product derived from grapes. Science 275:218–220

Jiang N, Dai Q, Su X et al (2020) Role of PI3K/AKT pathway in cancer: the framework of malignant behavior. In: Molecular biology reports. Springer, Cham, pp 4587–4629

Jose S, Anju SS, Cinu TA et al (2014) *In vivo* pharmacokinetics and biodistribution of resveratrol-loaded solid lipid nanoparticles for brain delivery. Int J Pharm 474:6–13

Juan ME, Wenzel U, Daniel H, Planas JM (2008) Resveratrol induces apoptosis through ROS-dependent mitochondria pathway in HT-29 human colorectal carcinoma cells. J Agric Food Chem 56:4813–4818

Keservani RK, Kesharwani RK, Sharma AK (2017) Introduction to nanotechnology in drug delivery. In: Drug delivery approaches and nanosystems. Apple Academic, pp 1–19. ISBN: 9781771885836

Keservani RK, Kesharwani RK, Sharma AK (2021) Nutraceuticals. In: Enhancing the therapeutic efficacy of herbal formulations. IGI Global, pp 241–266. ISBN13: 9781799844532

Kesharwani RK, Misra K (2010) Prediction of binding site for curcuminoids at human topoisomerase II α protein; an in silico approach. Curr Sci 101(8):1060–1064

Killip MJ, Fodor E, Randall RE (2015) Influenza virus activation of the interferon system. Virus Res 209:11–22

Kimura Y, Sumiyoshi M (2016) Resveratrol prevents tumor growth and metastasis by inhibiting lymphangiogenesis and M2 macrophage activation and differentiation in tumor-associated macrophages. Nutr Cancer 68:667–678

Kiskova T, Kubatka P, Büsselberg D et al (2020) The plant-derived compound resveratrol in brain cancer: a review. Biomol Ther 10(1):161–164

Kjær TN, Thorsen K, Jessen N et al (2015) Resveratrol ameliorates imiquimod-induced psoriasis-like skin inflammation in mice. PLoS One 10:e0126599

Ko JH, Sethi G, Um JY et al (2017) The role of resveratrol in cancer therapy. Int J Mol Sci 18:1–36

Kong D, Chen F, Sima N (2015) Inhibition of focal adhesion kinase induces apoptosis in bladder cancer cells via Src and the phosphatidylinositol 3-kinase/Akt pathway. Exp Ther Med 10: 1725–1731

Krawiec M, Westcott J, Hong WC et al (2001) Persistent wheezing in very young preschool children reflects lower respiratory inflammation. Am J Respir Crit Care Med 163:1290–1291

Kuršvietienė L, Stanevičienė I, Mongirdienė A, Bernatonienė J (2016) Multiplicity of effects and health benefits of resveratrol. Medicina 52:148–155

Lagouge M, Argmann C, Gerhart-Hines Z et al (2006) Resveratrol improves mitochondrial function and protects against metabolic disease by activating SIRT1 and PGC-1α. Cell 127:1109–1122

Lançon A, Frazzi R, Latruffe N (2016) Anti-oxidant, anti-inflammatory and anti-angiogenic properties of resveratrol in ocular diseases. Molecules 21(3):304

Larrosa M, García-Conesa MT, Espín JC, Tomás-Barberán FA (2010) Ellagitannins, ellagic acid and vascular health. Mol Asp Med 31:513–539

Latruffe N, Lançon A, Limagne E et al (2014) Bioreactivity of resveratrol toward inflammation processes. Recent advances in redox active plant and microbial products, pp 193–207

Latruffe N, Lancon A, Frazzi R et al (2015) Exploring new ways of regulation by resveratrol involving miRNAs, with emphasis on inflammation. Ann NY Acad Sci 1348:97–106

Lau SKP, Lau CCY, Chan KH et al (2013) Delayed induction of proinflammatory cytokines and suppression of innate antiviral response by the novel Middle East respiratory syndrome coronavirus: implications for pathogenesis and treatment. J Gen Virol 94:2679–2690

Lee M, Kim S, Kwon O et al (2009a) Anti-inflammatory and anti-asthmatic effects of resveratrol, a polyphenolic stilbene, in a mouse model of allergic asthma. Int Immunopharmacol 4:418–424

Lee JH, Song MY, Song EK et al (2009b) Overexpression of SIRT1 protects pancreatic β-cells against cytokine toxicity by suppressing the nuclear factor-κB signaling pathway. Diabetes 58: 344–351

Lee S-M, Yang H, Tartar DM et al (2011) Prevention and treatment of diabetes with resveratrol in a non-obese mouse model of type 1 diabetes. Diabetologia 54:1136–1146

Lee JH, Kim JS, Park SY, Lee YJ (2016) Resveratrol induces human keratinocyte damage via the activation of class III histone deacetylase, Sirt1. Oncol Rep 35:524–529

Li S, Zhao G, Chen L et al (2016) Resveratrol protects mice from paraquat-induced lung injury: the important role of SIRT1 and NRF2 antioxidant pathways. Mol Med Rep 13:1833–1838

Lin HY, Tang HY, Davis FB, Davis PJ (2011) Resveratrol and apoptosis. Ann NY Acad Sci 1215: 79–88

Lin CJ, Lin HJ, Chen TH et al (2015) Polygonum cuspidatum and its active components inhibit replication of the influenza virus through toll-like receptor 9-induced interferon beta expression. PLoS One 10:1–13

Lin SC, Ho CT, Chuo WH et al (2017) Effective inhibition of MERS-CoV infection by resveratrol. BMC Infect Dis 17:1–10

Liu Y, Ma W, Zhang P et al (2015) Effect of resveratrol on blood pressure: a meta-analysis of randomized controlled trials. Clin Nutr 34:27–34

Liu Q, Zhu D, Jiang P et al (2019) Resveratrol synergizes with low doses of L-DOPA to improve MPTP-induced Parkinson disease in mice. Behav Brain Res 367:10–18

Lu X, Ji C, Xu H et al (2009) Resveratrol-loaded polymeric micelles protect cells from Abeta-induced oxidative stress. Int J Pharm 375:89–96

Lu X, Ma L, Ruan L et al (2010) Resveratrol differentially modulates inflammatory responses of microglia and astrocytes. J Neuroinflammation 8:1–14

Lu XY, Hu S, Jin Y et al (2012) Application of liposome encapsulation technique to improve anti-carcinoma effect of resveratrol. Drug Dev Ind Pharm 38:314–322

Lu DL, Ding DJ, Yan WJ et al (2013) Influence of glucuronidation and reduction modifications of resveratrol on its biological activities. Chembiochem 14:1094–1104

Luca SV, Macovei I, Bujor A et al (2020) Bioactivity of dietary polyphenols: the role of metabolites. Crit Rev Food Sci Nutr 60:626–659

Ma C, Wang Y, Dong L et al (2015) Anti-inflammatory effect of resveratrol through the suppression of NF-κB and JAK/STAT signaling pathways. Acta Biochim Biophys Sin Shanghai 47:207–213

Malaguarnera L (2019) Influence of resveratrol on the immune response. Nutrients 11:1–24

Malhotra A, Bath S, Elbarbry F (2015) An organ system approach to explore the antioxidative, anti-inflammatory, and cytoprotective actions of resveratrol. Oxidative Med Cell Longev 2015

Mancuso R, del Valle J, Modol L et al (2014) Resveratrol improves motoneuron function and extends survival in $SOD1^{G93A}$ ALS mice. Neurotherapeutics 11:419–432

Mastromarino P, Capobianco D, Cannata F et al (2015) Resveratrol inhibits rhinovirus replication and expression of inflammatory mediators in nasal epithelia. Antivir Res 123:15–21

Meng X, Zhou J, Zhao CN et al (2020) Health benefits and molecular mechanisms of resveratrol: a narrative review. Foods 9:1–27

Mihaylova MM, Shaw RJ (2011) The AMPK signalling pathway coordinates cell growth, autophagy and metabolism. Nat Cell Biol 13:1016–1023

Miki H, Uehara N, Kimura A et al (2012) Resveratrol induces apoptosis via ROS-triggered autophagy in human colon cancer cells. Int J Oncol 40:1020–1028

Moller M, Fjeldborg K, Juul M et al (2015) Resveratrol and inflammation: challenges in translating pre-clinical findings to improved patient outcomes. Biochim Biophys Acta Mol basis Dis 1852: 1124–1136

Mootha VK, Lindgren CM, Eriksson K-F et al (2003) PGC-1α-responsive genes involved in oxidative phosphorylation are coordinately downregulated in human diabetes. Nat Genet 34: 267–273

Moshawih S, Mydin RBSMN, Kalakotla S et al (2019) Potential application of resveratrol in nanocarriers against cancer: overview and future trends. J Drug Deliv Sci Technol 53:101187

Moussa C, Hebron M, Huang X et al (2017) Resveratrol regulates neuro-inflammation and induces adaptive immunity in Alzheimer's disease. J Neuroinflammation 14:1

Narayanan NK, Nargi D, Randolph C et al (2009) Liposome encapsulation of curcumin and resveratrol in combination reduces prostate cancer incidence in PTEN knockout mice. Int J Cancer 125:1–8
Neves AR, Lucio M, Martins S et al (2013) Novel resveratrol nanodelivery systems based on lipid nanoparticles to enhance its oral bioavailability. Int J Nanomedicine 8:177–187
Nicol T, Roc A, Mendoza-n M, Garc C (2020) The role of resveratrol on heavy metal-induced oxidative stress. Nutr Hosp 37:374–383
Nitulescu GM, Van De Venter M, Nitulescu G et al (2018) The Akt pathway in oncology therapy and beyond (review). Int J Oncol 53:2319–2331
Nunes S, Danesi F, Rio DD, Silva P (2018) Resveratrol and inflammatory bowel disease: the evidence so far. Nutr Res Rev 31:85–97
Olszowy M (2019) What is responsible for antioxidant properties of polyphenolic compounds from plants? Plant Physiol Biochem 144:135–143
Oroian M, Escriche I (2015) Antioxidants: characterization, natural sources, extraction and analysis. Food Res Int J 74:10–36
Palsamy P, Subramanian S (2010) Ameliorative potential of resveratrol on proinflammatory cytokines, hyperglycemia mediated oxidative stress, and pancreatic β-cell dysfunction in streptozotocin-nicotinamide-induced diabetic rats. J Cell Physiol 224:423–432
Pandita D, Kumar S, Poonia N (2014) Solid lipid nanoparticles enhance oral bioavailability of resveratrol, a natural polyphenol. Food Res Int 62:1165–1174
Patel KR, Scott E, Brown VA et al (2011) Clinical trials of resveratrol. Ann NY Acad Sci 1215: 161–169
Pecyna P, Wargula P, Murias M et al (2020) More than resveratrol: new insights into stilbene-based compounds. Biomol Ther 10(8):1111–1113
Penalva R, Esparza I, Larraneta E et al (2015) Zein-based nanoparticles improve the oral bioavailability of resveratrol and its anti-inflammatory effects in a mouse model of endotoxic shock. J Agric Food Chem 63:5603–5611
Penumathsa SV, Thirunavukkarasu M, Zhan L et al (2008) Resveratrol enhances GLUT-4 translocation to the caveolar lipid raft fractions through AMPK/Akt/eNOS signalling pathway in diabetic myocardium. J Cell Mol Med 12:2350–2361
Petrovski G, Gurusamy N, Das DK (2011) Resveratrol in cardiovascular health and disease. Ann NY Acad Sci 1215:22–33
Prysyazhna O, Wolhuter K, Switzer C et al (2019) Blood pressure-lowering by the antioxidant resveratrol is counterintuitively mediated by oxidation of cGMP-dependent protein kinase. Circulation 140:126–137
Ramos PS, Shedlock AM, Langefeld CD (2015) Genetics of autoimmune diseases: insights from population genetics. J Hum Genet 60:657–664
Raskovic A, Cucuz V, Torovic L et al (2019) Resveratrol supplementation improves metabolic control in rats with induced hyperlipidemia and type 2 diabetes. Saudi Pharmaceutical J 27: 1036–1043
Rayalam S, Yang J-Y, Ambati S et al (2008) Resveratrol induces apoptosis and inhibits adipogenesis in 3T3-L1 adipocytes. Phytother Res 22:1367–1371
Ruderman NB, Xu XJ, Nelson L et al (2010) AMPK and SIRT1: a long-standing partnership? Am J Physiol Endocrinol Metab 298:751–760
Saha A, Robertson E (2019) Mechanisms of B-cell oncogenesis induced by Epstein-Barr virus. J Virol 93:e002381–e0023813
Salehi B, Mishra A, Nigam M, Sener B, Kilic M, Sharifi-Rad M, Fokou P, Martins N, Sharifi-Rad J (2018) Resveratrol: a double-edged sword in health benefits. Biomedicines 6(3):91. https://doi.org/10.3390/biomedicines6030091
Samsamikor M, Daryani NE, Asl PR, Hekmatdoost A (2016) Resveratrol supplementation and oxidative/anti-oxidative status in patients with ulcerative colitis: a randomized, double-blind, placebo-controlled pilot study. Arch Med Res 47:304–309

Sanna V, Siddiqui IA, Sechi M et al (2013) Resveratrol-loaded nanoparticles based on poly(epsilon-caprolactone) and poly(D,L-lactic-co-glycolic acid)-poly(ethylene glycol) blend for prostate cancer treatment. Mol Pharm 10:3871–3881

Senturk S, Yaman M, Aydin H et al (2018) Effects of resveratrol on inflammation and apoptosis after experimental spinal cord injury. Turk Neurosurg 28(6):889–896

Seo Y, Park J, Choi W et al (2019) Antiatherogenic effect of resveratrol attributed to decreased expression of ICAM-1 (intercellular adhesion molecule-1). Arterioscler Thromb Vasc Biol 39: 675–684

Sessa M, Balestrieri ML, Ferrari G et al (2014) Bioavailability of encapsulated resveratrol into nanoemulsion-based delivery systems. Food Chem 147:42–50

Sharma AK, Keservani RK, Kesharwani RK (eds) (2018) Nanobiomaterials: applications in drug delivery. CRC Press. ISBN: 9781771885911

Shi Y, Zhou J, Jiang B, Miao M (2017) Resveratrol and inflammatory bowel disease. Ann N Y Acad Sci 1403:38–47

Shidhaye SS, Vaidya R, Sutar S et al (2008) Solid lipid nanoparticles and nanostructured lipid carriers—innovative generations of solid lipid carriers. Curr Drug Deliv 5:324–331

Shukla SD, Mahmood MQ, Weston S et al (2017) The main rhinovirus respiratory tract adhesion site (ICAM-1) is upregulated in smokers and patients with chronic airflow limitation (CAL). Respir Res 18:1–10

Silva F, Figueiras A, Gallardo E et al (2014) Strategies to improve the solubility and stability of stilbene antioxidants: a comparative study between cyclodextrins and bile acids. Food Chem 145:115–125

Singh UP, Singh NP, Busbee B et al (2012) Alternative medicines as emerging therapies for inflammatory bowel diseases. Int Rev Immunol 31:66–84

Singh DB, Gupta M, Kesharwani R, Sagar M, Dwivedi S, Misra K (2014) Molecular drug targets and therapies for Alzheimer's disease. Transl Neurosci 5(3):203–217

Soares MV, Puntel RL, Ávila DS (2018) Resveratrol attenuates iron-induced toxicity in a chronic post-treatment paradigm in *Caenorhabditis elegans*. Free Radic Res 52:939–951

Song J, Liu L, Hao K et al (2021) Resveratrol elongates the lifespan and improves antioxidant activity in the silkworm, *Bombyx mori*. J Pharmaceut Anal 11:374–382

Soo E, Thakur S, Qu Z et al (2016) Enhancing delivery and cytotoxicity of resveratrol through a dual nanoencapsulation approach. J Colloid Interface Sci 462:368–374

Springer M, Moco S (2019) Resveratrol and its human metabolites—effects on metabolic health and obesity. Nutrients 11:1–17

Sun Y, Reddy P (2013) Intracellular sensors of immunity and allogeneic hematopoietic stem cell transplantation. In: Immune biology of allogeneic hematopoietic stem cell transplantation, 2nd edn. Academic, San Diego, pp 425–447

Sun L, Chen B, Jiang R et al (2017) Resveratrol inhibits lung cancer growth by suppressing M2-like polarization of tumor associated macrophages. Cell Immunol 311:86–93

Szkudelsk K, Okulicz M, Hertig I et al (2020) Resveratrol ameliorates inflammatory and oxidative stress in type 2 diabetic Goto-Kakizaki rats. Biomed Pharmacother 125:110026

Szkudelska K, Szkudelski T (2010) Resveratrol, obesity and diabetes. Eur J Pharmacol 635:1–8

Szkudelski T, Szkudelska K (2011) Anti-diabetic effects of resveratrol. Ann N Y Acad Sci 1215: 34–39

Takashina M, Inoue S, Tomihara K et al (2017) Different effect of resveratrol to induction of apoptosis depending on the type of human cancer cells. Int J Oncol 50:787–797

Taniguchi K, Karin M (2018) NF-B, inflammation, immunity and cancer: coming of age. Nat Rev Immunol 18:309–324

Teijaro JR (2016) Type I interferons in viral control and immune regulation. Curr Opin Virol 16:31–40

Tellone E, Galtieri A, Russo A et al (2014) Resveratrol: a focus on several neurodegenerative diseases. Oxidative Med Cell Longev 2015:392169

Thapa SB, Pandey RP, Il Park Y, Sohng JK (2019) Biotechnological advances in resveratrol production and its chemical diversity. Molecules 24:1–21

Timmers S, Konings E, Bilet L et al (2011) Calorie restriction-like effects of 30 days of resveratrol supplementation on energy metabolism and metabolic profile in obese humans. Cell Metab 14: 612–622

Tome-Carneiro J, Larrosa M, Gonzalez-Sarrias A et al (2013) Resveratrol and clinical trials: the crossroad from in vitro studies to human evidence. Am J Phys Med Rehabil 19:6064–6093

Torres J, Mehandru S, Colombel J-F, Peyrin-Biroulet L (2017) Crohn's disease. Lancet 389:1741–1755

Tsai M-H, Hsu L-F, Lee C-W et al (2017) Resveratrol inhibits urban particulate matter-induced COX-2/PGE2 release in human fibroblast-like synoviocytes via the inhibition of activation of NADPH oxidase/ROS/NF-κB. Int J Biochem Cell Biol 88:113–123

Vélez-Marín M, Souza LFU-V, Lenz MI (2012) Papel Del Resveratrol De Uva Como Antioxidante. Luna Azul 34:240–256

Venuti V, Cannava C, Cristiano MC et al (2014) A characterization study of resveratrol/sulfobutyl ether-beta-cyclodextrin inclusion complex and *in vitro* anticancer activity. Colloids Surf B: Biointerfaces 115:22–28

Walle T, Hsieh F, DeLegge MH et al (2004) High absorption but very low bioavailability of oral resveratrol in humans. Drug Metab Dispos 32:1377–1382

Wang P, Sang S (2018) Metabolism and pharmacokinetics of resveratrol and pterostilbene. Biofactors 44:16–25

Wang XX, Li YB, Yao HJ et al (2011) The use of mitochondrial targeting resveratrol liposomes modified with a dequalinium polyethylene glycol-distearoylphosphatidyl ethanolamine conjugate to induce apoptosis in resistant lung cancer cells. Biomaterials 32:5673–5687

Wang B, Sun J, Li X et al (2013) Resveratrol prevents suppression of regulatory T-cell production, oxidative stress, and inflammation of mice prone or resistant to high-fat diet–induced obesity. Nutr Res 33(11):971–981

Wang Z, Zhang J, Duan Y et al (2015) MicroRNA-214 participates in the neuroprotective effect of resveratrol via inhibiting α-synuclein expression in MPTP-induced Parkinson's disease mouse. Biomed Pharmacother 74:252–256

Wang G, Hu Z, Song X et al (2017a) Analgesic and anti-inflammatory activities of resveratrol through classic models in mice and rats. Evid Based Complement Alternat Med 2017:5197567

Wang X, Li T, Li J et al (2017b) The effects of resveratrol on inflammation and oxidative stress in a rat model of chronic obstructive pulmonary disease. Molecules 22(9):1529

Wark PAB, Johnston SL, Moric I et al (2002) Neutrophil degranulation and cell lysis is associated with clinical severity in virus-induced asthma. Eur Respir J 19:68–75

Wenzel E, Somoza V (2005) Metabolism and bioavailability of trans-resveratrol. Mol Nutr Food Res 49:472–481

WHO World Health Organization (2020a) Obesity and overweight. https://www.who.int/news-room/fact-sheets/detail/obesity-and-overweight. Accessed 28 Sept 2020

WHO World Health Organization (2020b) Diabetes. https://www.who.int/es/news-room/fact-sheets/detail/diabetes. Accessed 28 Sept 2020

Wiciński M, Malinowski B, Rajewski P et al (2020) Resveratrol's impact on vascular smooth muscle cells hyporeactivity: the role of rho-kinase inhibition. Biomed Res Int 9012071

Xie XH, Zang N, Min LS et al (2012) Resveratrol inhibits respiratory syncytial virus-induced IL-6 production, decreases viral replication, and downregulates TRIF expression in airway epithelial cells. Inflammation 35:1392–1401

Xing L, Casasnovas JM, Cheng RH (2003) Structural analysis of human rhinovirus complexed with ICAM-1 reveals the dynamics of receptor-mediated virus uncoating. J Virol 77:6101–6107

Yamamoto Y, Gaynor RB (2001) Therapeutic potential of inhibition of the NF-κB pathway in the treatment of inflammation and cancer. J Clin Investig 107:135–142

Yan B, Cheng L, Jiang Z et al (2018) Resveratrol inhibits ROS-promoted activation and glycolysis of pancreatic stellate cells via suppression of miR-21. Oxidative Med Cell Longev 2018:1–15

Yanez M, Jhanji M, Murphy K et al (2019) Nicotinamide augments the anti-inflammatory properties of resveratrol through PARP1 activation. Sci Rep 9:1–10
Yang X, Jiang T, Wang Y et al (2019) The role and mechanism of SIRT1 in resveratrol-regulated osteoblast autophagy in osteoporosis rats. Sci Rep 9:1–15
Yoon H, Dehart JP, Murphy JM, Lim STS (2015) Understanding the roles of FAK in cancer: inhibitors, genetic models, and new insights. J Histochem Cytochem 63:114–128
Yun CW, Lee SH (2018) The roles of autophagy in cancer. Int J Mol Sci 19:1–18
Zhang XS, Li W, Wu Q et al (2016) Resveratrol attenuates acute inflammatory injury in experimental subarachnoid hemorrhage in rats via inhibition of TLR4 pathway. Int J Mol Sci 17:1–14
Zhang L, Yu X, Ji M et al (2018) Resveratrol alleviates motor and cognitive deficits and neuropathology in the A53T α-synuclein mouse model of Parkinson's disease. Food Funct 9:6414–6426
Zhou Z, Mou S, Chen X et al (2018) Anti-inflammatory activity of resveratrol prevents inflammation by inhibiting NF-κB in animal models of acute pharyngitis. Mol Med Rep 17:1269–1274
Zhou X, Yang J, Zhou M et al (2019) Resveratrol attenuates endothelial oxidative injury by inducing autophagy via the activation of transcription factor EB. Nutr Metab 16:1–12

Chapter 10
Immunomodulation Impact of Curcumin and Its Derivative as a Natural Ingredient

Eknath D. Ahire, Khemchand R. Surana, Vijayraj N. Sonawane, Swati G. Talele, Sanjay J. Kshirsagar, Umesh D. Laddha, Nilima A. Thombre, and Gokul S. Talele

Abstract Curcumin is a phytochemical derived from the bulb of the plant *Curcuma longa* and is the major component of turmeric. The herb has been used for thousands of years in traditional Ayurvedic and Asian medicine, initially to heal wounds but also to treat diverse conditions. There is a definite need to develop effective, inexpensive, safe treatments for the millions of patients with chronic diseases, including cancer, autoimmune diseases, and degenerative conditions such as arthritis. From the current scenario and study of curcumin, it is believed that data from ongoing and future research will continuously boost researchers to enhance the use of natural, synthetic, and semi synthetic derivatives of the compound as both a primary and secondary therapeutic in a variety of disease states. It has also been reported that curcumin is capable of inhibiting the development of cancer cells and/or inducing many signaling pathways affecting chronic inflammation, including nuclear factor kappa beta (NF-κB) and cyclooxygenase-2 enzymes (COX-2). Curcumin has also shown antiviral and anti-inflammatory effects and to be helpful for both prevention and treatment of the new coronavirus. However, well-designed clinical trials are needed to demonstrate the potential efficacy of curcumin against SARS-CoV-2 infection and its ensuing complications.

Eknath D. Ahire and Khemchand R. Surana contributed equally with all other contributors.

E. D. Ahire (✉) · S. J. Kshirsagar · U. D. Laddha · N. A. Thombre
Department of Pharmaceutics, MET's Institute of Pharmacy, Bhujbal Knowledge City, Affiliated to Savitribai Phule Pune University, Nashik, Maharashtra, India

K. R. Surana · V. N. Sonawane
Department of Phamaceutical Chemistry, SSS's, Divine College of Pharmacy, Satana, Maharashtra, India

S. G. Talele
Department of Pharmaceutics, Sandip Institute of Pharmaceutical Sciences, Nashik, Maharashtra, India

G. S. Talele
Department of Pharmaceutics, Matoshri College of Pharmacy, Nashik, Maharashtra, India

R. K. Kesharwani et al. (eds.), *Nutraceuticals and Functional Foods as Immunomodulators*, https://doi.org/10.1007/978-981-19-2507-8_10

Keywords Curcumin · Turmeric · Immunomodulator · Curcuma longa · Natural Ingredients

10.1 Introduction

Immune response modulated by using medicinal plants and their products as an achievable remedial measure has become an accepted curative approach. Since antiquity, plants and minerals sources have been used for the treatment of many illnesses and disorders. When the defense mechanism of the host is triggered under conditions that impede immune acceptance or when a selective immunosuppressive drug has to be used in certain situations, such as in an autoimmune condition or organ transplant, has helped to realize that immunomodulation of immune response can be changed by conventional chemotherapy for a variety of disease conditions. The homeostatic process is a series of finely balanced complex, multicellular, and physiologic mechanisms that enable a person to distinguish foreign material from "self" and being able to neutralize and/or eliminate the foreign matter is called immunity (Šturdík 2013, Farooqui and Farooqui 2019).

10.2 Immunomodulation

Growth in the field of clinical and experimental immunology has shown that a number of infectious diseases and disorders emerge because of environmental situations associated with weakened immune responses. It is noticeable that some types of stress evoke physiological changes that affect sensitivity to infection and cancerous state. The capability to adapt the immune response in animals and humans emerges from a desire to obtain higher protection against infectious material through an entire understanding of the functioning of the immune system, and of the ways in which nonspecific and specific immune mechanisms developed. Natural products or synthetic compounds are able to alter those mechanisms allowing further prospects for modulation of the immune system. Nutraceuticals are food products that are known to provide health and medicinal benefits, including the prevention and treatment of diseases (Keservani et al. 2010; Keservani et al. 2017a, b). Diet has a major role in modulating the risk of developing a number of diseases. Edible plants, through dietary compounds, are important in that they contain phytochemicals which can control biochemical and physiological processes at the cellular level in human organisms. Among nutraceuticals, preparations of the turmeric rhizome, species Curcuma longa L., family *Zingiberaceae*, have been used medicinally and as spices for millennia in several Asian countries and traditional Chinese medicine to treat bacterial infections, digestive disorders, inflammation, and burns (Kesharwani and Misra 2011; Vyas et al. 2010; Kesharwani et al. 2015). Some of the medicinal properties are associated with curcumin (diferuloylmethane), a major polyphenolic compound of the rhizome (5–10% of dry weight). Biomedical evaluations of curcumin and curcuminoids have reported a wide range of molecular and cellular activities, most related to anti-inflammatory, signal transduction, and redox reactions. It is available in supplemental form for gastrointestinal discomfort as an antiseptic (Zahedipour et al. 2020; Catanzaro et al. 2018; Scartezzini and Speroni 2000; Mishra et al. 2019; Singh et al. 2016). Turmeric is traditionally used in the Middle East as a liver protector, a stimulant of bile duct secretion, antiflatulent, and diuretic, for curing catarrh, for the circulation, antifever, anti-inflammatory, and for rheumatisms or sprains. Safety assessment studies demonstrate that both curcumin and turmeric can be used at a very high dose without any toxicity. Curcumin, a safe, cheap, natural bioactive product of turmeric has shown efficacy against diverse human ailments and is one of the most broadly considered natural compounds. However, it is limited by its loss of aqueous solubility and poor bioavailability due to poor absorption, rapid metabolism, and rapid systemic elimination. To reduce the above limitations, different nanostructured water soluble delivery systems of curcumin and curcuminoids were recently developed to increase the solubility and bioavailability (Boroumand et al. 2018; Upadhyaya et al. 2009; Keservani et al. 2017a, b; Keservani et al. 2018).

10.3 Chemical Composition of Turmeric

Turmeric (Curcuma longa L.), a tropical herb of the *Zingiberaceae* family, indigenous to southern Asia, is chiefly used in the form of the powdered rhizome, which has a yellow pigment, and is one of the main natural spices used for centuries in Indian cuisine. The curcuminoids and volatile oils are the main secondary metabolites of turmeric, and they are mainly responsible for the biological activities of turmeric powder extracted from rhizomes. Its medicinal properties have been attributed mainly to the primary compound of this group, that is, curcumin – (1, 7-bis (4-hydroxy, 3-methoxyphenyl)-1, 6-heptadiene- 3, 5-dione), also called diferuloylmethane. Curcumin has an equilibrium of equivalent enol forms rather than a tautomeric keto–enol mixture. Nonetheless, the content of curcumin in turmeric is approximately 4–5%. Commercially, curcumin has a purity >95% and contains three major compounds: curcumin (77%), demethoxycurcumin (17%), and bisdemethoxycurcumin (3%), which are called curcuminoids. Curcumin is hydrophobic and based on calculated log P values ranged from 2.56 to 3.29, which explains the experimental conclusions about lipid membrane affinity, interaction with hydrophobic territory of proteins, and ability to cross the blood–brain barrier (Keservani et al. 2011; Rahimi et al. 2019; Kesharwani et al. 2018). Turmeric (Curcuma longa), also known as "Indian saffron" due to its brilliant yellow color, is a spice herb, member of the ginger family (Zingiberaceae) native to the Indian subcontinent and Southeast Asia, and has a more than a two centuries old scientific history (Cundell and Wilkinson 2014). The worldwide main producer of turmeric is

India, which has been used as an Ayurvedic remedy and flavoring agent since antiquity (more than 4000 years). Depending on its origin and growth conditions, turmeric obtained from the ground-dried root contains different percentages of volatile and non-volatile oils, proteins, fats, minerals, carbohydrates, curcuminoids, and moisture. Commercially available curcumin is a combination of three molecules, together called curcuminoids. Curcumin is the most represented (60–70%), followed by demethoxycurcumin (20–27%), and bisdemethoxycurcumin (10–15%). Curcuminoids differ in potency, efficacy and stability, with no clear supremacy of curcumin over the other two compounds or the whole mixture. In addition to curcuminoids, the other active components of turmeric are monoterpenes, diterpenes, and triterpenoids. To date, many limitations have been recognized for a therapeutic use of curcumin: its poor pharmacokinetic/pharmacodynamic properties, its chemical instability, its low efficacy in different in vitro and in vivo disease models, its toxic profile under certain experimental settings, and the very recent suggestion that curcumin may be part of a series of molecules recognized for their interference with biological assays called pan assay interference compounds (PAINS) (Sasikumar 2012; Singh et al. 2011).

Different delivery systems have ways to overcome the pharmaceutical limitations related to curcumin pharmacokinetics to enhance its therapeutic efficacy and give new goals for clinical applications of this natural product. Distinct preclinical and clinical data have indicated the efficacy of curcumin in the blockage and treatment of various human diseases, including cardiovascular, cancer, metabolic, inflammatory, neurological, and skin diseases. Among the distinct properties attributed to curcumin, one of the most studied is the anti-inflammatory profile that may be helpful in both acute and chronic inflammation. The immunomodulatory capabilities of curcumin activate from its interaction with several immunomodulators, including not only cellular components, such as macrophages, dendritic cells and both B and T lymphocytes, but also molecular components involved in the inflammatory processes, such as cytokines and various transcription factors with their downstream signaling pathways(Ahire et al. 2021; Jayaprakasha et al. 2001; Sasikumar 2012).

10.3.1 Curcumin and the Curcuminoids: Structure–Activity Relationships and Synthetic Analogs

Among the curcuminoids, diarylheptanoids, consisting of polyphenols in which a seven-carbon (7C) chain combines together two oxygen substituted aryl rings, constitute approximately 5% of the dried rhizome. Curcumin is the most predominant of these molecules, constituting around 77% of the available curcuminoids in spice preparations, with demethoxycurcumin (DMC) and bisdemethoxycurcumin (BDMC) at 17% and 3%, respectively, (Fig 10.1) (Amalraj et al. 2017).

As well as the three main curcuminoids from turmeric, distinct curcuminoid metabolites have also been shown to be active moieties of which tetrahydrocurcumin

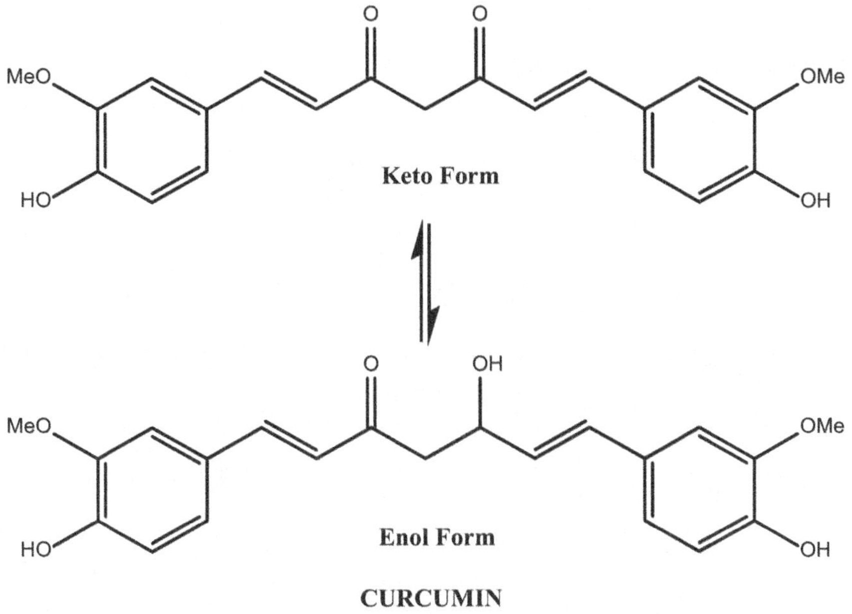

Fig. 10.1 Keto–enol conversion of curcumin

(THC) is indicated to give the majority of antioxidant activity to curcumin itself. Research on THC in animals has also shown it is cardio protective through modification of cholesterol and low density lipoproteins (LDL) (Fig. 10.2) (Amalraj et al. 2017).

The comparison of the structure–activity relationships of these curcuminoid molecules suggest that those with o-methoxy substitutions, such as curcumin, are more effective as oxygen radicals and antioxidants than those lacking this structure, e.g., DMC. Furthermore, it was indicated that NF-κB and COX-2 enzyme inhibitory activity, which is prominent in the THC molecule, is associated with the possession of an unsaturated ß-diketone in conjugation with the aromatic rings (Shirley et al. 2008).

The comparison of the salient structural features of the curcumin molecule, namely "a pharmacophore containing one aryl function with 3, 4 substitution, i.e., either a methoxylated phenol or catechol" and many natural analogs of the molecule exist. These include phytochemicals with a single aromatic ring such as caffeic acid, prominent in echinacea, gingerols from the related species ginger, and eugenols from clove as well as those with two aromatic rings such as oregonin from oregano. Computational chemistry studies for the curcumin molecule have suggested that its optimal structure is a planar molecule. Derivatives of the molecule have then been successfully obtained by a variety of reaction methods and several of these appear to have more immunological potency than the original curcumin molecule. Most originate from the altered assembly of the curcumin molecule using aldehydes and

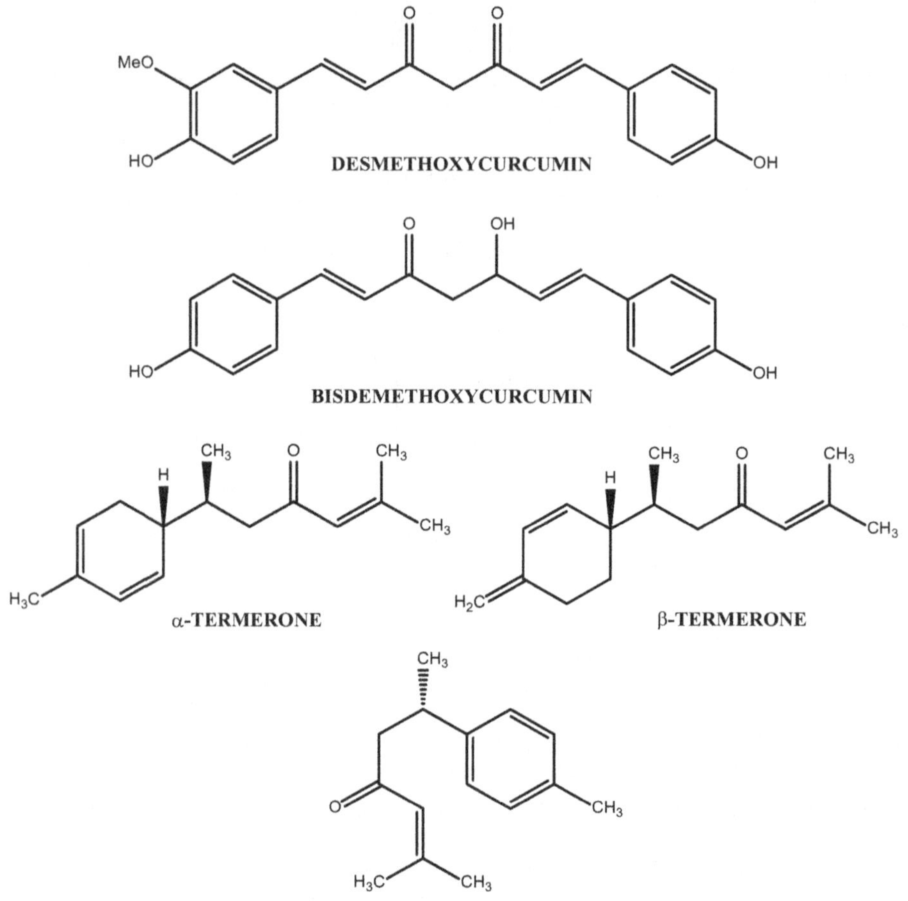

Fig. 10.2 Chemical structures of important constituents present in turmeric

acetyl acetone rather than modifications of the natural parent curcumin itself, and this has also yielded new information on the structural features needed for anti-tumor and anti-inflammatory (NF-κB and COX-2) activity. These grassroots studies are anticipated to ultimately provide more of a guide to the molecules that will be useful in clinical settings (Karbalaei and Keikha 2019; Kohyama et al. 2015).

As well as synthetic analogs, many scientists have been designing effective strategies aimed at improving the bioavailability of the curcumin molecule. First is the usage of adjuvants such as piperazine, which interfere with enzymes that would otherwise metabolize the phytochemical. In a small human study, six volunteers took either curcumin alone or with piperazine for 1 week, and it was found that piperazine doubled the absorption of the herb. Second, the use of lipid based delivery systems (nanoparticles) has been demonstrated as a successful way to deliver curcumin to rat macrophages through the bloodstream, as well as via topical treatments. Bisht and colleagues were able to synthesize and characterize a nanoparticular version of curcumin that they term "nanocurcumin. Nanocurcumin was found to possess similar efficacy as native curcumin at inhibiting NF-κB activation in pancreatic cell lines. Bisht and colleagues, demonstrated this was also associated with the ability of both nanocurcumin and curcumin to maintain levels of the inflammatory interleukins and tumor necrosis factor alpha (TNF-α) (Gupta et al. 2017; Jarouliya et al. 2018; Keservani et al. 2019; Sharma et al. 2018). A third method to deliver curcumin is via the use of liposomes as delivery systems. Murine studies by Ruby and colleagues, using liposomes loaded with albumin and curcuminoids (11–20 mg/ml), have demonstrated that this delivery system resulted in more curcuminoid bioavailability for lymphoma cells in Ehrlich ascites tumors than did dimethyl sulfoxide or serum albumin alone. These authors also reported that the lymphoma cells they used preferentially took up the liposomes compared with normal B lymphocytes, results indicative of a targeted delivery to tumor cells. A second study, by Bruzell and colleagues investigated the phototoxicity of liposomes containing cyclodextrin and curcumin (0.4–13.5 μM) on a salivary gland cell line (SM 10–12 cells). Bruzell and colleagues observed that the phototoxicity induced by natural and synthetic curcumin was similar. Significant effects on the cells were seen after 3 hours of incubation, using concentrations of curcumin below 1 μM and with low doses of light. From these studies Bruzell and colleagues have suggested that their liposome preparation is the most efficient delivery method for curcumin (Sahebkar et al. 2015; Bisoffi and O'neill 2021). Finally, as with other pharmaceuticals such as dolichol and silymarin, phospholipid complexing of curcumin has also been shown to improve bioavailability. Liu and colleagues administered a 100 mg/kg oral dose of curcumin alone or as a complex with phospholipid to Sprague-Dawley male rats. The authors observed that the complexed curcumin was able to maintain maximal plasma levels of the herb of 600 ng/ml 140 mins after ingestion, whereas for the free curcumin peaks of 267 ng/ml occurred only until 96 mins after oral dosing. Liu and colleagues also observed a 1.5-fold increase in curcumin half-life for the curcumin phospholipid complex over free curcumin. Complexes of curcumin with polyethylene glycol (PEG) have also been reported to enhance the bioavailability of curcumin. In 2005, Anand and colleagues patented Biocurcurmax® (BCM-95®; Arjuna Natural Extracts Ltd. (Patent No. 200430), Edayar, India), which is a combination of sesquiterpenoids and curcuminoids from turmeric that has demonstrated significantly improved bioavailability and blood half-life compared with curcumin itself. BCM-95® is stated in the patent to be a "reconstituted turmeric extract standardized with not less than 95% curcuminoids ("curcuminoid" is a mixture of curcumin, demethoxycurcumin, and bisdidemethoxycurcumin, wherein curcumin is the major component of the curcuminoid and comprises about 95% of the curcuminoid, and demethoxycurcumin and bisdidemethoxycurcumin are minor components of the curcuminoid)." Three human trials have already been performed using this product, which will be discussed later in this monograph (Campbell et al. 2018; Surana et al. 2021b; Singh et al. 2013b).

10.4 Immunomodulatory Effect

The immune system has an important role in the defense mechanism against infections. Immunomodulators guide to improve or normalize the immune responses by impairing or modulating the activity of the immune response, resulting in the decrease of the inflammatory response (Singh et al. 2015). A number of flavonoids have significant pharmacological and biochemical activity, influencing the normal functions of immune cells, such as B and T cells, neutrophils, basophils macrophages, and mast cells. Curcumin has been known to downregulate the expression of several pro-inflammatory cytokines such as TNF, IL-1, IL-2, IL-6, IL-8, IL-12, and chemokines, most likely through inactivation of the transcription factor NF-κB. However, at low doses, curcumin can also increase antibody responses, implicating that curcumin's favorable effects in curing of chronic diseases such as allergy, asthma, arthritis, atherosclerosis, heart disease, AD, diabetes, and cancer may be partially due to modulation of the immune responses (Boroumand et al. 2018). Curcumin has been established to exert an inhibitory response on the production of inflammatory cytokines by human monocytes (Mimche et al. 2011; Singh et al. 2013a, b). In an animal model of multiple sclerosis (MS) such as experimental auto immune encephalomyelitis, curcumin decreases IL-12 production, and signal transducer and activator of transcription 4 (STAT4) activation. The action of curcumin on STAT4 activation depends on the stimulus to which the T cells have been exposed. The molecular mechanisms underlying the targets of curcumin for treating distinct cancers and inflammation-mediated diseases are diverse and may involve combinations of multiple signaling pathways, including NF-κB and STAT3 signaling (De Geus and Vervelde 2013). Several preclinical and clinical trials have revealed immunomodulatory actions of curcumin, which arise from its effects on immune cells and mediators involved in the immune response (e.g., various T-lymphocyte subsets and dendritic cells, as well as different inflammatory cytokines), implying that curcumin can affect different immune cells, warranting that further consideration of curcumin is required as a therapy for immune diseases (Cundell and Wilkinson 2014). The dried powder of turmeric has the ability to modulate the antimicrobial peptides, mainly crustin and lysozyme, of the shrimp *M. rosenbergii* when boosted with *V. alginolyticus*. The combination of turmeric extract with commercial feeds also enhanced the survival rate of shrimp infected with a bacterial pathogen. The protective action of turmeric in viral infection will be worth pursuing, and the plant can be used by the shrimp aquaculture industry for shrimp feed formulation (Hosseinzade et al. 2019; Surana et al. 2021a).

It has also been reported that the antiviral action of curcumin is worth further investigation to determine the potency of curcumin against the novel SARS-CoV-2 (COVID-19) infection. The ability of curcumin to modulate a broad range of molecular targets makes it a suitable candidate for the treatment of COVID-19. Curcumin is involved in modulating cellular signaling pathways such as apoptosis, RNA replication, and inflammation. Curcumin may also suppress pulmonary edema and fibrosis associated pathways in COVID-19 infection. Despite the potential beneficial effects and safety profile of curcumin against various diseases, the limited bioavailability of this turmeric-derived compound, especially via oral administration, may be a problematic issue. Several clinical trials have shown that the issue regarding the bioavailability of curcumin can be mitigated by administering higher concentrations within non-toxic limits. Since curcumin is known to have strong inhibitory effects on NF-κB and several pro inflammatory cytokines, it can be particularly helpful as an adjunct in reversing the fatal cytokine storm that occurs in serious cases of COVID-19 (Alambra et al. 2012; Kim et al. 2005).

10.5 Immunomodulation Properties of Curcuma Derivatives

Curcuma species and their bioactive compounds were much investigated for their diverse organic and pharmacological activities, such as antioxidant, anti-inflammatory, anticancer, hepato-protective, antifungal, antihypertensive, neuroprotective, and immunomodulatory outcomes via in vitro and in vivo research. The six Curcuma species and their bioactive compounds mentioned in this text were documented to demonstrate diverse pharmacological activities, mainly via modulation of the immune system. There are in-depth mechanistic studies on the immunomodulation outcomes of some of those species in the literature (Yuandani et al. 2021).

10.5.1 Curcuma Longa

10.5.1.1 In Vitro Immunomodulating Effect of C. longa

Of all the Curcuma species investigated, the immunomodulatory consequences of C. longa have been the most studied. Interestingly, most of the experimental research studies involving the extracts of C. longa have used in vivo animal models, and there have been few in vitro studies. The followings are reviews of the few in vitro research studies that have been done to assess the immunomodulating consequences of C. longa. C longa fermented with the aid of using Aspergillus oryzae (FCL) exhibited immunomodulatory consequences in RAW 264.7 cells. The different extracts of FLC on phagocytic activity, TNF-α, NO formation, NK mobileular activity, and mRNA expression of LP-BM5 eco displayed the following results: warm water and 20% ethanol extracts expanded the phagocytic activity; however, there has been no considerable expansion in NO formation relative to the control (Yuandani et al. 2021).

10.5.1.2 In Vivo Immunomodulating Effect of C. longa

Most immunomodulating research has been performed using aqueous and alcoholic extracts. The ethanol extract of C. longa was stated to suppress immune function, and behavioral and neuroendocrine changes in a rat persistent slight stress (CMS) model. The enhancement of cytokine stage (TNF-α and IL-6) activity and NK mobileular activity inhibition in the CMS-prompted rat splenocytes was reversed with the aid of 35 mg/kg of C. longa ethanol extract and 7 mg/kg of fluoxetine as a control. The putative antidepressant properties of the extract had been due to suppressive results on cytokine biosynthesis. However, the extract accelerated the IL-6 stage in the nonstress group. There was no sizable distinction compared with the ones of the everyday group and triggered a mild but not sizable reduction in TNF-α levels. Although the extract is more suitable for splenic NK mobileular activity in CMS-treated rats, the NK mobileular activity of nonstressed rats no longer changed after treatment with C. longa. In another observation, treatment with C. longa methanol extract with an unmarried dose of 200 mg/kg for 14 days in mice induced innate and adaptive immunity. The impact of the extract on adaptive immunity was investigated via immunizing the mice with sheep purple blood cells (sRBCs) on days 7 and 14, respectively. C. longa was more suitable for the adaptive immunity with growing leukocyte number, antibody titer, spleen index, and delayed-type hypersensitive reaction. However, the outcomes of this observation are initial because more doses of the extract need to be administered to decide a dose–reaction relationship and the most effective dose for efficacy (Kumolosasi et al. 2018).

10.5.2 Curcuma Zanthorrhiza Roxb

10.5.2.1 In Vitro Immunomodulating Effect of C. zanthorrhiza

C. zanthorrhiza methanol extract has been stated to inhibit ROS technology in a luminol and lucigenin-enhanced chemiluminescence (CL) assay. C. zanthorrhiza rhizomes decreased ROS formation from whole human blood in an in vitro study. Moreover, the extract considerably inhibited the launch of ROS from zymosan-precipitated PMNs and macrophages. C. zanthorrhiza additionally confirmed strong inhibition on PMN migration, with an IC50 amount of 2.5 μg/ml. A previous study stated that the methanol extract of C. zanthorrhiza rhizomes confirmed strong inhibition at the expression of CD18/11a; meanwhile, the extract has a low effect on leukocyte phagocytosis. The mRNA levels of IL-1β, NF-κB p65, MMP-2, and MMP-eight on LPS-induced human gingival fibroblast-1 cells decreased after treatment with crude polysaccharide extract of C. zanthorrhiza. The extract of C. zanthorrhiza inhibited MAPK/activator protein-1 (AP-1) signaling pathways. C. zanthorrhiza has been documented to showcase anti-inflammatory activity in LPS-induced RAW264.7 monocytes and H_2O_2-treated HT22 hippocampal cells (Nogami et al. 2011; Kim et al. 2018).

10.5.2.2 In Vivo Immunomodulating Effect of C. zanthorrhiza

Curcuminoid cider, a conventional fermented product made through the addition of Acetobacter xylinum to curcuminoid extract isolated from C. zanthorrhiza, decreased the gene expression of IL1β, TNFα, and chemokine in hypercholesterolemia rats, according to a study that confirmed the inhibitory activity of curcuminoid cider and curcuminoid extract from C. zanthorrhiza at the gene expression of CD44, ICAM-1, iNOS, and LOX-1 in high cholesterol weight-reduction plan rats. Volatile oil from C. zanthorrhiza increased the lymphocyte proliferation from human male B blood type. C. zanthorrhiza extract administration reduced inflammatory lymphocytes in alcohol-precipitated hepatitis in mice. C. zanthorrhiza ethanol extracts strongly decreased cytokine gene expression, which consists of TNF-α, IL-6, IL-1β, and C-reactive protein (CRP) in the liver, adipose tissue, and muscle of high-fats weight-reduction plan-precipitated overweight mice (Kim et al. 2014; Mauren and Yanti 2016; Miksusanti 2012).

10.5.3 Curcuma Aeruginosa Roxb

10.5.3.1 In Vitro Immunomodulating of C. aeruginosa

The methanol extract of C. aeruginosa at 106.25 μg/ml confirmed mild inhibition on CD18/11a expression of phagocytes, which prompted the use of a flow cytometry method. The extract on the identical concentrations additionally established low inhibition on phagocytosis of leukocytes. A study on the impact of C. aeruginosa methanol extract on ROS technology from polymorph nuclear cells (PMNs) and peritoneal macrophages in human whole blood discovered that the extract possessed ROS inhibitory action for luminol-induced chemiluminescence (CL). C. aeruginosa rhizomes inhibited oxidative bursts of PMNs and macrophages, with IC50 values of 1.8 and 4.6 μg/ml, respectively. Interestingly, C. aeruginosa extract additionally possessed enormous ROS inhibitory action for lucigenin-enhanced CL. However, C. aeruginosa revealed low inhibition on PMN chemotaxis toward the chemoattractant, N-formylmethionyl-leucyl-phenylalanine (fMLP), with 49.9% inhibition (Harun et al. 2015; Jantan et al. 2011).

10.5.3.2 In Vivo Immunomodulating of C. aeruginosa

C. aeruginosa extract, acquired through steam distillation, has been indicated to boost the number of CD4+ and CD8+ cells. A study investigated if C. aeruginosa ethanol extract could boost IFN-γ, TNF-α, IL-2, and IL-12 tiers in 7,12-dimethylbenz [a]anthracene (DMBA)-precipitated Wistar rats. The maximum stimulation on cytokines launch was proven after treatment with the ethanol extract of C. aeruginosa at a dose of 80 mg/200 g frame weight. The aqueous extract of

C. aeruginosa in aggregate with Piper retrofractum and Curcuma zanthorrhiza supplemented in fish fed on the concentrations of 0.5, 1, and 1.5%, respectively, presented better nonspecific immunity of Epinephelus fuscoguttatus. The addition of C. aeruginosa extract precipitated enormous distinction in the general leukocyte matter of Epinephelus fuscoguttatus after being inflamed through Vibrio alginolyticus and V. parahaemolyticus at some stage in 15 days of observation. C. aeruginosa treatment multiplied the entire leukocyte expect day four and day eight. Moreover, C. aeruginosa at awareness of 1% confirmed the most powerful stimulation on phagocytosis action, which became known on day eight. In vitro and in vivo immunomodulating research on C. aeruginosa has been done on their crude aqueous and ethanol extracts. The bioactive metabolites contributing to the modulating results have not been identified. It is essential to chemically signify the extract to determine the bioactive compounds contributing to the immunomodulatory properties and mechanistic research to achieve the plant efficiency and results for the immune-associated disorders (Jantan et al. 2011; Yuandani et al. 2021; Subagiyo et al. 2019).

10.5.4 Curcuma Zedoaria (Christm.) Roscoe

10.5.4.1 In Vitro Immunomodulating Effect of C. zedoaria

C. zedoaria (Christm.) Roscoe rhizome extract has been mentioned to inhibit NO formation from LPS-induced RAW264.7 cells. It has additionally been shown to lessen iNOS and COX-2 expressions. In another study, C. zedoaria prevented ß-hexosaminidase launch in RBL-2H3 cells and confirmed passive cutaneous anaphylaxis response in mice. ß-Hexosaminidase is a marker of antigen-IgE-mediated degranulation. Essential oil from C. zedoaria reduced TNF-α launch from L. monocytogenes and S. aureus-induced RAW264.7 cells. The polysaccharide fraction of C. zedoaria rhizome improved phagocytosis activity and splenocyte proliferation. It additionally induced the number one and secondary titers as well as a delayed-type hypersensitive reaction (Faradilla and Iwo 2014; Lobo et al. 2009).

10.5.4.2 In Vivo Immunomodulating Effect of C. zedoaria

The impact of C. zedoaria extract on tumor development and peripheral blood cells in C57Bl/6 J mice injected with B16F10 murine cancer cells helped determine the distinct routes of administration. A reduction in peritoneal mobileular range and a big boost in general red and white blood mobileular counts has been observed. Oral administration of the extract showed a noteworthy boost in the general leukocyte count. C. zedoaria has additionally been mentioned to stimulate immune reactions in fish. Supplemented diets with C. zedoaria improved the phagocytic reduction and lysosome activity in Epinephelus coioedes fish. C. zedoaria boosted reactive oxygen formation, determined the use of distinct methods, NBT check, and chemiluminescent assay (Nan et al. 2015).

10.5.5 Curcuma Mangga Valeton & Zijp

10.5.5.1 In Vitro Immunomodulating Effect of C. mangga

A study investigated in vitro NO inhibition action of C. mangga, which may contribute to its anti-inflammatory effect (Liu and Nair 2011). Furthermore, C. mangga rhizome extract and its chloroform, hexane, and ethyl acetate fractions decreased NO formation from LPS-induced RAW 264.7 cells. Among the fractions, the chloroform fraction confirmed the very best NO inhibition, followed by hexane, and ethyl acetate fractions. A study investigating the methanol extract of C. mangga rhizomes on whole blood confirmed that the extract exhibited inhibitory action upon activation with zymosan. C. mangga rhizome extract possessed excessive ROS inhibitory action in PMNs and peritoneal macrophages as investigated in a luminol-enhanced CL assay. The extract also inhibited the discharge of ROS from PMNs and macrophages in a lucigenin-based CL assay, with IC50 values of 0.9 and 6.6 μg/ml, respectively (Jantan et al. 2011).

10.5.5.2 In Vivo Immunomodulating Effect of C. mangga

C. mangga Valeton & Zijp rhizome ethanol extract, its different natural fractions (hexane, chloroform, and ethyl acetate), and aqueous fraction have confirmed considerable anti-inflammatory and analgesic activity in mice and inflammatory activity using croton oil-induced mouse ear edema and carrageenan-induced rat paw edema. The plant extract and its fractions at 200 mg/kg demonstrated analgesic action using a decreased range of movement and also produced antinociception using a warm plate and formalin test. At 200 mg/kg, the hexane and chloroform fractions significantly extended the latency time; however, ethyl acetate and aqueous fractions were then no longer active. In addition, the ethanol extract of C. mangga rhizome and its fractions displayed a reduction of paw and ear edema in rats (Ruangsang et al. 2010). A study suggested that the n-hexane, ethyl acetate, and ethanol extracts of C. mangga rhizomes at the doses of 100, 200, and 400 mg/kg accelerated the carbon clearance rate, indicating the enhancement of carbon engulfment using cells in the reticuloendothelial system of mice, thus stimulating the phagocytosis action in mice (Yuliasmi et al. 2019).

10.5.6 Curcuma Amada Roxb

The ethanol, petroleum ether, chloroform, and acetone extracts of C. amada improve the phagocytosis activity of PMNs. The ethanol extract at a dose of 3 mg/ml confirmed the best stimulation on percent of phagocytosis. A further look at late allergic reaction toward sRBCs confirmed that the ethanol extract of C. amada accelerated the paw volume. Moreover, the ethanol extract at the doses of 100, 200, and 400 mg/kg improve the antibody titer dose-dependently (Karchuli and Pradhan n.d.). Supercritical carbon dioxide (CO_2) extract from C. amada rhizomes has the potential for use for the treatment of immune ailments such as autoimmune diseases. Specifically, the extract may be used to treat or prevent allergic reaction diseases, especially IgE mediated allergies in addition to autoimmune disorders (Weidner et al. 2001).

10.6 Conclusion

Curcumin is a biologically active product of turmeric, and it is known to show strong antioxidant, immunomodulatory, anti-inflammatory, anti-exitotoxic, anti-apoptotic actions, and cognitive-enhancing properties. Many of the above reported actions of curcumin may not be seen in target tissues in vivo with oral dosing. In spite of our statement about its poor oral bioavailability, curcumin has some known neuroprotective activity, and many of these may be realized later in vivo. Accumulating cell culture and animal model data advise that dietary curcumin is a potential molecule or compounds for utilization in the prevention or treatment of major age-related neurodegenerative diseases such as AD and PD. Therefore, more comprehensive and well-controlled clinical trials are required to check its bio-efficacy for treating neurodegenerative diseases in humans.

Acknowledgement We would like to thank the MET's, Institute of Pharmacy, BKC, affiliated under Savitribai Phule Pune University, Nashik, Divine College of Pharmacy Satana and Sandip Institute of Pharmaceutical Science for their constant support and providing all facilities to complete this work.

References

Ahire E, Sonawane V, Surana K, Talele G (2021) Drug discovery, drug-likeness screening, and bioavailability: development of drug-likeness rule for natural products. In: Applied pharmaceutical practice and nutraceuticals. Apple Academic Press

Alambra JR, Alenton RRR, Gulpeo PCR, Mecenas CL, Miranda AP, Thomas RC, Velando MKS, Vitug LD, Maningas MBB (2012) Immunomodulatory effects of turmeric, Curcuma longa (Magnoliophyta, Zingiberaceae) on macrobrachium rosenbergii (Crustacea, Palaemonidae)

against vibrio alginolyticus (Proteobacteria, Vibrionaceae). Aquaculture, Aquarium, Conservation & Legislation 5:13–17

Amalraj A, Pius A, Gopi S, Gopi S (2017) Biological activities of curcuminoids, other biomolecules from turmeric and their derivatives–a review. J Tradit Complement Med 7:205–233

Bisoffi M, O'neill JM (2021) Curcumin analogs, oxidative stress, and prostate cancer. In: Cancer. Elsevier

Boroumand N, Samarghandian S, Hashemy SI (2018) Immunomodulatory, anti-inflammatory, and antioxidant effects of curcumin. J Herbmed Pharmacol 7:211–219

Campbell NK, Fitzgerald HK, Malara A, Hambly R, Sweeney CM, Kirby B, Fletcher JM, Dunne A (2018) Naturally derived Heme-oxygenase 1 inducers attenuate inflammatory responses in human dendritic cells and T cells: relevance for psoriasis treatment. Sci Rep 8:1–15

Catanzaro M, Corsini E, Rosini M, Racchi M, Lanni C (2018) Immunomodulators inspired by nature: a review on curcumin and echinacea. Molecules 23:2778

Cundell RD, Wilkinson F (2014) Curcumin: powerful immunomodulator from turmeric. Curr Immunol Rev 10:122–132

De Geus ED, Vervelde L (2013) Regulation of macrophage and dendritic cell function by pathogens and through immunomodulation in the avian mucosa. Develop Comp Immunol 41: 341–351

Faradilla M, Iwo MI (2014) Immunomodulatory effect of polysaccharide from white turmeric [Curcuma zedoaria (Christm.) roscoe] rhizome. J Ilmu Kefarmasian Indonesia 12:273–278

Farooqui T, Farooqui AA (2019) Curcumin: historical background, chemistry, pharmacological action, and potential therapeutic value. In: Curcumin Neurological and Psychiatric Disorders, pp 23–44

Gupta A, Khan S, Manzoor M, Yadav A, Sharma G, Anand R, Gupta S (2017) Anticancer curcumin: natural analogues and structure-activity relationship. In: Studies in natural products chemistry. Elsevier

Harun NH, Septama AW, Jantan I (2015) Immunomodulatory effects of selected Malaysian plants on the CD18/11a expression and phagocytosis activities of leukocytes. Asian Pac J Trop Biomed 5:48–53

Hosseinzade A, Sadeghi O, Naghdipour Biregani A, Soukhtehzari S, Brandt GS, Esmaillzadeh A (2019) Immunomodulatory effects of flavonoids: possible induction of T CD4+ regulatory cells through suppression of mtor pathway signaling activity. Front Immunol 10:51

Jantan I, Harun NH, Septama AW, Murad S, Mesaik M (2011) Inhibition of chemiluminescence and chemotactic activity of phagocytes in vitro by the extracts of selected medicinal plants. J Nat Med 65:400–405

Jarouliya U, Kesharwani RK, Kesharwani RK (2018) Advances of nanotechnology in cancer therapy. In: Nanobiomaterials. Apple Academic Press, pp 291–328. ISBN: 9781771885911

Jayaprakasha GK, Negi PS, Anandharamakrishnan C, Sakariah KK (2001) Chemical composition of turmeric oil-a byproduct from turmeric oleoresin industry and its inhibitory activity against different fungi. Zeitschrift für Naturforschung C 56:40–44

Karbalaei M, Keikha M (2019) The curcumin effects on immuneresponse and its potential properties against tuberculosis. JSM Clin Cytol Pathol 4

Karchuli MS, Pradhan D Curcuma Amada Roxb. Rhizome extract modulates cellular ad humoral immu e system

Keservani RK, Kesharwani RK, Sharma AK (2017b) Introduction to Nanotechnology in Drug Delivery. In: Drug delivery approaches and nanosystems. Apple Academic Press, pp 1–19. ISBN: 9781771885836

Keservani RK, Kesharwani RK, Vyas N, Jain S, Raghuvanshi R, Sharma AK (2010) Nutraceutical and functional food as future food: a review. Pharm Lett 2(1):106–116

Keservani RK, Sharma AK, Ayaz MD, Kesharwani RK (2011) Novel drug delivery system for the vesicular delivery of drug by the niosomes. Int J Res Control Release 1(1):1–8

Keservani RK, Sharma AK, Kesharwani RK (eds) (2019) Nanocarriers for brain targeting: principles and applications. CRC Press. ISBN: 9781771887304

Keservani RK, Sharma AK, Kesharwani RK (2017a) An overview and therapeutic applications of nutraceutical and functional foods. Recent Advances in Drug Delivery Technology 160-201: ISBN:9781522507543

Keservani RK, Sharma AK, Kesharwani RK (2018) Nanoparticulate nanocarriers in drug delivery. In: Nanobiomaterials. Apple Academic Press, pp 3–22. ISBN: 9781771885911

Kesharwani RK, Misra K (2011) Prediction of binding site for curcuminoids at human topoisomerase II α protein; an in silico approach. Curr Sci 101(8):1060–1065

Kesharwani RK, Singh DB, Singh DV, Misra K (2018) Computational study of curcumin analogues by targeting DNA topoisomerase II: a structure-based drug designing approach. Netw Model Anal Health Inform Bioinform 7(1):1–7

Kesharwani RK, Srivastava V, Singh P, Rizvi SI, Adeppa K, Misra K (2015) A novel approach for overcoming drug resistance in breast cancer chemotherapy by targeting new synthetic curcumin analogues against aldehyde dehydrogenase 1 (ALDH1A1) and glycogen synthase kinase-3 β (GSK-3β). Appl Biochem Biotechnol 176(7):1996–2017

Kim G-Y, Kim K-H, Lee S-H, Yoon M-S, Lee H-J, Moon D-O, Lee C-M, Ahn S-C, Park YC, Park Y-M (2005) Curcumin inhibits immunostimulatory function of dendritic cells: mapks and translocation of NF-κb as potential targets. J Immunol 174:8116–8124

Kim OK, Yoo SA, Nam D-E, Kim Y, Kim E, Jun W, Hwan K, Lee J (2014) Immunomodulatory effects of Curcuma longa L. extract in LP-BM5 murine leukemia viruses-induced murine acquired immune deficiency syndrome. J Korean Soc Food Sci Nutr 43:1317–1324

Kim S, Kook KE, Kim C, Hwang J-K (2018) Inhibitory effects of Curcuma xanthorrhiza supercritical extract and xanthorrhizol on LPS-induced inflammation in HGF-1 cells and RANKL-induced osteoclastogenesis in RAW264. 7 cells. J Microbiol Biotechnol 28:1270–1281

Kohyama A, Yamakoshi H, Hongo S, Kanoh N, Shibata H, Iwabuchi Y (2015) Structure-activity relationships of the antitumor C5-curcuminoid GO-Y030. Molecules 20:15374–15391

Kumolosasi E, Ibrahim SNA, Shukri SMA, Ahmad W (2018) Immunostimulant activity of standardised extracts of Mangifera indica leaf and Curcuma domestica rhizome in mice. Trop J Pharm Res 17:77–84

Liu Y, Nair MG (2011) Labdane diterpenes in curcuma mangga rhizomes inhibit lipid peroxidation, cyclooxygenase enzymes and human tumour cell proliferation. Food Chem 124:527–532

Lobo R, Prabhu KS, Shirwaikar A, Shirwaikar A (2009) Curcuma zedoaria Rosc.(white turmeric): a review of its chemical, pharmacological and ethnomedicinal properties. J Pharm Pharmacol 61: 13–21

Mauren FM, Yanti BWL (2016) Efficacy of oral curcuminoid fraction from curcuma xanthorrhiza and curcuminoid cider in high-cholesterol fed rats. Pharm Res 8:153

Miksusanti M (2012) Lymphocyte Proliferation by Temu Lawak (Curcuma xanthorrhiza ROXB) Essential Oil. In: International Conference on Bioinformatics and Biomedical Technology? ICBBT 2012. CBEES

Mimche PN, Taramelli D, Vivas L (2011) The plant-based immunomodulator curcumin as a potential candidate for the development of an adjunctive therapy for cerebral malaria. Malar J 10:1–9

Mishra H, Kesharwani RK, Singh DB, Tripathi S, Dubey SK, Misra K (2019) Computational simulation of inhibitory effects of curcumin, retinoic acid and their conjugates on GSK-3 beta. Netw Model Anal Health Inform Bioinform 8(1):3

Nan F-H, Agus Putra A, Margie B, Lee M (2015) The effects of Curcuma zedoaria and Zingiber zerumbet on non-specific immune responses of grouper Epinephelus coioides. IJFS 14:598–611

Nogami A, Sakata Y, Uchida N, Yamaguchi K, Kawasaki C, Shindo T, Kubota K, Katsurabayashi S, Takasaki K, Mishima K (2011) Effects of yokukansan on anxiety-like behavior in a rat model of cerebrovascular dementia. J Nat Med 65:275–281

Rahimi K, Ahmadi A, Hassanzadeh K, Soleimani Z, Sathyapalan T, Mohammadi A, Sahebkar A (2019) Targeting the balance of T helper cell responses by curcumin in inflammatory and autoimmune states. Autoimmun Rev 18:738–748

Ruangsang P, Tewtrakul S, Reanmongkol W (2010) Evaluation of the analgesic and anti-inflammatory activities of curcuma mangga Val and Zijp rhizomes. J Nat Med 64:36–41

Sahebkar A, Serban M-C, Ursoniu S, Banach M (2015) Effect of curcuminoids on oxidative stress: a systematic review and meta-analysis of randomized controlled trials. J Funct Foods 18:898–909

Sasikumar B (2012) Turmeric. Elsevier, Handbook of herbs and spices

Scartezzini P, Speroni E (2000) Review on some plants of Indian traditional medicine with antioxidant activity. J Ethnopharmacol 71:23–43

Sharma AK, Keservani RK, Keseharwani RK (eds) (2018) Nanobiomaterials: applications in drug delivery. CRC Press. ISBN: 9781771885911

Shirley SA, Montpetit AJ, Lockey R, Mohapatra SS (2008) Curcumin prevents human dendritic cell response to immune stimulants. Biochem Biophys Res Commun 374:431–436

Singh DB, Gupta MK, Kesharwani RK, Misra K (2013a) Comparative docking and ADMET study of some curcumin derivatives and herbal congeners targeting β-amyloid. Network Modeling Analysis in Health Informatics and Bioinformatics 2(1):13–27

Singh DV, Agarwal S, Kesharwani RK, Misra K (2013b) 3D QSAR and pharmacophore study of curcuminoids and curcumin analogs: interaction with thioredoxin reductase. Interdiscip Sci 5(4):286–295

Singh P, Kesharwani RK, Misra K, Rizvi SI (2015) The modulation of erythrocyte Na+/K+-ATPase activity by curcumin. J Adv Res 6(6):1023–1030

Singh P, Kesharwani RK, Misra K, Rizvi SI (2016) Modulation of erythrocyte plasma membrane redox system activity bycurcu min. Biochem Res Int 2016:702039

Singh S, Sankar B, Rajesh S, Sahoo K, Subudhi E, Nayak S (2011) Chemical composition of turmeric oil (Curcuma longa L. cv. Roma) and its antimicrobial activity against eye infecting pathogens. J Essent Oil Res 23:11–18

Šturdík E (2013) Therapeutic potential of curcumin in medicinal chemistry. Acta Chimica Slovaca 6:89–99

Subagiyo S, Pramesti R, Pringgenies D (2019) Effectiveness of herbal extract (Piper retrofractum, Curcuma aeruginosa, and curcuma zanthorrhiza) as Immunomodulator in non-specific immunity system of Tiger grouper (Epinephelus fuscoguttatus) against infection from vibrio alginolyticus and vibrio parah. SciTechnol Indonesia 4:94–100

Surana KR, Ahire ED, Sonawane VN, Talele SG, Talele G (2021a) Informatics and methods in nutrition design and development. In: Natural food products and waste recovery. Apple Academic Press

Surana KR, Ahire ED, Sonawane VN, Talele SG, Talele G (2021b) Molecular modeling: novel techniques in food and nutrition development. In: Natural food products and waste recovery. Apple Academic Press

Upadhyaya J, Kesharwani RK, Misra K (2009) Metabolism, pharmacokinetics and bioavailability of ascorbic acid; synergistic effect with tocopherols and curcumin. J Comput Intell Bioinform 2(1):77–84

Vyas N, Keservani RK, Nayak A, Jain S, Singhal M (2010) Effect of Tamarindus Indica and curcuma longa on stress induced alopecia. Pharmacologyonline 1:377–384

Weidner, M. S., Petersen, M. J. & Jacobsen, N. 2001. Certain diterpenes and extracts or concentrates of curcuma amada containing them for use as medicaments. Google Patents

Yuandani IJ, Rohani AS, Sumantri IB (2021) Immunomodulatory effects and mechanisms of curcuma species and their bioactive compounds: a review. Front Pharmacol 12:643119

Yuliasmi S, Satria D, Dongoran R, Sinaga M, Marpaung N (2019) Correlation between the phytochemical constituents of curcuma mangga and its immunomodulatory effect. Rasāyan J Chem 12:1–6

Zahedipour F, Hosseini SA, Sathyapalan T, Majeed M, Jamialahmadi T, Al-Rasadi K, Banach M, Sahebkar A (2020) Potential effects of curcumin in the treatment of COVID-19 infection. Phytother Res 34:2911–2920

Chapter 11
Colchicine and Andrographolide as Natural Immunomodulators

S. Yasri and V. Wiwanitkit

Abstract Conceptually, a promotion of appropriate immune reaction to foreign body will help in the management of the medical disorder. The immunomodulating management is a specific approach. It is the use of other substances for modifying the immunological process. The immunomodulator is a specific agent that has main action in modifying of immunological reaction. This modulating process is helpful for managing immune-related disorder. There are many immunomodulatory agents. The examples of immunomodulating agents are methotrexate, alisporivir, tocilizumab, chloroquine, anakinra, steroid, colchicine, and andrographolide. Of the many immunomodulating substances, some are also naturally available. In this chapter, the authors summarize and discuss on some important natural immunomodulators. The summarization on colchicine and andrographolide as a natural immunomodulator is done. At present, both colchicine and andrographolide are well-known biochemies that are widely used in clinical practice. The immunomodulating role of both biosubstances is interesting. The immunomodulation effect of both biosubstances is already proposed in managing many medical disorders. The good examples are the application of colchicine and andrographolide in management of malignancies and infectious diseases. The applied usage of colchicine and andrographolide is an extraordinary use apart from its classic indication. The studies on the specific issues on immunomodulating action of both colchicine and andrographolide are interesting and should be performed for getting useful data for further application in the field of immunotherapy.

Keywords Immunomodulator · Colchicine · Andrographolide · Natural

S. Yasri (✉)
Private Academic Consultant, Bangkok, Thailand

V. Wiwanitkit
Dr. DY Patil University, Mumbai, India

Hainan Medical University, Haikou, China

R. K. Kesharwani et al. (eds.), *Nutraceuticals and Functional Foods as Immunomodulators*, https://doi.org/10.1007/978-981-19-2507-8_11

11.1 Introduction

Immunity is an important thing for homeostasis. The good immune system is required for maintenance of normal human health. The disorder of immunity is a common problem in clinical medicine. There are many possible medical disorders due to malfunction of the immune system. Physiologically, immune system is an important system that plays important role in defense against foreign bodies. When there is an invasion of foreign bodies, human body will create a defensive reaction. Of several defenses, immunoreaction is an important biological process. If there is an abnormality in immune reaction, there might be a possible medical problem.

Generally, there are many medical problems that are the results of the defect of immunity. The examples are inflammation, infection, malignancy, hypersensitivity, etc. There are many forms of abnormality of immune system. An appropriate function of immune system due to the disorder can lead to abnormal clinical presentation. Physically, an appropriate immunoactivity is required. A too much or a too little immune reactivities are considered important problems in medicine. The poor immune status or immune impairment or immunodefect can result in many clinical problems. The medical disorders in the group of immunodeficiency include immunodeficiency virus infection, opportunistic infection, and malignancy. On the other hand, a too much immune activity can also result in an unwanted clinical presentation. Hypersensitivity is the good example of inappropriate excessive immune response causing medical problems.

Based on clinical epidemiological concept, the triad of host, foreign body, and environment play a role in immune reaction process. The triad must exist at the same time and same place. When interrelationship among triad occurs, the immunological process will start. When a foreign body enters into human body, a defensive reaction will be triggered. The defensive mechanism plays its important function for counteracting the invader (Furman and Davis 2015). Immunological system is an important part of defensive mechanism in human body. Ideally, if there is a strong appropriate immunological system of the host, the clearance of any possible results from alien foreign bodies is expected. This concept is applicable for management of a disease. The clinical immunologist mainly focuses on the immune mechanism. The specific study to understand the exact immunological pathway that host responded to the different foreign bodies is useful. The studies can provide scientific data for further searching for proper diagnosis and treatment of the medical disorders. The knowledge on immunopathogenesis can help explain the pathobiology of the disease. For management of the medical problem, the immune system is an important target since immune system has its main function in counteracting with alien foreign bodies. Basically, a promotion of appropriate immune reaction to foreign body is a way for managing the medical disorder. An immune system is manipulation of the current concept of immunotherapy (Fig. 11.1).

For management of the disease, a promotion of appropriate immune reaction to foreign body can help managing the medical problem. To specifically manipulate on immune system is the core immunotherapeutic concept. There are many ways for

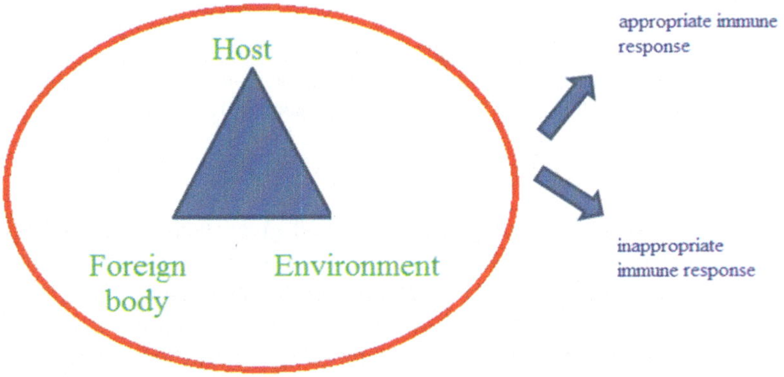

Fig. 11.1 Epidemiological triad, immunological reaction, and immune-related disorder

immunotherapy such as vaccine therapy, immunoglobulin therapy, and immunomodulation. To achieve a successful immunotherapeutic outcome, the basic requirements include (a) effective immunotherapeutic agent, (b) safe immunotherapeutic agent, (c) sufficient immunotherapeutic agent, (d) good immunotherapeutic agent administration, and (e) good biotransportation and biodistribution of the agent (Tiligada et al. 2015). As already mentioned, the immunotherapy is a new therapeutic approach for management of several medial diseases such as infectious diseases and cancers. The immunopharmacology regarding the immunotherapy for treatment is interesting and will be further mentioned. To provide a good immunotherapy, the knowledge on both immunopathogenesis and immunopharmacology are necessary. In immunopharmacology, the necessary knowledge include principles of immunopharmaceutic, immunopharmacostatic, and immunopharmacodynamic (Tiligada et al. 2015). At present, there are many new immunological treatments proposed for management of the infection. The specific researches on this area are necessary.

As earlier mentioned, there are many forms of immunotherapy at present. The good examples are vaccinotherapy, immunoglobulin therapy, and immunomodulation. Immunoglobulin therapy is classified as a passive immunotherapy, whereas vaccinotherapy is classified as an active immunotherapy. The immunomodulation is another specific group. Immunomodulation makes use of other substances for modifying the immunological process. The immunomodulator is a specific agent that has main action in modifying of immunological reaction. This modulating process is helpful for managing of immune-related disorder. There are many immunomodulatory agents. The examples of immunomodulating agents are methotrexate, alisporivir, tocilizumab, chloroquine, anakinra, steroid, colchicine, and andrographolide. Of many immunomodulating substances, some are also naturally available (Table 11.1).

The natural immunomodulating substance is a specific natural substance that has the ability for immunological modification. The natural immunomodulating substance is considered as a natural product. Hence, it is concordant with present

Table. 11.1 Examples of immunomodulatory agents for immunotherapy

Diseases	Examples
1. Non-natural immunomodulator	Methotrexate, alisporivir, tocilizumab, chloroquine, anakinra
2. Natural immunomodulator	Colchicine, andrographolide, beta-glucan

concept of promotion of natural product use and green technology. As a natural product, natural immunomodulating substance is usually available from natural sources, especially for plants. The natural product can reduce the problematic hazardous substance from chemical synthesis process. Therefore, it is within the scope of natural medicine, which is a contemporary concept for managing medical problem. In fact, phytomedicine is classical concept in alternative medicine. Several herbs have been used for many centuries, and the rooted local wisdom is an actual hidden gem in medicine. The natural immunomodulating substance will be a useful tool for immunotherapy if there is a good systematic evaluation of it property.

In this chapter, the authors summarize and discuss on some important natural immunomodulating agents. The authors hereby summarize on colchicine and andrographolide as a natural immunomodulaton. At present, both biosubstances are well-known biochemies that are widely used in clinical practice. The immunomodulating role of these biosubstances is interesting. The immunomodulation effect of the two biosubstances is already proposed in managing many medical disorders. The good examples are the application of colchicine and andrographolide in management of cancers and infections.

11.2 Natural Immunomodulator

As already mentioned, immunomodulation is an important kind of immunotherapy. Conceptually, immunomodulation therapy uses the modifying of the existed immune system for management of the disease. The modification might be a stimulant or suppression. Also, it might be an induction for a new alternative pathway. The use of substance that can affect the existing immunological process is the main practice. For immunomodulation therapy, the immunomodulating substance plays important key function in treatment. There are many kinds of immunomodulating substances. The natural immunomodulating substance is an important group.

A natural immunomodulating substance is derived from natural source. It is a kind of natural product. In fact, numerous substances existed in nature, and some are medically beneficial (Bratman et al. 2019; Franco et al. 2017). Therefore, it is a basic concept in natural medicine on use of natural product for management of a medical disorder (Falzon and Balabanova 2017). In fact, there is a long history of natural product use for management of medical disorder. It is the basic principle of many alternative medicine systems (such as Chinese medicine and Ayurveda). Many natural products are available from plants or animals and have been used for alternative medicine treatment. It is confirmed that herbal medicine is a useful classic wisdom for management of disease. Many new modern drugs are derived from well-known herbs. The good examples are artemisinin, an effective antimalarial drug (Lee 2007). It is no doubt that there are many present ongoing researches to find active components in classic herb for new drug search. Many herbs have their direct or indirect effects on immune system. Many herbs give the biological components that can be further used as immunomodulating substances.

To find a new herb-derived immunomodulating substance is very interesting. This might follow the standard guidelines for new drug search (Jiang et al. 2020; Leonti and Casu 2018). Briefly, it is necessary to gather the discrete hidden data on classic herbs. This might be by literature search on alternative medicine documents. Sometimes, it might require an in-depth field study to explore hidden alternative medicine practice within the community. The researcher has to follow the ethnopharmacological approach to get the data on ethnopharmacological herbs. Also, the manipulation of the herbal plant must base on the principles of medical botany. The horticultural studies might also be needed for getting a pure sample of herb for further studies. It should be noted that the active ingredient might be available from any parts of the plant such as stem, bark, leaf, flower, or root. The environmental contamination is possible.

The environmental contamination might be an important confounding factor for verification; the active components of the herbal plant are well controllable by a standard controlled horticulture.

When data are collected, a systematic arrangement for further analysis is required. The analysis might be by component analysis. The herbal sample collection might be done for further laboratory component analysis. Sometimes, the pharmaco-informatics might be applied for comparative analysis Oprea et al. 2012; Tanaka 2005). The comparison on the available database of active ingredients within plants might be done and can reduce required time for ingredient identification of the collected samples. After getting the data on active ingredient, it can further be considered for possible pharmacological usefulness. If it is evidenced that there is a possible pharmacological useful component, the herb might be further studied for extraction of active ingredient. If the active ingredients can be extracted and further purified, a further pharmaceutical process can be started. When a formulated form of active ingredients is successfully prepared, it has to pass further standard clinical trial before it can be finalized for clinical usage. The mentioned process is a strict requirement for scientific verification on the efficacy and safety of the new medical natural product (Singh et al. 2012). Without standardization and quality control, there might be a question on the quality and safety of the new natural product for clinical use.

Many new drugs passed the mentioned process and are already in use. Many drugs from natural source have its immunological modification property. This is an interesting issue for modern immunotherapy. Briefly, natural substances have immunological modification properties, and if those substances are well studied and developed as new drugs, it will be very useful. The natural-derived immunomodulating agent can promote the local research and local pharmaceutical system. In fact, many countries with their own alternative medicine system can develop new drugs including immunomodulating agents from local herbs. For example, China can successfully prepare many new drugs from classic Chinese herbs. The good examples are drugs made from ginseng (Shergis et al. 2013a, b) and ganoderma (Cao et al. 2018; Pan and Lin 2019; Sanodiya et al. 2009). This is a way to promote local wisdom and local pharmaceutic system. This can result in a cost-effective pharmaceutical management.

11.3 What Are Colchicine and Andrographolide?

As already mentioned, there are many biosubstances that might be used in immunotherapy. The application of immunotherapeutic agent is a new issue in clinical medicine. The use of immunotherapy to manage disease is a very interesting clinical issue. The data on this specific is little mentioned. Of several agents, the two important agents, colchicine and andrographolide, are widely mentioned as a natural immunomodulator. Here, the authors will summarize the important information on both substances.

11.3.1 Colchicine

Colchicine is an anti-inflammatory agent that is widely used in clinical practice for management of gout. This biosubstance that has been used for the treatment of various medical problems for many years. Examples of medical problem that might be managed by colchicine include gout, familial Mediterranean fever, atrial fibrillation pericarditis, chronic urticaria, cutaneous vasculitis, and psoriasis (Alkadi et al. 2018; Robinson and Chan 2018). Chemically, colchicine is a tricyclic, lipid-soluble alkaloid (Angelidis et al. 2018). Colchicine is an alkaloid extracted from natural source. Naturally, it is detected in the *Colchicum autumnale* plant. *Colchicum autumnale* has its common name as autumn crocus, meadow saffron, or naked ladies (Hermanns-Clausen et al. 2006). This plant is a toxic autumn-blooming flowering plant (Hermanns-Clausen et al. 2006). It has leaves up to 25 cm (10 in) long. Its flowers are solitary, about 5 cm, with six tepals and six stamens with orange anthers and three white styles (Hermanns-Clausen et al. 2006). Its health effects have been recognized for many centuries (Tsoucalas et al. 2018).

Regarding its alkaloids and colchicine, it was firstly isolated from colchicum autumnale in the nineteenth century (Karamanou et al. 2018). The molecular chemical formula of colchicine is C22H25NO6. This biosubstance consists of three rings (Alkadi et al. 2018). The plant source is the Lily family *Colchicum autumnale*, sometimes, or "autumn crocus" (Angelidis et al. 2018). Colchicine is predominantly metabolized in the gastrointestinal tract (Angelidis et al. 2018). Two proteins,

P-glycoprotein (P-gp) and CYP3A4, are main determinants for colchicine pharmacokinetic (Angelidis et al. 2018). Usefulness of colchicine is impaired by the impact of underlying comorbidities and drug interactions that can modify its pharmacokinetics and pharmacodynamics (Pascart and Richette 2018). The clinical use of colchicine may be limited by concerns over its unwanted adverse effects and the potential for toxicity (Robinson and Chan 2018). There are researches with structural modification of colchicine in order to increase their pharmacological effectiveness and reduce the side effects of toxicity arising from this substance (Alkadi et al. 2018). In clinical practice, colchicine is a widely used anti-inflammatory agent. This specific drug is used for the treatment of several rheumatological conditions, especially for gouty arthritis (Indraratna et al. 2018). Gout is a specific metabolic disorder that there is an abnormality of the nucleic acid biopathway (Akram et al. 2014; Desai et al. 2017; Sunkureddi 2011; Wilson and Saseen 2016; Yamanouchi 2003). The clinical presentations of gouty arthritis include hyperuricemia and arthritis (Akram et al. 2014; Desai et al. 2017; Sunkureddi 2011; Wilson and Saseen 2016; Yamanouchi 2003). In a severe case, it might cause renal problem. In gout therapy, colchicine action is blocking polymerization of tubulin that further results in preventing the activation of the inflammasome (Slobodnick et al. 2015). The primary mechanism of action of colchicine is tubulin disruption that further results in downregulation of multiple inflammatory pathways. This is the process of modulation of innate immunity (Leung et al. 2015). Colchicine also has various inhibitory effects on macrophages including the inflammasome, inhibition of pore formation activated by purinergic receptors P2X7 and P2X2, inhibition of the NACHT-LRRPYD-containing protein 3 (NALP3), and promotion of dendritic cell maturation and antigen presentation (Leung et al. 2015). Additionally, colchicine also has antifibrotic properties and endothelial functional regulatory effect (Leung et al. 2015).

Since microtubule polymerization affects a variety of cellular processes including maintenance of shape, signaling, division, migration, and cellular transport, colchicine has effects on microtubule polymerization; hence, it can result in change of cellular mechanisms (Angelidis et al. 2018). In brief, c colchicine interferes with several inflammatory pathways including adhesion and neutrophil induction, inflammasome activation, superoxide production, the RhoA/Rho effector kinase (ROCK) pathway and the tumor necrosis factor alpha (TNF-α)-induced nuclear factor κB (NF-κB) pathway (Angelidis et al. 2018). The mentioned change results in a direct effect on the inflammatory response (Angelidis et al. 2018). The efficacy of colchicine for the treatment of gout is confirmed from many randomized-controlled trials. The use of colchicine must be with a high cautions. In pregnant women, colchicine therapy did not cause fetal malformations or miscarriage when taken during pregnancy (Indraratna et al. 2018). Colchicine therapy should not be withheld in obstetric case (Indraratna et al. 2018). Additionally, Indraratna et al. (2018) found that the incidence of miscarriage was significantly lower in women who received colchicine therapy compared with those that did not.

The toxicity of colchicine is interesting. This drug has a narrow therapeutic index, with no clear-cut distinction between nontoxic, toxic, and lethal doses. This nature results in a substantial confusion among clinicians. The commonest side effects of colchicine are nausea, vomiting, and, particularly, diarrhea (Angelidis et al. 2018). The side effect is usually dose related, and it can occur in one-tenth of the patient receiving colchicine (Angelidis et al. 2018). Toxicity of colchicine is also important problem in clinical practice. Although colchicine poisoning is sometimes intentional; unintentional toxicity is not rare (Babu et al. 2012). The colchicine's toxicity resulted from excessive action – binding to tubulin and disrupting the microtubular network (Finkelstein et al. 2010). If the poisoning occurs, it is usually associated with a poor clinical outcome (Finkelstein et al. 2010). Impaired glomerular filtration rate also increases risk of colchicine toxicity (Solak and Sipahi 2016). Colchicine also has an interaction with many drugs including clarithromycin, erythromycin, ketoconazole, and ciclosporin (Finkelstein et al. 2010). It also has interaction with natural grapefruit juice (Finkelstein et al. 2010). If drug-drug interaction occurs, increased colchicine concentrations might occur (Finkelstein et al. 2010).

11.3.2 Andrographolide

Andrographolide is a natural compound extracted from *Andrographis paniculata* (Lu et al. 2019). *Andrographis paniculata* is a kind of plant. Its general name is creat or green chiretta. The plant has a height of 30 to 110 cm and is well grown in moist and shady places (Akbar 2011; Hossain et al. 2014). The stem color is dark green (Akbar 2011; Hossain et al. 2014). It is square in cross-section and has longitudinal furrows and wings along the angles (Akbar 2011; Hossain et al. 2014). The lance-shaped leaves have hairless blades. It has small pink solitary flowers. Its fruit is a capsule around 2 cm containing many yellow-brown seeds (Akbar 2011; Hossain et al. 2014). Andrographolide was successfully isolated from the herbal plant as a major bioactive in 1951 (Tan et al. 2017). The plant source is a well-known traditional herb that has been used in ancient Asia for a very long time (Dai et al. 2019). This herbal plant is indicated for management of many medical problems such as malignancy, rheumatoid arthritis, diarrhea, and upper respiratory tract infection (Lu et al. 2019). The part of the herbal plant that has high andrographolide content is its leaf. The pharmacological mechanisms of action of andrographolide include antioxidative stress, anti-inflammation, antiapoptosis, and proapoptosis (Lu et al. 2019). Therefore, andrographolide has a wide range of biological activities such as antioxidant, anti-inflammatory, and anticancer properties (Gupta et al. 2017; Mussard et al. 2019). Additionally, andrographolide is also proposed for its antibacterial, antimalarial, anti-hepatitis, anti-HIV, anti-atherosclerosis, hepatoprotective, and α-glucosidase inhibition properties (Kandanur et al. 2019). Andrographolide expresses immunomodulatory effects. The immunomodulation is by effectively enhancing natural killer (NK) cells, cytotoxic T cells, phagocytosis, and antibody-dependent cell-mediated cytotoxicity (ADCC) (Mussard et al. 2019). Andrographolide also has activities on cell cycle arrest, necrosis, and autophagy and poses anti-angiogenic activity (Islam et al. 2018). Regarding the effect on immunological system, inhibition of NF-κB activity is the prevailing anti-inflammatory mechanism of andrographolide (Tan et al. 2017) Nrf2 activation is also observed (Tan et al. 2017). Andrographolide and its analogs can provide anti-inflammatory benefits in inflammatory disease models. There are also some clinical trials conducted using andrographolide in fixed combination formulation. In those studies, anti-inflammatory property is reported (Tan et al. 2017). Therefore, it is the main focus for further research and development (Dai et al. 2019). Since the 1980s, andrographolide and its analogs have been focused on new anti-inflammatory drug search, and it is believed that the new drugs developed based on andrographolide will be a new useful anti-inflammatory agent (Tan et al. 2017; Zhang et al. 2020).

11.4 Colchicine and Andrographolide as Natural Immunomodulators

11.4.1 Colchicine as a Natural Immunomodulator

As already mentioned, colchicine poses the anti-inflammatory and immunomodulatory property. The application in immunotherapy is a very interesting issue. Colchicine is proposed for its role in immunomodification in many diseases. Regarding infectious disease, colchicine is widely studied for its efficacy in many infections. Of several infections, the role of colchicine in managing human immunodeficiency virus (HIV) should be mentioned. At present, HIV infection is the most common immune disorder. This disease results from a retrovirus infection. The impairment of lymphocyte function is observed, and it can lead to clinical problem. Opportunistic infection and opportunistic malignancy might occur in HIV-infected patient due to immune status. To manage HIV-infected patient, the adjustment of immunity is an important concept. It is questionable whether immunomodulatory effect of colchicine can help immune adjust in HIV cases. There are many new attempts for new drug design for managing HIV infection, and colchicine derivatives are under research and development (Worachartcheewan et al. 2019). Tatematsu et al. (1991) reported on inhibitory effects of colchicine derivatives on HIV replication in H9 lymphocyte cells. Tatematsu et al. (1991) found that colchicine derivatives might be useful for managing HIV infection. Zhang et al. (2011) performed an animal study and found that an unpolarized release of and HIV antigen by colchicine treatment could enhance intranasal HIV antigen expression and mucosal humoral responses. Zhang et al. (2011) noted that this is an important clue for HIV vaccine development. Nevertheless, some reports also present data that do not support the usefulness of colchicine in HIV infection management. Yoder et al. (2011) reported that although colchicine could disrupt microtubule integrity, almost no inhibition of HIV-1 infection was observed. Yoder et al. (2011) gave a conclusion against the use of colchicine for managing HIV infection.

Table 11.2 Examples of using colchicine for cancer therapy

Authors	Details
Bhattacharya et al. (2016)	Bhattacharya et al. (2016) reported that colchicine could induce autophagy and senescence in lung cancer cells at clinically admissible concentration. Bhattacharya et al. (2016) proposed for a potential use of colchicine in combination with autophagy inhibitor in cancer therapy
Cho et al. (2017)	Cho et al. (2017) studied on anticancer effects of colchicine on hypopharyngeal cancer. Cho et al. (2017) concluded that colchicine might be a useful additional agent for cancer treatment
Kumar et al. (2016)	Kumar et al. (2016) found that a colchicine analogue could trigger apoptosis and autophagy in HCT-116 colon cancer cells
Kuo et al. (2015)	Kuo et al. (2015) found that colchicine could reduce incident cancer in gout male patients
Lin et al. (2013)	Lin et al. (2013) studied on anticancer mechanisms of clinically acceptable colchicine concentrations on hepatocellular carcinoma
Lin et al. (2016)	Lin et al. (2016) studied on anticancer effects of clinically acceptable colchicine concentrations on human gastric cancer cell lines
Sun et al. (2016)	Sun et al. (2016) studied on proliferation inhibition and apoptosis processes of breast cancer MCF-7 cells under the influence of colchicine

Another interesting application is the proposed usefulness for management of coronavirus disease 2019 (COVID-19). COVID-19 is a new emerging coronavirus infection that can induce a severe respiratory viral syndrome (Vieira et al. 2020). This new viral respiratory infection causes pandemic in 2020. Many millions of world populations already get COVID-19. The immunopathological process in COVID-19 can lead to severe clinical presentation. The patient might have severe pneumonia and other lung complications (Liang et al. 2020). Abnormal immune responses in COVID-19 can result in immunopathological inflammatory disorder. The use of colchicine for managing abnormal immune response in COVID-19 is interesting. Mansouri et al. (2020) concluded that colchicine was a safe, inexpensive, and oral medication for the treatment of mild-to-moderate COVID-19.

Focusing on cancer therapy, the immunomodulatory effect of colchicine is widely studied, and it is a possible new application of the classic drug (Dasgeb et al. 2018) (Table 11.2). At present, there are many novel colchicine derivatives that are proposed for their useful anticancer activity (Johnson et al. 2017). It seems that there are many in vitro reports showing possible advantages of colchicine in cancer therapy. Since there is still no complete clinical trials in human subjects, it is difficult to conclude on the advantage of the colchicine. Further researches are required.

Colchicine is usually for managing not only immune defect problem but also hypersensitivity problem. The colchicine use for managing reactive vasculitis is another interesting application. Colchicine is useful for managing nodular vasculitis, a chronic relapsing lobular panniculitis that is believed to be a hypersensitivity reaction to antigenic triggers (Gilchrist and Patterson 2010; Wee and Kelly 2017). The main indication for using colchicine for managing cutaneous vasculitis is mild recurrent or persistent disease (Chen and Carlson 2008). At present, colchicine becomes a widely used immunological dermatologic therapeutic agent (Robinson and Chan 2018, b). In addition to cutaneous vasculitis, colchicine is also useful for managing visceral problem (Cocco et al. 2010). The management of pericarditis is a good example. Colchicine is a recommended drug for management of immune-related recurrent pericarditis (Imazio et al. 2016; Imazio et al. 2017). Morel et al. (2015) reported an observational study on patients with systematic lupus erythematosus. Morel et al. (2015) found that colchicine was safe and effective in treating lupus pericarditis and proposed that colchicine might be used as a steroids-sparing agent.

11.4.2 Andrographolide as a Natural Immunomodulator

Andrographolide is proposed for its role in immunomodification in many diseases. Regarding infectious disease, andrographolide is widely studied for its efficacy in many infections. Dengue is a good example that is widely studied. In general dengue is a mosquito-borne infection. This disease is an acute febrile illness and usually presents with a clinical triad, hemoconcentration, atypical lymphocytosis, and thrombocytopenia (Wiwanitkit 2010). The disease is endemic in tropical country, and it is still an important public health problem in many countries. For treatment of dengue, the fluid replacement therapy is the main management. For immunotherapy, it is an interesting concept. Panraksa et al. (2017) studied on activity of andrographolide against dengue virus. Panraksa et al. (2017) found that andrographolide had expressed a significant anti-dengue virus activity in both cell lines. Andrographolide could reduce both the levels of cellular infection and virus output (Panraksa et al. 2017). Similar observations are also reported in other studies by different investigators (Edwin et al. 2016; Paemanee et al. 2019). Apart from dengue virus, the inhibitory effects of Andrographolide on other virus infection such as hepatitis B virus are also reported (Mehrotra et al. 1990). Focusing on malaria, another important tropical mosquito-borne infection, there are also reports on clinical advantages of andrographolide. Zaid et al. (2015) found that andrographolide could effect on both *Plasmodium falciparum*-infected erythrocyte. Zaid et al. (2015) proposed that this observation might or might not associate with pathogenic parasite cellular invasion.

An interesting use of Andrographolide is for management of HIV infection. Andrographolide and its derivatives are tested for anti-HIV properties (Reddy et al. 2005). Indeed, HIV infection is a well-known immunodeficiency disorder. Since andrographolide has immunomodulatory effect, it might be useful for managing the HIV-infected patients. There are few reports on this specific issue. Uttekar et al. (2012) studied on anti-HIV activity of semisynthetic derivatives of andrographolide and concluded that andrographolide and its derivatives, 6 and 9, could inhibit gp120-mediated cell fusion. Additional molecular analysis showed that the tested molecules reacted by binding to the residues of V3 loop of gp120 (Uttekar et al. 2012). In another clinical trial by Calabrese et al. (2000), andrographolide showed inhibitory effect on HIV-induced cell cycle dysregulation, and this could result in a rise in CD4(+) lymphocyte levels in HIV-1-infected individuals.

Another interesting application is used for management of COVID-19. COVID-19 is caused by a new viral pathogen namely SARS-CoV2. The disease can result in a respiratory viral syndrome (Vieira et al. 2020). This new viral respiratory infection is an important global public health threat. After its first report from China, it already spreads to more than 200 countries around the world. More than 22 million of world populations already get COVID-19. The immunopathological process in COVID-19 can lead to severe clinical presentation. The patient might have severe pneumonia. A respiratory distress syndrome (ARDS) and pulmonary fibrosis (PF) might be important complications (Liang et al. 2020). Pathophysiologically, maladjusted, too longed, or exaggerated immune responses can cause immunopathological inflammatory disorder. Therefore, the immunotherapy might play important role in management of this new emerging infection. Murugan et al. (2020) recently performed a computational study on *Andrographis paniculata* phytochemicals to evaluate their potency against SARS-CoV-2 pathogen. Murugan et al. (2020) focused on the potency of the four selected phytochemicals namely andrographolide (AGP1), 14-deoxy andrographolide (AGP4), 14-deoxy 11,12-didehydro andrographolide (AGP2), and neoandrographolide (AGP3) and from the plant. The four biosubstances were tested against the four key targets including three nonstructural proteins (3 L main protease (3CLpro), Papain-like proteinase (PLpro) and RNA-directed RNA polymerase (RdRp)) and a structural protein (spike protein (S)) of the SARS-CoV-2 (Murugan et al. 2020). Murugan et al. (2020) concluded that AGP3 could be used for management of the infection, and note that this substance was as a cost-effective drug analog. Regarding molecular mechanism, Enmozhi et al. (2020) proposed that andrographolide acts as a potential inhibitor of SARS-CoV-2 main protease. From an in silico molecular docking experiment, Enmozhi et al. (2020) found that the andrographolide could successfully form a molecular binding in the binding site of SARS-CoV-2 Mpro. Banerjee et al. (2020) concluded that Andrographolide had many interesting pharmacobiological properties including immunomodulation and determining SARS-CoV-2 binding site.

Apart from andrographolide, Maurya et al. (2020) analyzed study on Indian Ayurveda regimen and proposed that withaferin A, piperine, curcumin, nimbin, mangiferin, thebaine, and berberine had significant binding affinity towards spike glycoprotein of SARS-CoV-2.

Apart from infectious disease, the usefulness of andrographolide for managing other medical disorders is also proposed. Regarding autoimmune disease, in animal model experiment, andrographolide could affect PI3K/Akt pathway, and this process is beneficial for management of autoimmune myocarditis (Zhang et al. 2019). In another animal model study, Iruretagoyena et al. (2005) found that andrographolide could interfere with T-cell activation, and this process is useful for managing autoimmune encephalomyelitis. Additionally, andrographolide is also reported for its usefulness in managing rheumatoid arthritis (Li et al. 2017). In an animal model study, Li et al. (2017) found that andrographolide could inhibit rheumatoid arthritis via inhibiting MAPK pathways.

Table 11.3 Examples of using andrographolide for cancer therapy

Authors	Details
Chen et al. (2014)	Chen et al. (2014) studied on antitumor effect of traditional Chinese herbal medicines against lung cancer. Chen et al. (2014) found that andrographolide was one of the five identified useful substances. The other four substances are polyphyllin I, tanshinone IIA, isochaihulactone, and 25-OCH3-PPD (Chen et al. 2014)
Gunn et al. (2011)	Gunn et al. (2011) found that andrographolide could exhibit anticancer stem cell activity in multiple myeloma
Nateewattana et al. (2014)	Nateewattana et al. (2014) found that andrographolide analogue could induce apoptosis in cholangiocarcinoma by topoisomerase II alpha inhibition
Wang et al. (2016)	Wang et al. (2016) proposed that andrographolide could reverse 5-FU resistance in human colorectal cancer by elevating BAX expression
Wang et al. (2020)	Wang et al. (2020) found that andrographolide could induce apoptosis in human osteosarcoma cells via the ROS/JNK pathway
Yang et al. (2016)	Yang et al. (2016) found that andrographolide could inhibit growth of human T-cell acute lymphoblastic leukemia. Jurkat cells by downregulation of PI3K/AKT and upregulation of p38 MAPK pathways
Zhang et al. (2014)	Zhang et al. (2014) found that andrographolide could suppress tumor growth by inhibiting TLR4/NF-κB signaling activation in insulinoma

Focusing on malignancy management, the immunomodulatory effect of andrographolide is widely studied (Table 11.3). Sheeja and Kuttan (2007a) studied on *Andrographis paniculata* extract and andrographolide and found that the tested substances could activate cytotoxic T-lymphocyte responses (Sheeja and Kuttan 2007a). This process is useful for controlling tumor growth (Sheeja and Kuttan 2007a). Based on a cancerous animal model study, Sheeja and Kuttan (2007b) found that the main pharmacological effect of andrographolide includes antibody-dependent cellular cytotoxicity, modulation of natural killer cell activity, and antibody-dependent complement-mediated cytotoxicity. In a cell line study, Khan et al. (2018) found that andrographolide could induce cell cycle arrest and programmed cell death via augmentation of intracellular reactive oxygen species level. Khan et al. (2018) noted that the mentioned biological process results in inhibitory activity against human colon cancer cells. Forestier-Román et al. (2019) found that andrographolide could induce DNA damage in prostate cancer cells. This process is useful for management of prostate cancer (Forestier-Román et al., 2019). Liu et al. (2020) found that andrographolide could potentiate PD-1 blockade immunotherapy by inhibiting COX2-mediated PGE2 release. Liu et al. (2020) noted that the COX2-mediated PGE2 release inhibition could promote antitumor activity of anti-PD-1 therapy. An interesting consideration of using andrographolide in clinical oncology is that there is still no complete clinical trials in human subjects for conclusion on the advantage of the substance. Most reports are experimental studies. It is an interesting research area for further studies.

11.5 Future Prospects

The immunotherapy will become more clinical therapeutic management in the future. The immunotherapy will be worldwide used. The immunopharmacological modification will play important role in therapy of medical disorders including cancer and infection. There will be new researches on this specific issue. There will be some new treatments based on immunomodulatory concepts. The natural immunomodulator will be the focused natural products for application on the immunomodulating therapy. The two substances, colchicine and andrographolide, are the examples of the natural immunomodulating agents that will be useful for management of disease. Based on the continuous improvement of immunopharmacological technology, there will be many new immunological agents, including immunomodulatory agents, based on these two classic biosubstances in the near future.

It is no doubt that the two biosubstances will be as therapeutic agents for many new indications. The studies on immunopathogenesis of the disease and immunopharmacological interrelationship with the two biosubstances will be required. Attempts to improve the formulations based on the two biosubstances will result in a more effective and safe new immunomodulating agents for management of the medical problem. The newly registered and proven immunomodulatory agents based on these two biosubstances will be new hopes for management of many diseases. Nevertheless, there will be also new reports on possible adverse effects of the biosubstances, and the closed monitoring of the unwanted adverse side effect will be necessary.

11.6 Conclusion

Immunological response plays important role in defense process against alien foreign bodies. The knowledge on immunopathology is very important in clinical immunology. There are many kinds of immunotherapy including active immunotherapy, passive immunotherapy, and immunomodulation. The immunotherapy by immunomodulation is the new concept that is usually proposed as an alternative new treatment for many diseases. There are many available immunomodulating agents, either natural or non-natural biosubstances.

The good examples of natural immunomodulating substances are colchicine and andrographolide. Both are proposed for usefulness in immunotherapy. The agents are applied for management of malignancies and infectious diseases. The applied usage of colchicine and andrographolide is an extraordinary clinical use. The studies on the specific issues on immunomodulating action of both biosubstances are interesting. There are many concerns on the cost effectiveness, safety, and adverse effects of the two natural immunomodulators. The assessment on effectiveness and safety of the two immunomodulatory agents is required. Further researches on both biosubstances are necessary for getting useful data for further application in the field of immunotherapy.

References

Akbar S (2011) Andrographis paniculata: a review of pharmacological activities and clinical effects. Altern Med Rev 16(1):66–77

Akram M, Usmanghani K, Ahmed I, Azhar I, Hamid A (2014) Comprehensive review on therapeutic strategies of gouty arthritis. Pak J Pharm Sci 27(5):1575–1582

Alkadi H, Khubeiz MJ, Jbeily R (2018) Colchicine: a review on chemical structure and clinical usage. Infect Disord Drug Targets 18(2):105–121

Angelidis C, Kotsialou Z, Kossyvakis C, Vrettou AR, Zacharoulis A, Kolokathis F, Kekeris V, Giannopoulos G (2018) Colchicine pharmacokinetics and mechanism of action. Curr Pharm Des 24(6):659–663

Babu YP, Manjunath S, Bakkannavar SM, Acharya J, Kumar GN (2012) Colchicine poisoning: a case report and review of literature. Med Sci Law 52(4):205–206

Banerjee A, Czinn SJ, Reiter RJ, Blanchard TG (2020) Crosstalk between endoplasmic reticulum stress and anti-viral activities: a novel therapeutic target for COVID-19. Life Sci 255:117842

Bhattacharya S, Das A, Datta S, Ganguli A, Chakrabarti G (2016) Colchicine induces autophagy and senescence in lung cancer cells at clinically admissible concentration: potential use of colchicine in combination with autophagy inhibitor in cancer therapy. Tumour Biol 37(8): 10653–10664

Bratman GN, Anderson CB, Berman MG, Cochran B, de Vries S, Flanders J, Folke C, Frumkin H, Gross JJ, Hartig T, Kahn PH Jr, Kuo M, Lawler JJ, Levin PS, Lindahl T, Meyer-Lindenberg A, Mitchell R, Ouyang Z, Roe J, Scarlett L, Smith JR, van den Bosch M, Wheeler BW, White MP, Zheng H, Daily GC (2019) Nature and mental health: an ecosystem service perspective. Sci Adv 5(7):eaax0903

Calabrese C, Berman SH, Babish JG, Ma X, Shinto L, Dorr M, Wells K, Wenner CA, Standish LJ (2000) A phase I trial of andrographolide in HIV positive patients and normal volunteers. Phytother Res 14(5):333–338

Cao Y, Xu X, Liu S, Huang L, Gu J (2018) Ganoderma: a cancer immunotherapy review. Front Pharmacol 9:1217

Chen KR, Carlson JA (2008) Clinical approach to cutaneous vasculitis. Am J Clin Dermatol 9(2): 71–92

Chen Y, Zhu J, Zhang W (2014) Antitumor effect of traditional Chinese herbal medicines against lung cancer. Anti-Cancer Drugs 25(9):983–991

Cho JH, Joo YH, Shin EY, Park EJ, Kim MS (2017) Anticancer effects of colchicine on Hypopharyngeal cancer. Anticancer Res 37(11):6269–6280

Cocco G, Chu DC, Pandolfi S (2010) Colchicine in clinical medicine. A guide for internists. Eur J Int Med 21(6):503–508

Dai Y, Chen SR, Chai L, Zhao J, Wang Y, Wang Y (2019) Overview of pharmacological activities of Andrographis paniculata and its major compound andrographolide. Crit Rev Food Sci Nutr 59(sup1):S17–S29

Dasgeb B, Kornreich D, McGuinn K, Okon L, Brownell I, Sackett DL (2018) Colchicine: an ancient drug with novel applications. Br J Dermatol 178(2):350–356

Desai J, Steiger S, Anders HJ (2017) Molecular pathophysiology of gout. Trends Mol Med 23(8): 756–768

Edwin ES, Vasantha-Srinivasan P, Senthil-Nathan S, Thanigaivel A, Ponsankar A, Pradeepa V, Selin-Rani S, Kalaivani K, Hunter WB, Abdel-Megeed A, Duraipandiyan V, Al-Dhabi NA (2016) Anti-dengue efficacy of bioactive andrographolide from Andrographis paniculata

(Lamiales: Acanthaceae) against the primary dengue vector Aedes aegypti (Diptera: Culicidae). Acta Trop 163:167–178
Enmozhi SK, Raja K, Sebastine I, Joseph J (2020) Andrographolide as a potential inhibitor of SARS-CoV-2 main protease: an in silico approach. J Biomol Struct Dyn 39(9):1–7
Falzon CC, Balabanova A (2017) Phytotherapy: an introduction to herbal medicine. Prim Care 44(2):217–227
Finkelstein Y, Aks SE, Hutson JR, Juurlink DN, Nguyen P, Dubnov-Raz G, Pollak U, Koren G, Bentur Y (2010) Colchicine poisoning: the dark side of an ancient drug. Clin Toxicol (Phila) 48(5):407–414
Forestier-Román IS, López-Rivas A, Sánchez-Vázquez MM, Rohena-Rivera K, Nieves-Burgos G, Ortiz-Zuazaga H, Torres-Ramos CA, Martínez-Ferrer M (2019) Andrographolide induces DNA damage in prostate cancer cells. Oncotarget 10(10):1085–1101
Franco LS, Shanahan DF, Fuller RA (2017) A review of the benefits of nature experiences: more than meets the eye. Int J Environ Res Public Health 14(8):864
Furman D, Davis MM (2015) New approaches to understanding the immune response to vaccination and infection. Vaccine 33(40):5271–5281
Gilchrist H, Patterson JW (2010) Erythema nodosum and erythema induratum (nodular vasculitis): diagnosis and management. Dermatol Ther 23(4):320–327
Gunn EJ, Williams JT, Huynh DT, Iannotti MJ, Han C, Barrios FJ, Kendall S, Glackin CA, Colby DA, Kirshner J (2011) The natural products parthenolide and andrographolide exhibit anti-cancer stem cell activity in multiple myeloma. Leuk Lymphoma 52(6):1085–1097
Gupta S, Mishra KP, Ganju L (2017) Broad-spectrum antiviral properties of andrographolide. Arch Virol 162(3):611–623
Hermanns-Clausen M, Schindler F, Stedtler U, Zilker T, Felgenhauer N (2006) Poisoning by the autumn crocus plant. MMW Fortschritte der Medizin 148(12):45–47
Hossain, M.S., Urbi, Z., Sule, A., Hafizur Rahman, K.M. (2014) Andrographis paniculata (Burm. F.) wall. Ex Nees: a review of ethnobotany, phytochemistry, and pharmacology, Sci World J, 2014,274905
Imazio M, Battaglia A, Gaido L, Gaita F (2017) Recurrent pericarditis. La *Revue de médecine interne* 38(5):307–311
Imazio M, Lazaros G, Brucato A, Gaita F (2016) Recurrent pericarditis: new and emerging therapeutic options. Nat Rev Cardiol 13(2):99–105
Indraratna PL, Virk S, Gurram D, Day RO (2018) Use of colchicine in pregnancy: a systematic review and meta-analysis. Rheumatology (Oxford) 57(2):382–387
Iruretagoyena MI, Tobar JA, González PA, Sepúlveda SE, Figueroa CA, Burgos RA, Hancke JL, Kalergis AM (2005) Andrographolide interferes with T cell activation and reduces experimental autoimmune encephalomyelitis in the mouse. J Pharmacol Exp Ther 312(1):366–372
Islam MT, Ali ES, Uddin SJ, Islam MA, Shaw S, Khan IN, Saravi SSS, Ahmad S, Rehman S, Gupta VK, Găman MA, Găman AM, Yele S, Das AK, de Castro E, Sousa JM, de Moura Dantas SM, Rolim M, de Carvalho HML, Melo-Cavalcante AA, Mubarak MS, Yarla NS, Shilpi JA, Mishra SK, Atanasov AG, Kamal MA (2018) Andrographolide, a diterpene lactone from Andrographis paniculata and its therapeutic promises in cancer. Cancer Lett 420:129–145
Jiang M, Zhao S, Yang S, Lin X, He X, Wei X, Song Q, Li R, Fu C, Zhang J, Zhang Z (2020) An "essential herbal medicine"-licorice: a review of phytochemicals and its effects in combination preparations. J Ethnopharmacol 249:112439
Johnson L, Goping IS, Rieger A, Mane JY, Huzil T, Banerjee A, Luduena R, Hassani B, Winter P, Tuszynski JA (2017) Novel colchicine derivatives and their anti-cancer activity. Curr Top Med Chem 17(22):2538–2558
Kandanur SGS, Tamang N, Golakoti NR, Nanduri S (2019) Andrographolide: a natural product template for the generation of structurally and biologically diverse diterpenes. Eur J Med Chem 176:513–533
Karamanou M, Tsoucalas G, Pantos K, Androutsos G (2018) Isolating colchicine in 19th century: an old drug revisited. Curr Pharm Des 24(6):654–658

Khan I, Khan F, Farooqui A, Ansari IA (2018) Andrographolide exhibits anticancer potential against human colon cancer cells by inducing cell cycle arrest and programmed cell death via augmentation of intracellular reactive oxygen species level. Nutr Cancer 70(5):787–803

Kumar A, Singh B, Sharma PR, Bharate SB, Saxena AK, Mondhe DM (2016) A novel microtubule depolymerizing colchicine analogue triggers apoptosis and autophagy in HCT-116 colon cancer cells. Cell Biochem Funct 34(2):69–81

Kuo MC, Chang SJ, Hsieh MC (2015) Colchicine significantly reduces incident cancer in gout male patients: a 12-year cohort study. Medicine (Baltimore) 94(50):e1570

Lee S (2007) Artemisinin, promising lead natural product for various drug developments. Mini-Rev Med Chem 7(4):411–422

Leonti M, Casu L (2018) Ethnopharmacology of love. Front Pharmacol 9:567

Leung YY, Hui LLY, Kraus VB (2015) Colchicine--update on mechanisms of action and therapeutic uses. Semin Arthritis Rheum 45(3):341–350

Li ZZ, Tan JP, Wang LL, Li QH (2017) Andrographolide benefits rheumatoid arthritis via inhibiting MAPK pathways. Inflammation 40(5):1599–1605

Liang Y, Wang ML, Chien CS, Yarmishyn AA, Yang YP, Lai WY, Luo YH, Lin YT, Chen YJ, Chang PC, Chiou SH (2020) Highlight of immune pathogenic response and Hematopathologic effect in SARS-CoV, MERS-CoV, and SARS-Cov-2 infection. Front Immunol 11:1022

Lin ZY, Kuo CH, Wu DC, Chuang WL (2016) Anticancer effects of clinically acceptable colchicine concentrations on human gastric cancer cell lines. Kaohsiung J Med Sci 32(2):68–73

Lin ZY, Wu CC, Chuang YH, Chuang WL (2013) Anti-cancer mechanisms of clinically acceptable colchicine concentrations on hepatocellular carcinoma. Life Sci 93(8):323–328

Liu W, Fan T, Li M, Zhang G, Guo W, Yang X, Jiang C, Li X, Xu X, Tang A, Liu K, Liu L, Kong L, Xu Q, Sun Y (2020) Andrographolide potentiates PD-1 blockade immunotherapy by inhibiting COX2-mediated PGE2 release. Int Immunopharmacol 81:106206

Lu J, Ma Y, Wu J, Huang H, Wang X, Chen Z, Chen H, He H, Huang C (2019) A review for the neuroprotective effects of andrographolide in the central nervous system. Biomed Pharmacother 117:109078

Mansouri N, Marjani M, Tabarsi P, von Garnier C, Mansouri D (2020) Successful treatment of Covid-19 associated cytokine release syndrome with colchicine. A case report and review of literature. *Immunol Investig* 50:1–7

Maurya VK, Kumar S, Prasad AK, Bhatt MLB, Saxena SK (2020) Structure-based drug designing for potential antiviral activity of selected natural products from Ayurveda against SARS-CoV-2 spike glycoprotein and its cellular receptor. Virus disease 31(2):179–193

Mehrotra R, Rawat S, Kulshreshtha DK, Patnaik GK, Dhawan BN (1990) In vitro studies on the effect of certain natural products against hepatitis B virus. Indian J Med Res 92:133–138

Morel N, Bonjour M, Le Guern V, Le Jeunne C, Mouthon L, Piette JC, Costedoat-Chalumeau N (2015) Colchicine: a simple and effective treatment for pericarditis in systemic lupus erythematosus? A report of 10 cases. Lupus 24(14):1479–1485

Murugan NA, Pandian CJ, Jeyakanthan J (2020) Computational investigation on Andrographis paniculata phytochemicals to evaluate their potency against SARS-CoV-2 in comparison to known antiviral compounds in drug trials. J Biomol Struct Dyn 50:1–12

Mussard E, Cesaro A, Lespessailles E, Legrain B, Berteina-Raboin S, Toumi H (2019) Andrographolide, a natural antioxidant: an update. Antioxidants (Basel) 8(12):571

Nateewattana J, Dutta S, Reabroi S, Saeeng R, Kasemsook S, Chairoungdua A, Weerachayaphorn J, Wongkham S, Piyachaturawat P (2014) Induction of apoptosis in cholangiocarcinoma by an andrographolide analogue is mediated through topoisomerase II alpha inhibition. Eur J Pharmacol 723:148–155

Oprea TI, Taboureau O, Bologa CG (2012) Of possible cheminformatics futures. J Comput Aided Mol Des 26(1):107–112

Paemanee A, Hitakarun A, Wintachai P, Roytrakul S, Smith DR (2019) A proteomic analysis of the anti-dengue virus activity of andrographolide. Biomed Pharmacother 109:322–332

Pan Y, Lin Z (2019) Anti-aging effect of Ganoderma (Lingzhi) with health and fitness. Adv Exp Med Biol 1182:299–309

Panraksa P, Ramphan S, Khongwichit S, Smith DR (2017) Activity of andrographolide against dengue virus. Antivir Res 139:69–78

Pascart T, Richette P (2018) Colchicine in gout: an update. Curr Pharm Des 24(6):684–689

Reddy VL, Reddy SM, Ravikanth V, Krishnaiah P, Goud TV, Rao TP, Ram TS, Gonnade RG, Bhadbhade M, Venkateswarlu Y (2005) A new bis-andrographolide ether from Andrographis paniculata nees and evaluation of anti-HIV activity. Nat Prod Res 19(3):223–230

Robinson KP, Chan JJ (2018) Colchicine in dermatology: a review. Australas J Dermatol 59(4): 278–285

Sanodiya BS, Thakur GS, Baghel RK, Prasad GB, Bisen PS (2009) Ganoderma lucidum: a potent pharmacological macrofungus. Curr Pharm Biotechnol 10(8):717–742

Sheeja K, Kuttan G (2007a) Activation of cytotoxic T lymphocyte responses and attenuation of tumor growth in vivo by Andrographis paniculata extract and andrographolide. Immunopharmacol Immunotoxicol 29(1):81–93

Sheeja K, Kuttan G (2007b) Modulation of natural killer cell activity, antibody-dependent cellular cytotoxicity, and antibody-dependent complement-mediated cytotoxicity by andrographolide in normal and Ehrlich ascites carcinoma-bearing mice. Integr Cancer Ther 6(1):66–73

Shergis JL, Zhang AL, Zhou W, Xue CC (2013a) Panax ginseng in randomised controlled trials: a systematic review. Phytother Res 27(7):949–965

Shergis JL, Zhang AL, Zhou W, Xue CC (2013b) Quality and risk of bias in Panax ginseng randomized controlled trials: a review. Am J Chin Med 41(2):231–252

Singh D, Gupta R, Saraf SA (2012) Herbs-are they safe enough? An overview. Crit Rev Food Sci Nutr 52(10):876–898

Slobodnick A, Shah B, Pillinger MH, Krasnokutsky S (2015) Colchicine: old and new. Am J Med 128(5):461–470

Solak Y, Sipahi S (2016) Impact of glomerular filtration rate in colchicine toxicity. Semin Arthritis Rheum 45(5):e15

Sun Y, Lin X, Chang H (2016) Proliferation inhibition and apoptosis of breast cancer MCF-7 cells under the influence of colchicine. *J BOUN* 21(3):570–575

Sunkureddi P (2011) Gouty arthritis: understanding the disease state and management options in primary care. Adv Ther 28(9):748–760

Tan WSD, Liao W, Zhou S, Wong WSF (2017) Is there a future for andrographolide to be an anti-inflammatory drug? Deciphering its major mechanisms of action. *Biochem Pharmacol* 139:71–81

Tanaka T (2005) Pharmacoinformatics and pharmacogenomics. Nihon Yakurigaku Zasshi 126(2): 113–116

Tatematsu H, Kilkuskie RE, Corrigan AJ, Bodner AJ, Lee KH (1991) Anti-AIDS agents, 3. Inhibitory effects of colchicine derivatives on HIV replication in H9 lymphocyte cells. J Nat Prod 54(2):632–637

Tiligada E, Ishii M, Riccardi C, Spedding M, Simon HU, Teixeira MM, Cuervo ML, Holgate ST, Levi-Schaffer F (2015) The expanding role of immunopharmacology: IUPHAR review 16. Br J Pharmacol 172(17):4217–4227

Tsoucalas G, Papaioannou T, Panayiotakopoulos G, Saridaki Z, Vrachatis DA, Karamanou M (2018) Colchicum genus in the writings of ancient Greek and byzantine physicians. Curr Pharm Des 24(6):648–653

Uttekar MM, Das T, Pawar RS, Bhandari B, Menon V, Nutan, Gupta SK, Bhat SV (2012) Anti-HIV activity of semisynthetic derivatives of andrographolide and computational study of HIV-1 gp120 protein binding. Eur J Med Chem 56:368–374

Vieira JM, Ricardo OMP, Hannas CM, Kanadani TCM, Prata TDS, Kanadani FN (2020) What do we know about COVID-19? A review article. Rev Assoc Med Bras (1992) 66(4):534–540

Wang S, Li H, Chen S, Wang Z, Yao Y, Chen T, Ye Z, Lin P (2020) Andrographolide induces apoptosis in human osteosarcoma cells via the ROS/JNK pathway. Int J Oncol 56(6):1417–1428

Wang W, Guo W, Li L, Fu Z, Liu W, Gao J, Shu Y, Xu Q, Sun Y, Gu Y (2016) Andrographolide reversed 5-FU resistance in human colorectal cancer by elevating BAX expression. Biochem Pharmacol 121:8–17

Wee E, Kelly RI (2017) Treatment of nodular vasculitis with colchicine. Australas J Dermatol 58(3):e79–e82

Wilson L, Saseen JJ (2016) Gouty arthritis: a review of acute management and prevention. Pharmacotherapy 36(8):906–922

Wiwanitkit V (2010) Dengue fever: diagnosis and treatment. Expert Rev Anti-Infect Ther 8(7): 841–845

Worachartcheewan A, Songtawee N, Siriwong S, Prachayasittikul S, Nantasenamat C, Prachayasittikul V (2019) Rational Design of Colchicine Derivatives as anti-HIV agents via QSAR and molecular docking. Med Chem 15(4):328–340

Yamanouchi T (2003) Gouty arthritis. Nihon Rinsho 61(Suppl 1):213–218

Yang T, Yao S, Zhang X, Guo Y (2016) Andrographolide inhibits growth of human T-cell acute lymphoblastic leukemia. Jurkat cells by downregulation of PI3K/AKT and upregulation of p38 MAPK pathways. Drug Des Devel Ther 10:1389–1397

Yoder A, Guo J, Yu D, Cui Z, Zhang XE, Wu Y (2011) Effects of microtubule modulators on HIV-1 infection of transformed and resting CD4 T cells. J Virol 85(6):3020–3024

Zaid OI, Abd Majid R, Sabariah MN, Hasidah MS, Al-Zihiry K, Yam MF, Basir R (2015) Andrographolide effect on both plasmodium falciparum infected and non-infected RBCs membranes. Asian Pac J Trop Med 8(7):507–512

Zhang L, Bao M, Liu B, Zhao H, Zhang Y, Ji XY, Zhao N, Zhang C, He X, Yi J, Tan Y, Li L, Lu C (2020) Effect of Andrographolide and its analogs on bacterial infection: a review. Pharmacology 105(3–4):123–134

Zhang Q, Hu LQ, Li HQ, Wu J, Bian NN, Yan G (2019) Beneficial effects of andrographolide in a rat model of autoimmune myocarditis and its effects on PI3K/Akt pathway. Korean J Physiol Pharmacol 23(2):103–111

Zhang QQ, Ding Y, Lei Y, Qi CL, He XD, Lan T, Li JC, Gong P, Yang X, Geng JG, Wang LJ (2014) Andrographolide suppress tumor growth by inhibiting TLR4/NF-κB signaling activation in insulinoma. Int J Biol Sci 10(4):404–414

Zhang Y, Yang J, Bao R, Chen Y, Zhou D, He B, Zhong M, Li Y, Liu F, Li Q, Yang Y, Han C, Sun Y, Cao Y, Yan H (2011) Unpolarized release of vaccinia virus and HIV antigen by colchicine treatment enhances intranasal HIV antigen expression and mucosal humoral responses. *PLoS One* 6(9):e24296

Chapter 12
Immune Booster Property of Epigallocatechin-3-Gallate and Catechin

Leidy Johana Valencia-Hernández, Juan Alberto Ascacio-Valdés, Jorge Enrique Wong-Paz, Haroon Khan, and Cristóbal N. Aguilar

Abstract The immune system plays a key role in the defense and protection of the human being, and its protection and stability is fundamental to a good functioning of it. The regulation of this system is complex but highly structured and synchronized. Several phytomolecules can act as booster of this complex system promoting the proper functioning and performance of the immune system. New trends in this research topic are focused on finding bioactive molecules and phytochemicals to modulate the biologic immune response and enhance the ability to resist disease. Among these relevant bioactive molecules, catechin and epigallocatechin-3-gallate stand out for their bioactivity, specificity, and reactivity, being the point of analysis, and discussion of this chapter that aims to describe the important biological activity of both compounds as booster of the immune system.

Keywords Immune system · Booster property · Catechin · Epigallocatechin-3-gallate

12.1 Introduction

Human health could be considered as the absence of diseases in a person. In this sense, the nutrition and immune system play an important role in promotion and maintenance of an adequate state of health. Immune system involves the interaction

L. J. Valencia-Hernández · J. A. Ascacio-Valdés · C. N. Aguilar (✉)
Departamento de Investigación en Alimentos, Facultad de Ciencias Químicas, Universidad Autónoma de Coahuila, Saltillo, Coahuila, Mexico
e-mail: cristobal.aguilar@uadec.edu.mx

J. E. Wong-Paz
Tecnológico Nacional de México, Instituto Tecnológico de Ciudad Valles, San Luis Potosí, Mexico

H. Khan
Department of Pharmacy, Abdul Wali Khan University, Mardan, Pakistan

R. K. Kesharwani et al. (eds.), *Nutraceuticals and Functional Foods as Immunomodulators*, https://doi.org/10.1007/978-981-19-2507-8_12

of a large number of metabolites and cellular components (Nair et al. 2019). Regulation of the immune system implies any modification in the induction, expression, amplification, and inhibition in the immune response including boosting, anti-inflammatory, and antiviral efficacy (Khanal et al. 2020). Also, several cell receptors and molecules implicated in signal transduction have been well identified in the immune cells. Actually, the research has been focused on finding compounds and substances that can modulate the biologic immune response and enhance the host's ability to resist disease (Ngcobo and Gqaleni 2016). Here the plants and phytochemicals seem to be a promising alternative in healthcare.

Plant-derived compounds are widely used as supplement for maintenance of health (Ngcobo and Gqaleni 2016). These compounds are considered as a safer alternative to replace synthetic immunomodulators which have been reported to possess adverse effects on human health (Nair et al. 2019). In fact, near to the 70–80% of the people belonging to the developing countries use herbal drugs as primary healthcare in comparison to the modern synthetic drugs (Hamilton 2004; Khanna et al. 2020; Kesharwani et al. 2019). Actually, the plant extracts and their phytochemicals have play the prime therapeutic option mainly attributed to the immunomodulatory effects. Its high use in the traditional medicine has originated that the people use it as antiviral therapy in the Pandemic Covid-19 Battle. With this, findings allow the development of deep and rigorous studies in order to determine efficacy and safety of these products (Ngcobo and Gqaleni 2016).

Plants and phytochemicals with medicinal properties act through a nonspecific widespread modulation of the immune response with effects on various immune cells (Nair et al. 2019). These are considered as immune booster which decreases the risk of getting sick because they improve the immune system. Polyphenols are phytochemicals widely mentioned as a class of compounds with high potential to be used as immune booster.

Polyphenols must be activated to different metabolites before reaching the biological targets (i.e., glucuronide, sulfate, and methylated). Some reports showed the immune booster potential of conjugated catechins owing to the antioxidant and immunomodulatory roles (González-Manzano et al. 2011). Epigallocatechin-3-gallate and catechin are some phytochemicals with high potential to be used as immune booster (Fatima et al. 2013; Singh et al. 2014).

They have been reported as good alternative to control both obesity and autoinflammatory arthritis aggravated by obesity (Byun et al. 2014). Also, catechins have antioxidant, anti-inflammatory, and immunomodulatory activities resulting in hepatic- and immune-protective properties against restraint stress (Tang et al. 2020), may exert anticancer effect (Jung et al. 2001), and can be a potent natural inhibitor of leukocyte elastase that may be used to reduce elastase-mediated progression (Sartor et al. 2002). Litchi extracts (with contains catechins and procyanidins) (Miranda-Hernández et al. 2019) diminish prostate cancer progression (Guo et al. 2017).

12.2 Phytochemicals with Immunomodulatory Activity

An immunomodulatory compound or agent (natural or synthetic) is one that has the ability to modify or alter the immune system and help the body fight infection or other diseases (Akram et al., 2020). Immunomodulators participate in dual phenomena, such as immunosuppression and immunostimulation that control the function of the immune system (Ganeshpurkar and Saluja 2018a, b). Natural plant-based immunomodulatory agents can serve as an alternative to conventional therapies related to the immune system and its response to infection or disease (Srikumar et al. 2006). For a long time and throughout human history, medicinal plants have played a very important role in healthcare, as alternative treatments. Due to this, synthetic compounds have been developed and used as drugs that have a high similarity to compounds of natural origin (Tilbur and Kaptchuk 2008). One of the main groups of phytochemicals found in plants is phenolic compounds, and it has been reported that these compounds have very relevant biological properties. Among these biological properties is the immunomodulatory property (Yahfoufi et al. 2018; Kesharwani et al. 2020). These polyphenolic compounds are classified based on their chemical structure into different groups: the flavonoids, such as flavones, flavonols, chalcones, anthocyanidins, and proanthocyanidins, and the nonflavonoids, such as phenolic acids, stilbenes, etc. (Tsao 2010). Flavonoids are reported to be the main group of natural phytochemicals with different roles in promoting human health (Formica and Regelson 1995). Catechin and catechin-containing compounds in their composition (e.g., epigallocatechin-3-gallate) are relevant compounds that possess important biological properties. In the specific case of the immunomodulatory property, it has been evaluated the anti-inflammatory and analgesic effect in different studies, and derived from these studies, it has been demonstrated that flavonoids as the catechin have the capacity to inhibit interleukins as the IL-1β and cytokines as COX-2 that are directly related to inflammation processes (Ganeshpurkar and Saluja 2018a, b).

Another important biological effect that catechin has and that has been evaluated in in vivo studies is the effect on cellular and humoral immunity. It has been demonstrated that the administration of catechin in biological models causes an increase in the total levels of antibodies and this is reflected in a positive immunomodulatory effect which demonstrates that phytochemical compounds such as catechin and compounds that contain it have the capacity to promote this immunomodulatory effect (Ganeshpurkar and Saluja 2019).

12.2.1 Catechin

Catechins are important polyphenols in plants and fruits, also called flavan-3-ols, that belong to the most studied polyphenol group, the flavonoids, which are representative in green wine and tea plant *Camellia sinensis* (L.), representing in this last vegetal material between 70 and 80% approximately of the total polyphenols (Liu et al. 2014; Wang et al. 2015); research has reported that they contribute to the bitter taste and black color to green tea in fermentation of tea leaves. Catechins can be oxidized to higher molecular weight molecules such as thearubigins and theaflavins in green tea and, in addition, are part of the basic structure of procyanidins.

In fresh tea leaves, there are eight commonly reported catechins; these can be grouped into esterified and non-esterified: the first group includes (+)-catechin (C), (−)-epicatechin (EC), (−)-gallocatechin (GC), and (−)-epigallocatechin (ECG); the second group includes (−)-catechin gallate (CG), (−)-epicatechin-3-gallate (EGC), (−)-gallocatechin gallate (GCG), and (−)-epigallocatechin-3-gallate (EGCG). In general, EGCG, ECG, and EGC are reported to be present with relatively high contents in leaves of many tea cultivars (Chen et al. 2018a, b; Theppakorn, 2016). EGCG is the most abundant catechin. Various studies have shown its benefit in human health; some pharmacological actions have been reported including antioxidant, anti-inflammatory, anticancer, antifungal, antibacterial, antiproliferative, and proapoptotic analgesic (Ma et al. 2019; Singh et al. 2014). Catechin can be associated in processes of angiogenesis, the regulation of cell death, and drug resistance (Zanwar et al. 2013). However, catechins have some disadvantages such as unstable at physiological pH, low bioavailability, and poor solubility; under alkaline conditions can be degraded, sensitivity to oxygen, light, and heat, and its taste is bitter and astringent (Cuevas-Valenzuela et al. 2015; Okabe et al. 1999). These characteristics limit the development of food and pharmaceutical products (Ho et al. 2019; Wang et al. 2015). In the food industry, catechins have been incorporated into oral supplements, nutraceutical products, functional foods, and food additives, although they are considered safe (Keservani et al. 2010; Mahindrakar and Rathod 2020; Chen et al., 2019; Ho et al. 2019; Xiang et al., 2016; Keservani et al. 2010). Some studies have evaluated the incorporation of catechin in food matrices such as cheese, yogurt, and milk, allowing the study of its stability during storage (Ho et al. 2019; Rashidinejad et al. 2016).

12.2.1.1 Chemistry, Sources, and Properties

Chemistry

Catechins (flavan-3-ols) are monomeric polyphenols with molecular structure of 2-phenylchromone (de Pascual-Teresa et al., 2010), known as a polyhydroxylated compound. The basic structure of the catechin consists 3 types of rings, two benzene rings and a dihydropyran heterocycle; the first ones are called A and B rings and the third C ring (dihydropyran). These rings are hydroxylated by the regulation of two enzymes flavonoid 3′,5′-hydroxylase (F3′5′H) and flavonoid 3′- hydroxylase (F3′H) (Cuevas-Valenzuela et al. 2015). This flavan-3-ols contains two chiral centers (carbons) on the ring B on C2 and C3. From these chiral carbons, cis and trans configurations can be formed, equivalent to (+) catechins and (−) epicatechin, respectively. These structural characteristics of catechin are like epicatechin, so both are isomers. The resorcinol and catechol moieties are similar on benzene rings A and B, respectively (Braicu et al. 2013).

Catechins can be divided into two groups according to the number of hydroxyl groups present in the B ring, dihydroxyl, and trihydroxyl (Wang et al., 2014). Liu et al. (2012) classified the monomeric catechins in galloylated that are galloylated in position 3 of the C ring and the nongalloylated; the catechin belongs to the group of those nongalloylated (Wang et al. 2018a, 2018b; Jiang et al., 2017; Madhan et al. 2005). The ability to quench free radicals and antioxidant action of the catechins is due to these hydroxyl groups that allow delocalization of electrons. However, these structures can increase their susceptibility (autoxidation) to neutral and alkaline pH. Controversially, reports have indicated that the stability of the catechin was altered to pH 7.4; the color of the solution changed after 96 h of incubation (Chobot et al. 2009). Catechins can undergo epimerization under hydrothermal conditions and at pH above 7, contributing to chemical changes and loss of these compounds in food matrices (Chen et al. 2018a, b). These compounds can be degraded, polymerized, and easily oxidized, which allows the generation of more complex structures and other oxidized molecules such as quinones and semiquinone radicals (Ashihara et al. 2010). Another characteristic of catechins is their susceptibility to illumination (Shi et al. 2016) with visible light and UV, which promotes their epimerization and degradation. Catechin can chelate metal ions, which avoid the formation of ROS (Khan et al. 2011). They can also be degraded in the presence of oxygen and cuprous or ferric ion (Yang et al. 2014; Sang et al. 2011).

The biosynthetic pathways have been widely studied; however, more research is still needed; enzymes such as dihydroflavonol 4-reductase (EC 1.1.1.219), anthocyanidin reductase (EC 1.3.1.77), anthocyanidin synthase (EC 1.3.1.77), and leucoanthocyanidin reductase (EC 1.3.1.77) have been involved in the pathways (Liu et al. 2012; Ashihara et al. 2010; Eungwanichayapant and Popluechai 2009; Pang et al. 2007; Punyasiri et al. 2004; Xie et al. 2003, 2004; Stafford & Lester, 1984, 1985, 1980). In another study, it was shown that these compounds are biosynthesized through the phenylpropanoid and flavonoid pathways; in tea plants, some genes and enzymes were involved in these pathways, among them are phenylalanine ammonia-lyase (PAL), 4-coumarate-CoA ligase (4CL), flavonoid 3-′-hydroxylase (F3′H), flavonoid 3′5′-hydroxylase (F3′5′H), cinnamate 4-hydroxylase (C4H), chalcone isomerase (CHI), chalcone synthase (CHS), dihydroflavonol 4-reductase (DFR), anthocyanidin reductase (ANR), flavonol synthase (FLS), and leucoanthocyanidin reductase (LAR) (Wang et al. 2018a, b).

Sources

Catechins are flavonoids with wide distribution in nature; they are detected in some fruits such as apples, pears, grapes (black, seeds), strawberries blueberries, blackberries, raspberries, cherries, gooseberries, cider, lemon, orange, kiwi, and peach (Zanwar et al. 2013). One of the most consumed fruits worldwide is the apple; it contains significant amounts of catechin, epicatechin, and chlorogenic acid

(Napolitano et al. 2004). In obtaining apple juice, large quantities of apple pomace are generated, from which catechins, epicatechin, chlorogenic acid, and other compounds have been extracted (Casazza et al. 2020; W. Li et al. 2020). The recovery of these compounds has been reported for the formulation of food products such dietary supplements, food ingredient, cosmetics, and drugs (Dhadge et al. 2019; Oliveira et al. 2015).

Native plant sources in the Amazon have been studied, such as guarana seeds, which contain catechins, epicatechin, and procyanidins. Experiments have been carried out on the gastrointestinal bio-accessibility and stability of catechin seeds obtained from different geographical locations. Catechin concentrations ranged from 30.0 to 26.8 mg catechin/g; moreover, the gastrointestinal bio-accessibility was high (86–95%), proving that it varies according to the geographical origin of the samples. Additionally, the results showed that the macronutrients did not affect the bio-accessibility and permeability of catechins (Mendes et al. 2019). These results can be compared with those obtained in food products derived from cocoa and achieved gastrointestinal bio-accessibility between 77% and 91% (Neilson et al. 2009).

These compounds are considered one of the most studied polyphenols in green tea and have received enormous attention due to its health-promoting effects, mainly antioxidant, cardioprotective, antidiabetic, and anticarcinogenic (Pastoriza et al. 2017; Gadkari and Balaraman 2015). In recent experiments, catechins have been isolated from *Rosa chinensis* flower buds, which had antiproliferative activity against five cancer cell lines; in addition, they did not present cytotoxic activity (Liu et al. 2020). Other points to note are the role of the catechin in stabilizing the skin protein collagen and the quality of the tea leaves (Wang et al. 2018a, 2018b; Madhan et al. 2005).

In recent years, new sources of phytochemicals have been evaluated, which are characterized as little-known plants with antioxidant potential. Some examples of catechin sources are those studied by Kah Hui et al. (2020) who recently reported extractions of bioactive compounds from *Morinda citrifolia* leaves (MCL), which is shrub commonly called as noni. *Morinda citrifolia* leaf extracts have been reported to contain (+)-catechin, (−)-epicatechin, and types of catechins. These results can be compared with data obtained from green tea leaf catechins. Other vegetable sources have been used to obtain catechins, among them *Azorean Camellia sinensis* flowers (Baptista et al., 2019) as well as off-season fresh leaves (Baptista et al., 2019).

Properties

Catechins have received great interest due to its biological and pharmacological properties, including antioxidant, anticancer, antiproliferative, proapoptotic analgesic, neuroprotective, antihypertensive antidiabetic, antimutagenic, antimicrobial, antifungal, antibacterial, antiviral, anti-inflammatory, anti-allergic, antiarthritic, antiplatelet, anti-obesity, antiaging, hypocholesterolemic, and chemopreventive (Jiang et al. 2019; Ma et al. 2019; Liu et al. 2014; Geetha et al., 2004). In addition, they can protect against ulcerative colitis and have the ability to reduce lipid content. Also, they can also catalyze oxidation reactions through the bonding of metal ions and have related to the environmental adaptability of the plant (Pheomphun et al. 2019; Zhang et al. 2017; Lambert and Elias 2010). Other properties have been found in green tea catechins, which are potential in the inhibition of various types of cancer such as the skin, stomach colon, mammary glands, prostate lung, liver, esophagus, small intestine, and bladder (C. S. Yang et al. 2002). In addition, these also improve the adaptability of the plants to the environment due to their ability to reduce reactive oxygen species (ROS) (Pheomphun et al. 2019; Zhang et al. 2017).

Numerous investigations have focused on the evaluation of the antioxidant activity of catechins from different plant sources, as well as compare this property with the other natural and synthetic antioxidants such as flavonols (morin and rutin), flavanones (hesperidin, naringenin and naringin), hydroxycinnamic acids (caffeic acid, chlorogenic acid, ferulic acid), and other polyphenolic compounds (Grzesik et al. 2018), being most effective the catechins. This catechin has a strong chelating and antioxidant activity, due to two gallocatechol rings which can directly remove free radicals with high efficacy (Braicu et al. 2013). The catechins structural characteristics confer antioxidant properties and the ability of ROS on them; this property is attributed the benefits in human health. Several studies have evaluated the effectiveness of catechin in all three stages of cancer (initiation, promotion, and progression); some lines such as breast, lung, and skin have been tested (Seeram and Nair 2002).

Several papers report the effect of catechin consumption on some diseases in in vitro and in vivo assays. Yoo et al. (2020) indicate that the consumption of catechin-rich green tea can prevent accumulation of lipids and their intestinal absorption. Delgado et al. (2014) showed that the use of catechin concentration over 10 M could exert antiproliferative effect on cellular models. It was also effective in regulating insulin signaling.

Catechin has been shown to have immunomodulatory effects on the cellular and humoral system in the rat model. In this study, catechin administration increased delayed type hypersensitivity reaction. Some treatments increased levels of white blood cells, immunoglobulins, and antibody titer. All these effects can be due to the biological properties of catechins such as cytoprotective, antioxidant, and anti-inflammatory. The authors suggest that these results may be due to the effect on "chemotaxis", which is dependent of leucocytes migration (Ganeshpurkar and Saluja 2018a, b).

12.2.1.2 Immune Booster Property of Catechin

Catechins can modulate biological processes related to the development of chronic diseases (Kaliora et al., 2006). The mechanisms of action by which these compounds act need to be investigated. One of the best-known modulatory effects is the inhibition of the activation of NF-κB, associated with inflammation processes. There are several cell signaling pathways that modulate the catechins, including inhibition of B-lymphocyte activation; T-cell proliferation; and inflammatory processes involving enzymes and proteins such as COX-2, MAPK, protein kinase-C, cytokines, tyrosine, serine-threonine, and protein kinases; other enzyme of importance in physiological processes is inducible nitric oxide synthase (iNOS) (Hussain et al. 2016). Catechins have a role in the expression of apoptosis regulating genes like caspase (Mandel et al. 2004), cyclogenase activity (Kapoor et al. 2004), and growth factor signaling (Yang et al. 2004) and induced superoxide dismutase and catalase activities (Meng et al. 2008). Furthermore, catechin has been found to modulate the activation of neutrophils, macrophages, and dendritic cells (Kamalden et al. 2012).

Catechins can have potential antiproliferative role in cancer cells. The effect from *Ligaria cuneifolia* catechin in the proliferation of cell lines associated with leukemia (murine lymphoma cell line LB02) was evaluated by Papademetrio et al. (2013). In this study was reported antiproliferative activity of catechin at concentrations of 50, 100, and 200 μg/mL; it was more effective at 100 and 200 μg/mL with reductions in their rate of proliferation in a 33 and 97%, respectively. These results were explained by the ability of the catechins to modulate the action of antiapoptotic proteins, which is related to loss of mitochondrial membrane potential ($\Delta\Psi m$) and cell death in LB02 cells (Khiewkamrop et al. 2018).

Catechins are natural agents with potential in the treatment and prevention of neurodegenerative disorders such as Parkinson's (PD) and Alzheimer's diseases (AD). This progressive neurodegenerative disorder affects approximately ten million people worldwide (Ball et al. 2019). The main characteristic is the degeneration of dopaminergic neurons in the substantia nigra pars compacta (SNc) and can be presented with motor symptoms, including bradykinesia, slowness of movement, resting tremors, muscle rigidity, gait instability, and impaired postural reflexes (Dickson 2018); other non-motor symptoms are olfactory dysfunction, constipation, pain, sleep disturbances, anxiety, depression, and cognitive dysfunction. These symptoms could occur due to the incorrect misfolding of proteins, which results in the formation of β-sheet-rich amyloid fibrils (Thanikachalam et al. 2020). Alpha-synuclein (SNCA) is an example of the proteins that cause this disease, which are part of neuronal inclusions or intracellular deposits called Lewy bodies (LBs). Some authors have reported that catechin could inhibit some signaling pathways associated with disruption of protein degradation, activation of apoptosis by c-Jun N-terminal kinase (JNK), mitochondrial dysfunction, and microglia-triggered inflammation (Bossy-Wetze et al. 2004).

This disease can affect people over 60 years of age (Lashuel et al. 2013) and is likely to be associated with genetic and environmental factors that can generate oxidative stress (Ball et al. 2019; Brouwer et al. 2017; Del Rey et al. 2018). However, the exact mechanisms of this disease remain to be fully elucidated; so far studies have been conducted to understand it.

On the other hand, Alzheimer's disease (AD) is the most common degenerative disorder, which causes irreversible loss of neurons in the cerebral cortex and hippocampus, characterized by chronic neuroinflammation, oxidative stress, reasoning impairment, loss of memory, and learning, swallowing, or walking problems, which can eventually lead to death. Alterations in cholesterol homeostasis and insulin signaling may also occur (Anand et al. 2014; Ying Liu et al. 2011). These characteristics can occur by the formation intracellular neurofibrillary tangles of tau protein and extracellular amyloid beta (Aβ peptide) plaques (Ittner and Götz 2011). The Aβ peptide can accumulate senile plaques that can be oligomers or aggregates of higher molecular weight (Sharman et al. 2019), which can lead to neuronal damage and cause dementia (Ballard et al. 2011). Authors have proposed catechin as a promising candidate for inhibiting the mechanisms of action on microglia-triggered inflammation, mitochondrial alterations, activation of glutamate receptors, generation of nitric oxide species and free radicals, and increased intracellular calcium levels, among others. (Bossy-Wetze et al. 2004).

12.2.2 Epigallocatechin-3-Gallate

Epigallocatechin-3-gallate also known as epigallocatechin gallate (EGCG) is one of the most abundant catechins of green tea (*Camellia sinensis)*, as well as the most biologically active, and has been reported with the highest antioxidant activity with respect to the other types of catechins (N. Khan and Mukhtar 2007). EGCG has the ability to inhibit peroxidative reactions, improve the function of mitochondria, increase endogenous antioxidants, and increase catalase activity (Chen et al. 2016; N. A. Singh et al. 2016; Tseng et al. 2015).

Reportes recientes han evaluado el efecto de EGCG en diferentes enfermedades degenerativas y actualmente en nuevos trastornos como el coronavirus (SARS-CoV-2). EGCG isolation is low cost and safe for food and pharmaceutical use. Despite these advantages, the EGCG present low intestinal bioavailability, unstable at physiological pH, under alkaline conditions can be degraded, sensitivity to oxygen, light, and heat (Cuevas-Valenzuela et al. 2015; Okabe et al. 1999). Therefore, strategies to improve bioavailability have been carried out, as is the case of nanoencapsulation (Aggarwal et al. 2020; Keservani et al. 2017; Keservani et al. 2018).

12.2.2.1 Chemistry, Sources, and Properties

Chemistry

EGCG which is a gallate-type catechin has been studied as a natural drug due to its potential in the prevention of chronic and degenerative diseases. These beneficial effects on human health are attributed to their structural characteristics. EGCG has a catechin-like backbone structure with an m-5,7-dihydroxy substitution on ring A and a di-hydroxyl or trihydroxy substitution on rings B and D; in ring C, at the 3′ position presents a gallate moiety (Braicu et al. 2013). This compound is part of the group of the esterified and galloylated catechins (Davinelli et al. 2012). The polyphenolic structure and the number of hydroxyl and galloyl groups of these compounds allow electron delocalization and display biological effectiveness in in vitro and in vivo models. Among all catechin derivatives and complex tannins, the EGCG is the most effective at scavenging free radical and inhibits reactive oxygen species (ROS) formation. Moreover, they make them good donors for hydrogen bonding and important feature in the absorption and bioavailability of EGCG.

These compounds can form dimers through autoxidation in cell culture conditions, a process in which superoxide anion and hydrogen peroxide are produced. On the other hand, EGCG has a catechol group, corresponding to a benzene ring, which has two hydroxyl groups in positions 3 and 4. This catechol group causes the EGCG to undergo methylation by the catecol-O-metiltransferasa enzyme. Other reactions catalyzed by the enzymes glucuronosiltransferasas and sulfotransferasas can occur as a defense mechanism (detoxification) (Lambert et al. 2003; Lu et al. 2003).

EGCG has a carbonyl group on the gallate moiety, which could avoid its oxidation when dissolved in alkaline solutions under blue light illumination conditions, giving the molecule stability. This polyphenol can suppress this type of oxidation in other molecules such as epicatechins (Shi et al. 2016). Other authors report low stability in cell culture media and pH above 7. These molecules not only oxidize, but they can also degrade during storage at temperatures above 50 °C. Polymerization and epimerization are other reactions that can occur in the presence of oxidative agents and under specific conditions of temperature and pH (Chen et al. 2015). These reactions can modify EGCG structure and generate oxidation products, such as dimer and dimer quinone (Ashihara et al. 2010; Wang et al., 2006). Despite the potentialities of the EGCG, these characteristics presented make the EGCG have low bioavailability.

Studies on the biosynthesis of EGCG from the phenylalanine pathway reported that some enzymes are involved with this process including flavanol-3-O-gallate synthase, dihydroflavanol 4-reductase, anthocyanidin reductase, and anthocyanidin synthase (Ashihara et al. 2010). The synthesis of EGCG depends on the amount of gallic acid, which is a precursor in the biosynthesis of galloylated catechins (Liu et al. 2012), which in the presence of enzymes such as lipoxygenase, trypsin, and pepsin can form precipitates; this phenomenon is attributed to the ester bonding of these compounds (Sekiya et al., 1984). Other authors investigated enzymatic reactions in in vitro assays to identify and purify the enzymes associated with the biosynthesis of this type of polyphenols (Liu et al. 2012).

Sources

EGCG is a class of antioxidant compounds that are most found in cacao, grapes, red wine, green tea, white tea, black tea, fruits, hazelnuts, pecans, and onions. EGCG is one of the catechins with the greatest bioactive potential; it has been found abundantly in green tea, representing contents between 32 and 75% of total catechins in the plant (Xu et al., 2018). The biosynthesis of EGCG in tea seedlings was evaluated, finding much greater concentrations in young leaves than the other organs. In addition, this compound was predominant over other polyphenols (Mamati et al., 2006).

Several investigations have reported the bioactivity of EGCG obtained from dry leaves of tea plants, reporting values between 3% and 13%. Other authors determined a content of EGCG of 3.51 y 6.78% in dried green tea leaves, corresponding to the varieties *Camellia sinensis* and *C. assamica* var. *kucha*, respectively (X. R. Yang et al. 2007).

Good dietary sources of EGCG are vegetables such as broad beans, apricots, onions, lettuce, and tomatoes; also they have been found in legumes such as green beans (Andersen and Markham 2005). High levels of total catechins are found in products like red wine, beer, cacao liquor, chocolate, cocoa, and coffee. Recent studies have reported that tea leaves contain high concentrations of catechin among them monomeric and polymeric, accounting approximately 12% on dry basis (Zhang et al. 2019). Although the content of catechins varies according species, variety, to the distribution of plant tissues, season, geographical location, and growth conditions such as altitude and illumination (Grzesik et al. 2018; Jiang et al. 2013). Green tea is mostly made from the shoots and young leaves of the tea plant *Camellia sinensis* L. (Liu et al. 2012).

Properties

EGCG is one of the catechins with the highest antioxidant potential and presents important benefits for human health due to different biological activities such as antitumor, anticarcinogenic, antimicrobial, anti-inflammatory, anti-cardiovascular antiproliferative, antiangiogenic, anti-metastatic, antimutagenic, antibacterial, antidiabetic, and antihypertensive activities (Liu et al. 2012; Oz et al., 2013, Wang et al. 2018a, 2018b). Other disorders such as Parkinson's disease, Alzheimer's disease, and obesity have been treated with EGCG, finding satisfactory results. Some authors report that galloylated catechins have greater pharmaceutical potential than nongalloylated catechins (Grzesik et al. 2018; Kao et al. 2006). The anticancer role of EGCG has been widely studied; in vitro and in vivo trials have been conducted to evaluate the benefits on different types of cancer including breast (Zeng et al. 2014), gastric (Yang et al. 2016), prostate (Johnson et al. 2010), and liver (Ni et al. 2017), among others.

Interesting results have been published on mechanisms of EGCG modulation in cellular processes such as proliferation, differentiation, angiogenesis, apoptosis, invasion, and cell death (Mondal et al. 2012). EGCG can inhibit cutaneous tumors in mice, as well as malignant transformation of papillomas, which were induced by UV light (Mittal et al. 2003). This catechin has been suggested to inhibit protein kinases involved in cell growth and apoptosis, osteoclast differentiation processes in stem cells, myeloperoxidase activity, metastasis of cancer cells, and infiltration of leukocytes and suppress matrix metalloproteinase activity (Wang et al. 2018a, 2018b; Jin et al. 2014; Oz et al., 2013).

The anticancer effects of EGCG have been investigated focusing on various mechanisms of action, such as electron transfer, hydrogen atom transfer, and the chelation of catalytic metals, which are associated with antioxidant activity. It has been proposed that EGCG has strong antioxidant activity due to the presence of hydroxyl groups and a gallate group in its structure. Other studies have evaluated the effect of EGCG on some mechanisms of action related to obesity, such as adipocyte proliferation, preadipocyte differentiation, induction of thermogenesis, and lipid accumulation, among others (Carrasco-Pozo et al. 2019). Moreover, EGCG can chelate metal ions, which are involved in the oxidation reactions of lipid peroxidation, also prevent oxidative DNA damage, and inhibit signaling pathway associated with generation of pro-inflammatory cytokines (Lakshmi et al. 2020; Zhao et al. 2017).

Recent studies have reported the role of EGCG on mechanisms related to mitochondrial function improvement, due to the importance of mitochondria in the body homeostasis, redox status, and apoptotic cell death regulation (Bhargava and Schnellmann 2017). Some organs are sensitive to cellular redox imbalance that leads to a wide range of pathologies associated to increase in ROS (Pedraza-Chaverri et al. 2016). Authors have demonstrated the potential of EGCG on the decrease of mitochondrial ROS in HK-2 cells and improvement of mitochondrial function of the kidney (Pan et al. 2015). Other studies revealed that EGCG was able to inhibit the formation of free radicals, oxidative stress, and lipid peroxidation; pro-inflammatory gene expression in cell treatments is stimulated with cigarette smoke extract (Lakshmi et al. 2020).

Currently, SARS-CoV-2, commonly known as coronavirus, has claimed more than 750,000 people worldwide as of August 2020, so assays of the role of EGCG in combination with other natural substances have been conducted. This study evaluated the effect of EGCG administration and theaflavin on the activity of 3CL-proteasa of SARS-CoV-2 in HEK293T cells, finding a minimum inhibitory concentration of (IC50) 7.58 μg/mL (Jang et al. 2020). This study reported the potential of antioxidant compounds for the treatment of SARS-CoV-2. However, more research is needed for prevention and treatment.

12.2.2.2 Immune Booster Property of Epigallocatechin-3-Gallate

EGCG and catechins have shown antibacterial, antioxidant, antimutagenic, anticarcinogenic, cardioprotective, and neuroprotective potential using in vitro and in vivo assays. These compounds can intervene in several pathways due to its ability to interact with different proteins (Shukla et al. 2018). They can be more easily accessible and less harmful to human health because of their side effects. EGCG is another natural compound with immunological properties, which can prevent the deterioration and oxidation of proteasome. However, the study of these regulation pathways has yet to be clarified. The use of these compounds in cancer therapies is of great interest, since it has been proven that they can regulate the functionality of the system that degrades proteins (oxidized, unfolded, and misfolded) in prokaryotic and eukaryotic cells, known as proteasomes. Other important functions are the regulation of apoptosis, cell proliferation, and drug resistance in human tumors, neuropathological disorders, signal transduction, and antigen processing (Bonfili et al. 2011).

These compounds are involved in the modulation of inflammatory signal transduction pathways, suppression of chemokines and cytokines and interactions with their receptors, and neutrophil responses to inflammatory stimuli (Li et al. 2012; Santilli et al. 2013). The activity of the transcription factors (nuclear factor kB) and other proteins (kinases and activator protein-1) is altered during the modulation process (Pae and Wu 2013). Kasper et al. (2016) evaluated in vivo the effect of intraoperative administration of EGCG on cardiopulmonary bypass-associated lung injury, finding decrease of neutrophil infiltration, interstitial edema, apoptosis-inducing factor translocation, and poly(ADP-ribose) polymerase 1 activation (PARP-1). The PARP-1 is a protein activated by oxidative stress and is associated with DNA repair. EGCG can also modulate MAPK pathway and inhibits the expression of cyclooxygenase 2 (COX2), the epidermal growth factors such as factor receptor-2 (HER2) in cancer cell lines (Shimizu et al. 2005), and monocyte migration (Santilli et al. 2013; Pae and Wu 2013).

Epigallocatechin-3-gallate has been found to modulate the T helper 17 /Treg cell ratio in animal with arthritis; these T cells play an important role in inflammation and autoimmunity. The effects of the administration of EGCG on T-cell activation and activation of transcription factors related to inflammation and inhibition of osteoclastogenesis were evaluated. This study showed a decrease in the production of cytokines IL-6, IL-17, and VEGF (vascular endothelial growth factor) in EGCG treatments with concentrations of 50 mg/kg. These proteins and the growth factor are involved in the processes of inflammation and have effect on the communications and interactions between cells. Oral EGCG administration inhibited the osteoclastogenesis and suppressed the p-STAT3 transcription factor, which intervenes in the activation and differentiation of Th17 cells. EGCG showed increase in Nrf2 gene (nuclear respiratory factor 2) expression in splenic cells in mice. Moreover, they could activate phosphorylate-extracellular signal-regulated kinase (p-ERK), which activates the STAT5 transcription factor, and finally, it occurs in the suppression of Th17 cells (Lee et al. 2016).

Current EGCG has demonstrated cellular mechanisms including activation and inhibition of signaling pathways (PKC, MAPK, and PI3K/Akt); enhancement of antioxidant action (radical scavenging, lipid peroxidation and production of endogenous defenses); modulation of cell survival genes and cell death genes (antiapoptotic action); neurite growth and bioenergetic action (stabilization of the mitochondrial potential), induction of iron-chelating effects on Aβ, tau, and α-synuclein; elevation of synaptic DA (via COMT activity inhibition and DAT internalization); production of non-amyloidogenic sAPPα (by increasing α secretase levels for preferential APP processing); and inhibition of Aβ fibrillation, plaques, tau accumulation, NFT generation, and α-synuclein fibrillation (Singh et al. 2016).

12.3 Conclusions

Catechin and epigallocatechin-3-gallate are two molecules with very outstanding functionality for their immune booster properties; their high effectiveness makes them target molecules in the design of new drugs and functional foods. However, there are scientific and technological challenges for the recovery, extraction, separation, purification, and stabilization of these molecules that are highly reactive and whose efficiency and function are highly dependent on the structure and its degree of oxidation and polymerization.

Acknowledgments Authors thank CONACYT-Mexico for the financial support in the frame of the project PROYECTO FOSEC_CB 2017-18 A1-S-42515 and for the fellowship given to LJVH.

Competing Interest The authors declare that there are no competing interests.

References

Aggarwal V, Tuli HS, Tania M, Srivastava S, Ritzer EE, Pandey A, Aggarwal D, Barwal TS, Jain A, Kaur G, Sak K, Varol M, Bishayee A (2020) Molecular mechanisms of action of epigallocatechin gallate in cancer: recent trends and advancement. Semin Cancer Biol. Academic Press. https://doi.org/10.1016/j.semcancer.2020.05.011

Akram M, Riaz M, Wadood AWC, Hazrat A, Mukhtiar M, Ahmad Zakki S et al (2020) Medicinal plants with anti-mutagenic potential. Biotechnol Biotechnol Equip 34(1):309–318

Anand R, Gill KD, Mahdi AA (2014) Therapeutics of Alzheimer's disease: past, present and future. Neuropharmacology 76(PART A):27–50. Pergamon. https://doi.org/10.1016/j.neuropharm.2013.07.004

Andersen OM, Markham KR (eds) (2005) Flavonoids: chemistry, biochemistry and applications. CRC press

Ashihara H, Deng WW, Mullen W, Crozier A (2010) Distribution and biosynthesis of flavan-3-ols in Camellia sinensis seedlings and expression of genes encoding biosynthetic enzymes. Phytochemistry 71(5–6):559–566. https://doi.org/10.1016/j.phytochem.2010.01.010

Ball N, Teo W-P, Chandra S, Chapman J (2019) Parkinson's disease and the environment. Front Neurol 10:218. https://doi.org/10.3389/fneur.2019.00218

Ballard C, Gauthier S, Corbett A, Brayne C, Aarsland D, Jones E (2011) Alzheimer's disease. The Lancet 377(9770):1019–1031). Lancet Publishing Group. https://doi.org/10.1016/S0140-6736(10)61349-9

Baptista J, Paiva LS, Lima E, Motta M (2019) The surplus value of Azorean Camellia sinensis flowers as an important contributor affecting the nutraceutical benefits of green tea quality. Pharm Pharmacol Int J 7(6)):327–332

Bhargava P, Schnellmann RG (2017) Mitochondrial energetics in the kidney. Nat Rev Nephrol 13(10):629–646). Nature Publishing Group. https://doi.org/10.1038/nrneph.2017.107

Bonfili L, Cuccioloni M, Mozzicafreddo M, Cecarini V, Angeletti M, Eleuteri AM (2011) Identification of an EGCG oxidation derivative with proteasome modulatory activity. Biochimie 93(5):931–940. https://doi.org/10.1016/j.biochi.2011.02.003

Bossy-Wetze E, Schwarzenbacher R, Lipton SA (2004) Molecular pathways to neurodegeneration. Nat Med 10(7):S2. https://doi.org/10.1038/nm1067

Braicu C, Ladomery MR, Chedea VS, Irimie A, Berindan-Neagoe I (2013) The relationship between the structure and biological actions of green tea catechins. Food Chem 141(3): 3282–3289). Elsevier Ltd. https://doi.org/10.1016/j.foodchem.2013.05.122

Brouwer M, Huss A, van der Mark M, Nijssen PCG, Mulleners WM, Sas AMG, van Laar T, de Snoo GR, Kromhout H, Vermeulen RCH (2017) Environmental exposure to pesticides and the risk of Parkinson's disease in the Netherlands. Environ Int 107:100–110. https://doi.org/10.1016/j.envint.2017.07.001

Byun JK, Yoon BY, Jhun JY, Oh HJ, Kim EK, Min JK, Cho ML (2014) Epigallocatechin-3-gallate ameliorates both obesity and autoinflammatory arthritis aggravated by obesity by altering the balance among CD4+ T-cell subsets. Immunol Lett 157(1–2):51–59

Carrasco-Pozo C, Cires MJ, Gotteland M (2019) Quercetin and epigallocatechin gallate in the prevention and treatment of obesity: from molecular to clinical studies. J Med Food 22(8): 753–770). Mary Ann Liebert Inc. https://doi.org/10.1089/jmf.2018.0193

Casazza AA, Pettinato M, Perego P (2020) Polyphenols from apple skins: a study on microwave-assisted extraction optimization and exhausted solid characterization. Sep Purif Technol 240: 116640. https://doi.org/10.1016/j.seppur.2020.116640

Chen CN, Chang KC, Lin RF, Wang MH, Shih RL, Tseng HC, Soung HS, Tsai CC (2016) Nitric oxide pathway activity modulation alters the protective effects of (−)Epigallocatechin-3-gallate on reserpine-induced impairment in rats. Behav Brain Res 305:198–211. https://doi.org/10.1016/j.bbr.2016.02.038

Chen LY, Cheng CW, Liang JY (2015) Effect of esterification condensation on the Folin-Ciocalteu method for the quantitative measurement of total phenols. Food Chem 170:10–15. https://doi.org/10.1016/j.foodchem.2014.08.038

Chen L-Y, Wu J-Y, Liang J (2018a) Using chromatography and mass spectrometry to monitor isomerization of catechin in alkaline aqueous with thermal processing. J Food Process Preserv 42(1):e13365. https://doi.org/10.1111/jfpp.13365

Chen Y, Zhou Y, Zeng L, Dong F, Tu Y, Yang Z (2018b) Occurrence of functional molecules in the flowers of tea (Camellia sinensis) plants: evidence for a second resource. Molecules 23(4):790. https://doi.org/10.3390/molecules23040790

Chen X, Han Y, Meng H, Li W, Li Q, Luo Y et al (2019) Characteristics of the emulsion stabilized by polysaccharide conjugates alkali-extracted from green tea residue and its protective effect on catechins. Ind Crop Prod 140:111611

Chobot V, Huber C, Trettenhahn G, Hadacek F (2009) (±)-catechin: chemical weapon, antioxidant, or stress regulator? J Chem Ecol 35(8):980–996. https://doi.org/10.1007/s10886-009-9681-x

Cuevas-Valenzuela J, González-Rojas Á, Wisniak J, Apelblat A, Pérez-Correa JR (2015) Solubility of (+)-catechin in water and water-ethanol mixtures within the temperature range 277.6–331.2K: fundamental data to design polyphenol extraction processes. Fluid Phase Equilib 382:279–285. https://doi.org/10.1016/j.fluid.2014.09.013

Davinelli S, Sapere N, Zella D, Bracale R, Intrieri M, Scapagnini G (2012) Pleiotropic protective effects of phytochemicals in Alzheimer's disease. Oxidative Med Cell Longev. https://doi.org/10.1155/2012/386527

Del Rey NLG, Quiroga-Varela A, Garbayo E, Carballo-Carbajal I, Fernández-Santiago R, Monje MHG, Trigo-Damas I, Blanco-Prieto MJ, Blesa J (2018) Advances in Parkinson's disease: 200 years later. Front Neuroanat 12:113. Frontiers Media S.A. https://doi.org/10.3389/fnana.2018.00113

Delgado L, Fernandes I, González-Manzano S, De Freitas V, Mateus N, Santos-Buelga C (2014) Anti-proliferative effects of quercetin and catechin metabolites. Food Funct 5(4):797–803. https://doi.org/10.1039/c3fo60441a

de Pascual-Teresa S, Moreno DA, García-Viguera C (2010) Flavanols and anthocyanins in cardiovascular health: a review of current evidence. Int J Mol Sci 11(4):1679–1703

Dhadge VL, Changmai M, Purkait MK (2019) Purification of catechins from Camellia sinensis using membrane cell. Food Bioprod Process 117:203–212. https://doi.org/10.1016/j.fbp.2019.07.010

Dickson DW (2018) Neuropathology of Parkinson disease. Parkinsonism Relat Disord 46:S30–S33. https://doi.org/10.1016/j.parkreldis.2017.07.033
Eungwanichayapant PD, Popluechai S (2009) Accumulation of catechins in tea in relation to accumulation of mRNA from genes involved in catechin biosynthesis. Plant Physiol Biochem 47(2):94–97. https://doi.org/10.1016/j.plaphy.2008.11.002
Fatima M, Kesharwani RK, Misra K, Rizvi SI (2013) Protective effect of theaflavin on erythrocytes subjected to in vitro oxidative stress. Biochem Res Int 2013
Formica JV, Regelson W (1995) Review of the biology of quercetin and related bioflavonoids. Food Chem Toxicol 33:1061–1080
Gadkari PV, Balaraman M (2015) Catechins: sources, extraction and encapsulation: a review. Food Bioprod Process 93:122–138. Institution of Chemical Engineers. https://doi.org/10.1016/j.fbp.2013.12.004
Ganeshpurkar A, Saluja A (2018b) In silico interaction of catechin with some immunomodulatory targets: a docking analysis. Indian J Biotechnol 17:626–631
Ganeshpurkar A, Saluja A (2019) Immunomodulatory effect of rutin, catechin, and hesperidin on macrophage function. Indian J Biochem Biophys 57:58–63
Ganeshpurkar A, Saluja AK (2018a) Protective effect of catechin on humoral and cell mediated immunity in rat model. Int Immunopharmacol 54:261–266. https://doi.org/10.1016/j.intimp.2017.11.022
Geetha T, Garg A, Chopra K, Kaur IP (2004) Delineation of antimutagenic activity of catechin, epicatechin and green tea extract. Mutat Res/Fundam Mol Mech Mutagen 556(1–2):65–74
González-Manzano S, Dueñas M, Surco-Laos F, Hidalgo M, González-Paramas A, Santos-Buelga-C, de Pascual-Teresa S (2011) In vitro evaluation of the antioxidant and anti-inflammatory activities of sulphated metabolites of catechins Evaluación in vitro de las actividades antioxidante y antiinflamatoria de metabolitos sulfatados de catequinas. CyTA J Food 9(4):257–264. https://doi.org/10.1080/19476337.2011.594523
Grzesik M, Naparło K, Bartosz G, Sadowska-Bartosz I (2018) Antioxidant properties of catechins: comparison with other antioxidants. Food Chem 241:480–492. https://doi.org/10.1016/j.foodchem.2017.08.117
Guo H, Luo H, Yuan H, Xia Y, Shu P, Huang X, Lu Y, Liu X, Keller ET, Sun D, Deng J, Zhang J (2017) Litchi seed extracts diminish prostate cancer progression via induction of apoptosis and attenuation of EMT through Akt/GSK-3β signaling. Sci Rep 7(1):41656. https://doi.org/10.1038/srep41656
Hamilton AC (2004) Medicinal plants, conservation and livelihoods. Biodivers Conserv 13(8):1477–1517. https://doi.org/10.1023/B:BIOC.0000021333.23413.42
Ho S, Thoo YY, Young DJ, Siow LF (2019) Stability and recovery of cyclodextrin encapsulated catechin in various food matrices. Food Chem 275:594–599. https://doi.org/10.1016/j.foodchem.2018.09.117
Hussain T, Tan B, Yin Y, Blachier F, Tossou MCB, & Rahu N (2016) Oxidative stress and inflammation: what polyphenols can do for us? https://doi.org/10.1155/2016/7432797
Ittner LM, Götz J (2011) Amyloid-β and tau—a toxic pas de deux in Alzheimer's disease. Nat Rev Neurosci 12(2):67–72. https://doi.org/10.1038/nrn2967
Jang M, Park Y-I, Cha Y-E, Park R, Namkoong S, Lee JI, Park J (2020) Tea polyphenols EGCG and theaflavin inhibit the activity of SARS-CoV-2 3CL-protease in vitro. Evid Based Complement Alternat Med 2020:1–7. https://doi.org/10.1155/2020/5630838
Jiang X, Liu Y, Li W, Zhao L, Meng F, Wang Y, Tan H, Yang H, Wei C, Wan X, Gao L, Xia T (2013) Tissue-specific, development-dependent phenolic compounds accumulation profile and gene expression pattern in tea plant [Camellia sinensis]. PLoS One 8(4):e62315. https://doi.org/10.1371/journal.pone.0062315
Jiang Z, Zhang J, Cai Y, Huang J, You L (2017) Catechin attenuates traumatic brain injury-induced blood–brain barrier damage and improves longer-term neurological outcomes in rats. Exp Physiol 102(10):1269–1277

Jiang Y, Ding S, Li F, Zhang C, Sun-Waterhouse D, Chen Y, Li D (2019) Effects of (+)-catechin on the differentiation and lipid metabolism of 3T3-L1 adipocytes. J Funct Foods 62:103558. https://doi.org/10.1016/j.jff.2019.103558

Jin P, Wu H, Xu G, Zheng L, Zhao J (2014) Epigallocatechin-3-gallate (EGCG) as a pro-osteogenic agent to enhance osteogenic differentiation of mesenchymal stem cells from human bone marrow: an in vitro study. Cell Tissue Res 356(2):381–390. https://doi.org/10.1007/s00441-014-1797-9

Johnson JJ, Bailey HH, Mukhtar H (2010) Green tea polyphenols for prostate cancer chemoprevention: a translational perspective. Phytomedicine 17(1):3–13. https://doi.org/10.1016/j.phymed.2009.09.011

Jung YD, Kim MS, Shin BA, Chay KO, Ahn BW, Liu W, Bucana CD, Gallick GE, Ellis LM (2001) EGCG, a major component of green tea, inhibits tumour growth by inhibiting VEGF induction in human colon carcinoma cells. Br J Cancer 84(6):844–850. https://doi.org/10.1054/bjoc.2000.1691

Kah Hui C, Majid NI, Mohd Yusof H, Mohd Zainol K, Mohamad H, Mohd Zin Z (2020) Catechin profile and hypolipidemic activity of Morinda citrifolia leaf water extract. Heliyon 6(6):e04337. https://doi.org/10.1016/j.heliyon.2020.e04337

Kaliora AC, Dedoussis GVZ, Schmidt H (2006) Dietary antioxidants in preventing atherogenesis. Atherosclerosis 187(1):1–17

Kamalden TA, Ji D, Osborne NN (2012) Rotenone-induced death of RGC-5 cells is caspase independent, involves the JNK and p38 pathways and is attenuated by specific green tea flavonoids. Neurochem Res 37(5):1091–1101. https://doi.org/10.1007/s11064-012-0713-5

Kao YH, Chang HH, Lee MJ, Chen CL (2006) Tea, obesity, and diabetes. Mol Nutr Food Res 50(2):188–210). John Wiley & Sons, Ltd. https://doi.org/10.1002/mnfr.200500109

Kapoor M, Howard R, Hall I, Appleton I (2004) Effects of epicatechin gallate on wound healing and scar formation in a full thickness incisional wound healing model in rats. Am J Pathol 165(1):299–307. https://doi.org/10.1016/S0002-9440(10)63297-X

Kasper B, Salameh A, Krausch M, Kiefer P, Kostelka M, Mohr FW, Dhein S (2016) Epigallocatechin gallate attenuates cardiopulmonary bypass-associated lung injury. J Surg Res 201(2):313–325. https://doi.org/10.1016/j.jss.2015.11.007

Keservani RK, Kesharwani RK, Sharma AK (2017) Introduction to nanotechnology in drug delivery. Drug Deliv Approaches Nanosyst:1–19. Apple Academic Press. ISBN: 9781771885836

Keservani RK, Kesharwani RK, Vyas N, Jain S, Raghuvanshi R, Sharma AK (2010) Nutraceutical and functional food as future food: a review. Pharm Lett 2(1):106–116

Keservani RK, Sharma AK, Kesharwani RK (2018) Nanoparticulate nanocarriers in drug delivery. Nano:3–22. Apple Academic Press. ISBN: 9781771885911

Kesharwani RK, Misra K, Singh DB (2019) Perspectives and challenges of tropical medicinal herbs and modern drug discovery in the current scenario. Asian Pac J Trop Med 12(1):1

Kesharwani RK, Vishwakarma VK, Keservani RK, Singh P, Katiyar N, Tripathi S (2020) Role of ADMET tools in current scenario: application and limitations. Comput-Aided Drug Des:71–87. Springer, Singapore. ISBN: 978-981-15-6814-5

Khan HY, Zubair H, Ullah MF, Ahmad A, Hadi SM (2011) Oral administration of copper to rats leads to increased lymphocyte cellular DNA degradation by dietary polyphenols: implications for a cancer preventive mechanism. Biometals 24(6):1169–1178. https://doi.org/10.1007/s10534-011-9475-9

Khan N, Mukhtar H (2007) Tea polyphenols for health promotion. Life Sci 81(7):519–533). Pergamon. https://doi.org/10.1016/j.lfs.2007.06.011

Khanal P, Patil BM, Chand J, Naaz Y (2020) Anthraquinone derivatives as an immune booster and their therapeutic option against COVID-19. Nat Prod Bioprospect 10(5):325–335. https://doi.org/10.1007/s13659-020-00260-2

Khanna K, Kohli SK, Kaur R, Bhardwaj A, Bhardwaj V, Ohri P, Sharma A, Ahmad A, Bhardwaj R, Ahmad P (2020) Herbal immune-boosters: substantial warriors of pandemic Covid-19 battle. Phytomedicine 153361. https://doi.org/10.1016/j.phymed.2020.153361

Khiewkamrop P, Phunsomboon P, Richert L, Pekthong D, Srisawang P (2018) Epistructured catechins, EGCG and EC facilitate apoptosis induction through targeting de novo lipogenesis pathway in HepG2 cells. Cancer Cell Int 18:1. https://doi.org/10.1186/s12935-018-0539-6

Lakshmi SP, Reddy AT, Kodidhela LD, Varadacharyulu NC (2020) Epigallocatechin gallate diminishes cigarette smoke-induced oxidative stress, lipid peroxidation, and inflammation in human bronchial epithelial cells. Life Sci 259:118260. https://doi.org/10.1016/j.lfs.2020.118260

Lambert JD, Elias RJ (2010) The antioxidant and pro-oxidant activities of green tea polyphenols: a role in cancer prevention. Arch Biochem Biophys 501(1):65–72. https://doi.org/10.1016/j.abb.2010.06.013

Lambert JD, Lee MJ, Lu H, Meng X, Hong JJJ, Seril DN, Sturgill MG, Yang CS (2003) Epigallocatechin-3-gallate is absorbed but extensively glucuronidated following oral administration to mice. J Nutr 133(12):4172–4177. https://doi.org/10.1093/jn/133.12.4172

Lashuel HA, Overk CR, Oueslati A, Masliah E (2013) The many faces of α-synuclein: from structure and toxicity to therapeutic target. Nat Rev Neurosci 14(1):38–48). Nature Publishing Group. https://doi.org/10.1038/nrn3406

Lee S-Y, Jung YO, Ryu J-G, Oh H-J, Son H-J, Lee SH, Kwon J-E, Kim E-K, Park M-K, Park S-H, Kim H-Y, Cho M-L (2016) Epigallocatechin-3-gallate ameliorates autoimmune arthritis by reciprocal regulation of T helper-17 regulatory T cells and inhibition of osteoclastogenesis by inhibiting STAT3 signaling. J Leukoc Biol 100(3):559–568. https://doi.org/10.1189/jlb.3a0514-261rr

Li J, Ye L, Wang X, Liu J, Wang Y, Zhou Y, Ho W (2012) (−)-epigallocatechin gallate inhibits endotoxin-induced expression of inflammatory cytokines in human cerebral microvascular endothelial cells. J Neuroinflammation 9(1):579. https://doi.org/10.1186/1742-2094-9-161

Li W, Yang R, Ying D, Yu J, Sanguansri L, Augustin MA (2020) Analysis of polyphenols in apple pomace: a comparative study of different extraction and hydrolysis procedures. Ind Crop Prod 147:112250. https://doi.org/10.1016/j.indcrop.2020.112250

Liu J, Lu J f, Kan J, Wen X y, Jin C h (2014) Synthesis, characterization and in vitro anti-diabetic activity of catechin grafted inulin. Int J Biol Macromol 64:76–83. https://doi.org/10.1016/j.ijbiomac.2013.11.028

Liu Y, Gao L, Liu L, Yang Q, Lu Z, Nie Z, Wang Y, Xia T (2012) Purification and characterization of a novel galloyltransferase involved in catechin galloylation in the tea plant (Camellia sinensis). J Biol Chem 287(53):44406–44417. https://doi.org/10.1074/jbc.M112.403071

Liu Y, Liu F, Grundke-Iqbal I, Iqbal K, Gong CX (2011) Deficient brain insulin signalling pathway in Alzheimer's disease and diabetes. J Pathol 225(1):54–62. https://doi.org/10.1002/path.2912

Liu Y, Lu Z, Wang M, Chen Y, Zhan R, Shao LD, Chen YG (2020) Catechins and lignan from the flower buds of Rosa chinensis Jacq. Phytochem Lett 38:46–48. https://doi.org/10.1016/j.phytol.2020.05.004

Lu H, Meng X, Yang CS (2003) Enzymology of methylation of tea catechins and inhibition of catechol-O-methyltransferase by (−)-epigallocatechin gallate. Drug Metab Dispos 31(5):572–579. https://doi.org/10.1124/dmd.31.5.572

Ma Y, Ding S, Fei Y, Liu G, Jang H, Fang J (2019) Antimicrobial activity of anthocyanins and catechins against foodborne pathogens Escherichia coli and salmonella. Food Control 106:106712. https://doi.org/10.1016/j.foodcont.2019.106712

Madhan B, Subramanian V, Rao JR, Nair BU, Ramasami T (2005) Stabilization of collagen using plant polyphenol: role of catechin. Int J Biol Macromol 37(1–2):47–53. https://doi.org/10.1016/j.ijbiomac.2005.08.005

Mahindrakar KV, Rathod VK (2020) Ultrasonic assisted aqueous extraction of catechin and gallic acid from Syzygium cumini seed kernel and evaluation of total phenolic, flavonoid contents and antioxidant activity. Chem Eng Process Process Intensif 149:107841. https://doi.org/10.1016/j.cep.2020.107841

Mamati GE, Liang Y, Lu J (2006) Expression of basic genes involved in tea polyphenol synthesis in relation to accumulation of catechins and total tea polyphenols. J Sci Food Agric 86(3):459–464

Mandel S, Weinreb O, Amit T, Youdim MBH (2004) Cell signaling pathways in the neuroprotective actions of the green tea polyphenol (−)-epigallocatechin-3-gallate: implications for neurodegenerative diseases. J Neurochem 88(6):1555–1569). Blackwell Publishing Ltd. https://doi.org/10.1046/j.1471-4159.2003.02291.x

Mendes TMN, Murayama Y, Yamaguchi N, Sampaio GR, Fontes LCB, Torres EAF d S, Tamura H, Yonekura L (2019) Guaraná (Paullinia cupana) catechins and procyanidins: gastrointestinal/colonic bioaccessibility, Caco-2 cell permeability and the impact of macronutrients. J Funct Foods 55:352–361. https://doi.org/10.1016/j.jff.2019.02.026

Meng Q, Velalar CN, Ruan R (2008) Effects of epigallocatechin-3-gallate on mitochondrial integrity and antioxidative enzyme activity in the aging process of human fibroblast. Free Radic Biol Med 44(6):1032–1041. https://doi.org/10.1016/j.freeradbiomed.2007.11.023

Miranda-Hernández AM, Muñiz-Márquez DB, Wong-Paz JE, Aguilar-Zárate P, de la Rosa-Hernández M, Larios-Cruz R, Aguilar CN (2019) Characterization by HPLC–ESI–MS 2 of native and oxidized procyanidins from litchi (Litchi chinensis) pericarp. Food Chem 291:126–131. https://doi.org/10.1016/j.foodchem.2019.04.020

Mittal A, Piyathilake C, Hara Y, Katiyar SK (2003) Exceptionally high protection of photocarcinogenesis by topical application of (−)-Epigallocatechin-3-gallate in hydrophilic cream in SKH-1 hairless mouse model: relationship to inhibition of UVB-induced global DNA hypomethylation. Neoplasia 5(6):555–565. https://doi.org/10.1016/s1476-5586(03)80039-8

Mondal S, Bandyopadhyay S, Ghosh MK, Mukhopadhyay S, Roy S, Mandal C (2012) Natural products: promising resources for cancer drug discovery. Asian Pac J Cancer Prev Indian J Pharmacol 12

Nair A, Chattopadhyay D, Saha B (2019) In: Ahmad Khan MS, Ahmad I, D. B. T.-N. L., Chattopadhyay P (eds) Chapter 17—plant-derived immunomodulators. Academic Press, pp 435–499. https://doi.org/10.1016/B978-0-12-814619-4.00018-5

Napolitano A, Cascone A, Graziani G, Ferracane R, Scalfi L, Di Vaio C, Ritieni A, Fogliano V (2004) Influence of variety and storage on the polyphenol composition of apple flesh. J Agric Food Chem 52(21):6526–6531. https://doi.org/10.1021/jf049822w

Neilson AP, George JC, Janle EM, Mattes RD, Ralf R, Matusheski NV, Ferruzzi MG (2009) Influence of chocolate matrix composition on cocoa flavan-3-ol bioaccessibility in vitro and bioavailability in humans. J Agric Food Chem 57(20):9418–9426. https://doi.org/10.1021/jf902919k

Ngcobo M, Gqaleni N (2016) Evaluation of the immunomodulatory effects of a south African commercial traditional immune booster in human peripheral blood mononuclear cells. BMC Complement Altern Med 16(1):300. https://doi.org/10.1186/s12906-016-1294-7

Ni CX, Gong H, Liu Y, Qi Y, Jiang CL, Zhang JP (2017) Green tea consumption and the risk of liver cancer: a meta-analysis. Nutr Cancer 69(2):211–220. https://doi.org/10.1080/01635581.2017.1263754

Okabe S, Ochiai Y, Aida M, Park K, Kim S-J, Nomura T, Suganuma M, Fujiki H (1999) Mechanistic aspects of green tea as a cancer preventive: effect of components on human stomach cancer cell lines. Jpn J Cancer Res 90(7):733–739. https://doi.org/10.1111/j.1349-7006.1999.tb00808.x

Oliveira A, Alexandre EMC, Coelho M, Lopes C, Almeida DPF, Pintado M (2015) Incorporation of strawberries preparation in yoghurt: impact on phytochemicals and milk proteins. Food Chem 171:370–378. https://doi.org/10.1016/j.foodchem.2014.08.107

Oz HS, Chen T, de Villiers WJ (2013) Green tea polyphenols and sulfasalazine have parallel antiinflammatory properties in colitis models. Front Immunol 4:132

Pae M, Wu D (2013) Immunomodulating effects of epigallocatechin-3-gallate from green tea: mechanisms and applications. Food Funct 4(9):1287–1303. https://doi.org/10.1039/c3fo60076a

Pan H, Chen J, Shen K, Wang X, Wang P, Fu G, Meng H, Wang Y, Jin B (2015) Mitochondrial modulation by epigallocatechin 3-gallate ameliorates cisplatin induced renal injury through decreasing oxidative/nitrative stress, inflammation and NF-kB in mice. PLoS One 10(4): e0124775. https://doi.org/10.1371/journal.pone.0124775

Pang Y, Peel GJ, Wright E, Wang Z, Dixon RA (2007) Early steps in proanthocyanidin biosynthesis in the model legume Medicago truncatula. Plant Physiol 145(3):601–615. https://doi.org/10.1104/pp.107.107326

Papademetrio DL, Trabucchi A, Cavaliere V, Ricco R, Costantino S, Wagner ML, Álvarez E (2013) The catechin flavonoid reduces proliferation and induces apoptosis of murine lymphoma cells LB02 through modulation of antiapoptotic proteins. Brazilian J Pharmacogn 23(3):455–463. https://doi.org/10.1590/S0102-695X2013005000025

Pastoriza S, Mesías M, Cabrera C, Rufián-Henares JA (2017) Healthy properties of green and white teas: an update. Food Funct 8(8):2650–2662). Royal Society of Chemistry. https://doi.org/10.1039/c7fo00611j

Pedraza-Chaverri J, Sánchez-Lozada LG, Osorio-Alonso H, Tapia E, Scholze A (2016) New pathogenic concepts and therapeutic approaches to oxidative stress in chronic kidney disease. Oxidative Med Cell Longev 2016. Hindawi Limited. https://doi.org/10.1155/2016/6043601

Pheomphun P, Treesubsuntorn C, Thiravetyan P (2019) Effect of exogenous catechin on alleviating O3 stress: the role of catechin-quinone in lipid peroxidation, salicylic acid, chlorophyll content, and antioxidant enzymes of Zamioculcas zamiifolia. Ecotoxicol Environ Saf 180:374–383. https://doi.org/10.1016/j.ecoenv.2019.05.002

Punyasiri PAN, Abeysinghe ISB, Kumar V, Treutter D, Duy D, Gosch C, Martens S, Forkmann G, Fischer TC (2004) Flavonoid biosynthesis in the tea plant Camellia sinensis: properties of enzymes of the prominent epicatechin and catechin pathways. Arch Biochem Biophys 431(1): 22–30. https://doi.org/10.1016/j.abb.2004.08.003

Rashidinejad A, Birch EJ, Sun-Waterhouse D, Everett DW (2016) Effect of liposomal encapsulation on the recovery and antioxidant properties of green tea catechins incorporated into a hard low-fat cheese following in vitro simulated gastrointestinal digestion. Food Bioprod Process 100:238–245. https://doi.org/10.1016/j.fbp.2016.07.005

Sang S, Lambert JD, Ho CT, Yang CS (2011) The chemistry and biotransformation of tea constituents. Pharmacol Res 64(2):87–99. https://doi.org/10.1016/j.phrs.2011.02.007

Santilli G, Anderson J, Thrasher AJ, Sala A (2013) Catechins and antitumor immunity. OncoImmunology 2(6):e24443. https://doi.org/10.4161/onci.24443

Sartor L, Pezzato E, Garbisa S (2002) (−)Epigallocatechin-3-gallate inhibits leukocyte elastase: potential of the phyto-factor in hindering inflammation, emphysema, and invasion. J Leukoc Biol 71(1):73–79. https://doi.org/10.1189/jlb.71.1.73

Seeram NP, Nair MG (2002) Inhibition of lipid peroxidation and structure-activity-related studies of the dietary constituents anthocyanins, anthocyanidins, and catechins. J Agric Food Chem 50(19):5308–5312. https://doi.org/10.1021/jf025671q

Sekiya J, Kajiwara T, Monma T, Hatanaka A (1984) Interaction of tea catechins with proteins: formation of protein precipitate. Agric Biol Chem 48(8):1963–1967

Sharman MJ, Gyengesi E, Liang H, Chatterjee P, Karl T, Li QX, Wenk MR, Halliwell B, Martins RN, Münch G (2019) Assessment of diets containing curcumin, epigallocatechin-3-gallate, docosahexaenoic acid and α-lipoic acid on amyloid load and inflammation in a male transgenic mouse model of Alzheimer's disease: are combinations more effective? Neurobiol Dis 124:505–519. https://doi.org/10.1016/j.nbd.2018.11.026

Shi M, Nie Y, Zheng X-Q, Lu J-L, Liang Y-R, Ye J-H (2016) Ultraviolet B (UVB) photosensitivities of tea catechins and the relevant chemical conversions. Molecules 21(10):1345. https://doi.org/10.3390/molecules21101345

Shimizu M, Deguchi A, Lim JTE, Moriwaki H, Kopelovich L, Weinstein IB (2005) (−)-epigallocatechin gallate and polyphenon E inhibit growth and activation of the epidermal growth factor receptor and human epidermal growth factor receptor-2 signaling pathways in human colon cancer cells. Clin Cancer Res 11(7):2735–2746. https://doi.org/10.1158/1078-0432.CCR-04-2014

Shukla AS, Jha AK, Kumari R, Rawat K, Syeda S, & Shrivastava A (2018) Role of catechins in chemosensitization. Role of Nutraceuticals in Chemoresistance to Cancer 169–198. Elsevier. https://doi.org/10.1016/b978-0-12-812373-7.00009-7

Singh NA, Mandal AKA, Khan ZA (2016) Potential neuroprotective properties of epigallocatechin-3-gallate (EGCG). Nutr J 15(1):1–17). BioMed Central Ltd. https://doi.org/10.1186/s12937-016-0179-4

Singh P, Kesharwani RK, Misra K, Rizvi SI (2014) In silico validation for the modulatory effect of tea catechins on erythrocyte Na+/K+-ATPase. Nat Prod J 4(3):173–182

Srikumar R, Parthasarathy NJ, Manikandan S, Narayanan G, Sheeladevi R (2006) Effect of tripala on oxidative stress and on cell-mediated immune respond against noise stress in rats. Mol Cell Biochem 283:67–74

Stafford HA, Lester HH (1980) Procyanidins (condensed tannins) in green cell suspension cultures of Douglas fir compared with those in strawberry and avocado leaves by means of C18-reversed-phase chromatography. Plant Physiol 66(6):1085–1090

Stafford HA, Lester HH (1984) Flavan-3-ol biosynthesis: the conversion of (+)-dihydroquercetin and flavan-3, 4-cis-diol (leucocyanidin) to (+)-catechin by reductases extracted from cell suspension cultures of douglas fir. Plant Physiol 76(1):184–186

Stafford HA, Lester HH (1985) Flavan-3-ol biosynthesis: the conversion of (+)-dihydromyricetin to its flavan-3, 4-diol (leucodelphinidin) and to (+)-gallocatechin by reductases extracted from tissue cultures of Ginkgo biloba and Pseudotsuga menziesii. Plant Physiol 78(4):791–794

Tang H, Hao S, Chen X, Li Y, Yin Z, Zou Y, Song X, Li L, Ye G, Zhao L, Guo H, He R, Lv C, Lin J, Shi F (2020) Epigallocatechin-3-gallate protects immunity and liver drug-metabolism function in mice loaded with restraint stress. Biomed Pharmacother 129:110418. https://doi.org/10.1016/j.biopha.2020.110418

Thanikachalam PV, Ramamurthy S, Kumar A, Gupta M, Bansal G (2020) Protective role of epigallocatechin gallate, a dietary antioxidant against oxidative stress in various diseases. Pathology:213–223. Elsevier. https://doi.org/10.1016/b978-0-12-815972-9.00021-4

Theppakorn T (2016) Stability and chemical changes of phenolic compounds during Oolong tea processing. Int Food Res J 23(2)

Tilbur JC, Kaptchuk TJ (2008) Herbal medicine research and global health: an ethical analysis. Bull World Health Organ 86:594–599

Tsao R (2010) Chemistry and biochemistry of dietary polyphenols. Nutrients 2:1231–1246

Tseng HC, Wang MH, Soung HS, Chang Y, Chang KC (2015) (−)Epigallocatechin-3-gallate prevents the reserpine-induced impairment of short-term social memory in rats. Behav Pharmacol 26(8):741–747. https://doi.org/10.1097/FBP.0000000000000171

Wang S, Noh SK, Koo SI (2006) Green tea catechins inhibit pancreatic phospholipase A2 and intestinal absorption of lipids in ovariectomized rats. J Nutr Biochem 17(7):492–498

Wang YS, Xu YJ, Gao LP, Yu O, Wang XZ, He XJ et al (2014) Functional analysis of flavonoid 3′, 5′-hydroxylase from tea plant (Camellia sinensis): critical role in the accumulation of catechins. BMC Plant Biol 14(1):1–14

Wang D, Wang Y, Wan X, Yang CS, Zhang J (2015) Green tea polyphenol (−)-epigallocatechin-3-gallate triggered hepatotoxicity in mice: responses of major antioxidant enzymes and the Nrf2 rescue pathway. Toxicol Appl Pharmacol 283(1):65–74

Wang W, Zhou Y, Wu Y, Dai X, Liu Y, Qian Y, Li M, Jiang X, Wang Y, Gao L, Xia T (2018a) Insight into catechins metabolic pathways of Camellia sinensis based on genome and transcriptome analysis. J Agric Food Chem 66(16):4281–4293. https://doi.org/10.1021/acs.jafc.8b00946

Wang Y, Meng G, Chen S, Chen Y, Jiang J, Wang YP (2018b) Correlation analysis of phenolic contents and antioxidation in yellow- and black-seeded Brassica napus. Molecules 23(7):1815. https://doi.org/10.3390/molecules23071815

Xiang LP, Wang A, Ye JH, Zheng XQ, Polito CA, Lu JL et al (2016) Suppressive effects of tea catechins on breast cancer. Nutrients 8(8):458

Xie DY, Jackson LA, Cooper JD, Ferreira D, Paiva NL (2004) Molecular and biochemical analysis of two cDNA clones encoding dihydroflavonol-4-reductase from Medicago truncatula. Plant Physiol 134(3):979–994. https://doi.org/10.1104/pp.103.030221

Xie DY, Sharma SB, Paiva NL, Ferreira D, Dixon RA (2003) Role of anthocyanidin reductase, encoded by BANYULS in plant flavonoid biosynthesis. Science 299(5605):396–399. https://doi.org/10.1126/science.1078540

Xu YQ, Zhang YN, Chen JX, Wang F, Du QZ, Yin JF (2018) Quantitative analyses of the bitterness and astringency of catechins from green tea. Food Chem 258:16–24

Yahfoufi N, Alsadi N, Jambi M, Matar C (2018) The immunomodulatory and anti-inflamatory role of polyphenols. Nutrients 10:1–23

Yang C, Du W, Yang D (2016) Inhibition of green tea polyphenol EGCG((−)-epigallocatechin-3-gallate) on the proliferation of gastric cancer cells by suppressing canonical wnt/β-catenin signalling pathway. Int J Food Sci Nutr 67(7):818–827. https://doi.org/10.1080/09637486.2016.1198892

Yang CS, Maliakal P, Meng X (2002) Inhibition of carcinogenesis by tea. Annu Rev Pharmacol Toxicol 42(1):25–54. https://doi.org/10.1146/annurev.pharmtox.42.082101.154309

Yang CS, Wang H, Chen JX, Zhang J (2014) Effects of tea catechins on cancer signaling pathways. Enzyme 36:195–221. Academic Press. https://doi.org/10.1016/B978-0-12-802215-3.00010-0

Yang XR, Ye CX, Xu JK, Jiang YM (2007) Simultaneous analysis of purine alkaloids and catechins in Camellia sinensis, Camellia ptilophylla and Camellia assamica var. kucha by HPLC. Food Chem 100(3):1132–1136. https://doi.org/10.1016/j.foodchem.2005.11.021

Yang YC, Lu FH, Wu JS, Wu CH, Chang CJ (2004) The protective effect of habitual tea consumption on hypertension. Arch Intern Med 164(14):1534–1540. https://doi.org/10.1001/archinte.164.14.1534

Yoo SH, Lee YE, Chung JO, Rha CS, Hong YD, Park MY, Shim SM (2020) Enhancing the effect of catechins with green tea flavonol and polysaccharides on preventing lipid absorption and accumulation. LWT 134:110032. https://doi.org/10.1016/j.lwt.2020.110032

Zanwar AA, Badole SL, Shende PS, Hegde MV, Bodhankar SL (2013) Antioxidant role of catechin in health and disease. Polyphenol Hum Health Dis 1:267–271. Elsevier Inc. https://doi.org/10.1016/B978-0-12-398456-2.00021-9

Zeng L, Holly JMP, Perks CM (2014) Effects of physiological levels of the green tea extract epigallocatechin-3-gallate on breast cancer cells. Front Endocrinol 5:61. https://doi.org/10.3389/fendo.2014.00061

Zhang L, Ho CT, Zhou J, Santos JS, Armstrong L, Granato D (2019) Chemistry and biological activities of processed Camellia sinensis teas: a comprehensive review. Compr Rev Food Sci Food Saf 18(5):1474–1495). Blackwell Publishing Inc. https://doi.org/10.1111/1541-4337.12479

Zhang Q, Liu M, Ruan J (2017) Metabolomics analysis reveals the metabolic and functional roles of flavonoids in light-sensitive tea leaves. BMC Plant Biol 17(1):1–10. https://doi.org/10.1186/s12870-017-1012-8

Zhao J, Xu L, Liang Q, Sun Q, Chen C, Zhang Y, Ding Y, Zhou P (2017) Metal chelator EGCG attenuates Fe(III)-induced conformational transition of α-synuclein and protects AS-PC12 cells against Fe(III)-induced death. J Neurochem 143(1):136–146. https://doi.org/10.1111/jnc.14142

Chapter 13
Recent Advances of Nutraceutical and Functional Foods in Immune Health

Saumya Das, Manas Kumar Das, and Rajesh K. Kesharwani

Abstract The recent phytopharmacological studies of nutritional immunology are part of a rapidly developing field. People's tendency to consume non-nutritive foods that reduce their fighting capacity against many diseases is a characteristic of the modern age. Diseases such as obesity, acute pancreatitis, osteoporosis, cardiovascular diseases (CVD), kidney dysfunction, cancer, diabetes, hepatitis, autoimmune disorders, allergies, and dental problems are directly related to nutraceutical and functional foods mediated immune health. In the current scenario, people worldwide are facing similar problems related to ageing and consumption of a high-energy and unbalanced diet. The nutritional components are very important and affect the intestinal micro flora and certain probiotic bacteria directly or indirectly, which play an important role in the modulation of the immune system. The present chapter describes the role of nutraceuticals/functional foods/food supplements in mitigating health problems, especially in the immune inflammatory diseases and gastrointestinal (GI) tract-related diseases.

Keywords Nutritional Immunology · Intestinal Microflora · Essential Proteins · Omega-3 Fatty Acids · Vitamins

S. Das (✉)
Noida Institute of Engineering and Technology (Pharmacy Institute), Greater Noida, UP, India

M. K. Das
Orlean College of Pharmacy, Greater Noida, India

R. K. Kesharwani
Department of Computer Application, Nehru Gram Bharati (Deemed to be University), Prayagraj, Uttar Pradesh, India

R. K. Kesharwani et al. (eds.), *Nutraceuticals and Functional Foods as Immunomodulators*, https://doi.org/10.1007/978-981-19-2507-8_13

13.1 Introduction

13.1.1 Nutraceuticals

Depending upon the property of nutraceuticals, therapists' interests vary widely, for example, neurologists are more interested in substances that boost or enhance brain function and cardiologists may be interested in those functional foods that reduce heart disease, hypertension, hypercholesterolemia, free radical-influenced disorders, and platelet-dependent thrombotic activity (Keservani et al. 2016). The interest in nutraceuticals and functional foods continues as research shows diseases can be prevented by diet and lifestyle modifications. Supplementation and fortified foods can optimize the health promoting capabilities of a person's diet. It should be noted that the term nutraceutical, as commonly used in marketing, has no regulatory definition (Keservani et al. 2010; Williamson et al. 2020). Nutraceuticals differ from dietary supplements in the following aspects: conventional foods use or as part of a meal or diet depending on the trends in food production and their consumption can have harmful health, environmental, and social impacts, and therefore have a direct negative impact on human health. Dietary supplements play a crucial role apart from normal daily food during some health issues.

Foods with nutraceutical factors such as omega-3 fatty acids, phytosterols, quercetin, and grape flavonoids are some dietary components of choice. Several naturally derived food substances with vitamin E, selenium, vitamin D, green tea, soy, and lycopene have been studied in immuno therapies. These compounds contribute to good human health, and many of these "natural" compounds have been found to have high therapeutic potential for human health. Polyunsaturated fatty acids (PUFAs) are abundant in flaxseed oil and fish oil. Based on the health benefits of PUFA, a balanced consumption of both is promoted to enhance immunity and metabolism.

13.2 Nutritional Food and Health

The understanding of nutraceutical and functional foods and their usability helps people to live life with good immunity. Pregnant women and adults over 50 years of age need vitamins such as vitamin D, folic acid, and minerals, especially calcium and iron (Galindo et al. 2021). Non-nutritional and processed foods are sodium rich, and are very dangerous for people with high blood pressure (Siscovick et al. 2017). The USFDA advises adults to consume less than 300 mg/day. According to the WHO, it is important that children 2 years or older should be fed a balanced diet with good nutritional value to help reduce the chances of developing chronic diseases (Martin and Li 2017). Vegetables and fruits are good sources of antioxidants, minerals, fiber, etc., with nutritional values, some of which have anticancerours properties (Springmann et al. 2018).

13.2.1 Cardiovascular Diseases

Cardiovascular diseases and tumors are the main cause of death worldwide and contribute more than 60% in economically developed countries. Recent studies indicate free radical species (FRS) are more reactive in the pathogenesis of both acute and chronic heart diseases as a result of cumulative oxidative stress. In particular, oxidation of low density lipoproteins (LDL), the latter emanating from saturated, trans fats and meat products, increases the chance of atherosclerosis and cardiovascular heart diseases through the initiation of the plaque formation process. The risk factor of CVD is increased not only by the consumption of poor diets without any nutrients but also by lifestyle habits such as smoking and alcohol intake. A reduced risk of CVD, leading to elevated serum total cholesterol, LDL-cholesterol, and triacylglycerol concentrations, while leading to reduced HDL cholesterol concentrations, leads to reduced risk of heart disease. Treatment of hypercholesterolemia has focused on increasing fecal excretion of cholesterol and bile acids, and reducing hepatic cholesterol synthesis through diet modification. Blood pressure control is important in the prevention of coronary artery disease, kidney disease, and stroke.

13.2.1.1 Blood Pressure

Blood pressure is an alarming indicator of human health that is influenced by many factors including food habits and internal body imbalance caused by angiotensin, insulin level, atherosclerosis, etc. Consequently, a general nutritional plan to minimize hypertension risk includes attaining and maintaining a healthy body weight; consuming a diet rich in calcium, phosphorus, and magnesium; and consuming non-alcoholic beverages and sodium in moderation (Barham et al. 2000).

13.2.1.2 Obesity

Obesity is a medical condition characterized by accumulation of excess body fat or increased cholesterol level. As a condition, obesity is associated with reduced life expectancy and increased health problems. The food additives have relevant effects on cells of the immune system that could contribute to immune-mediated metabolic dysregulation. Numerous studies indicate that higher levels of body fat are associated with an increased risk of many adverse health conditions (Nijhawan and Behl 2020). Food habits and a sedentary lifestyle contribute to overweight and obesity, especially in children of the western world over the past 20 years. Worldwide, over 20 million children under the age of six are obese or overweight. Some guidelines to being healthier include making appropriate dietary choices, embarking on good eating behavior, and having an active lifestyle (Joyce et al. 2020) (Fig. 13.1).

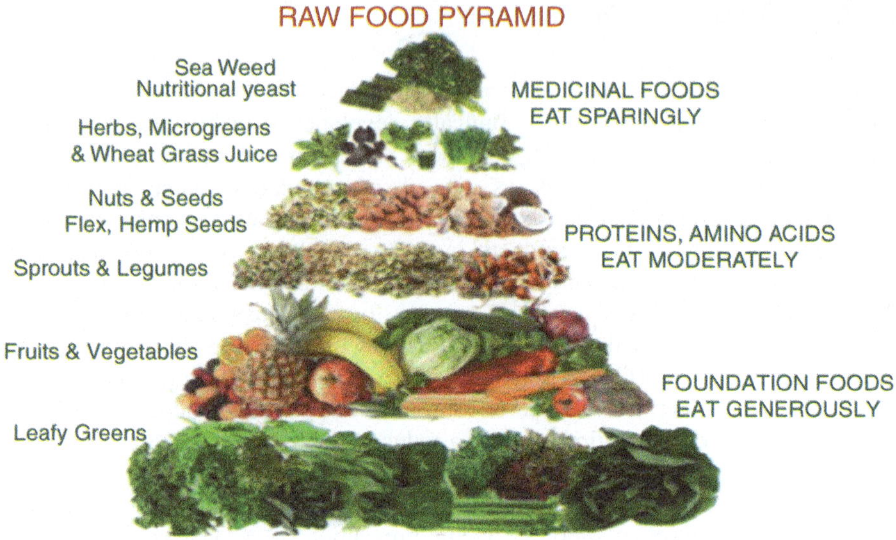

Fig. 13.1 Raw food pyramid (Barham et al. 2000)

13.2.2 Acute Pancreatitis and Immunity

Acute pancreatitis (AP) is a common type of gastrointestinal disorder in humans, and many functional foods or nutritional supplements can help in the management of AP (Gupta et al. 2021). It has been well proven that nutritional components are directly related to the immune system of the body. Thus, nutritional foods are very important for the enhancement of the immune system to defend against pathogens and also play a key role in the maintenance of homeostasis (Moynihan and Irvine 2017). A balance of innate and acquired immunity is desirable for good health (Yu et al. 2018) (Figs. 13.2 and 13.3).

13.2.3 Hypersensitivity

Hypersensitivity is mainly due to genetic causes, but environmental factors, including air pollution, dietary components and residential conditions, also play an important role. An allergic reaction is a sequential immune response involving the processing and presentation of the allergen, activation of allergen specific T and B cells, production of IgE against the allergen, and activation of mast cells and eosinophils triggered by the allergen.

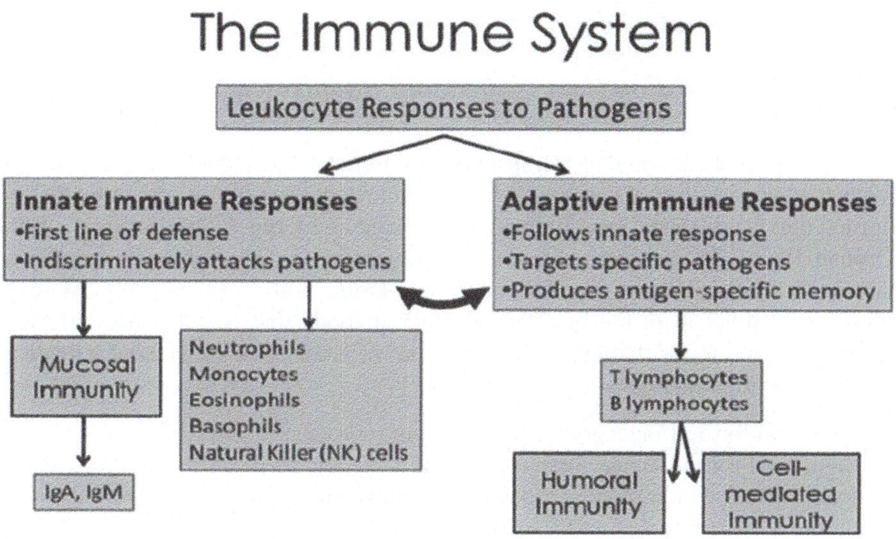

Fig. 13.2 Immune system

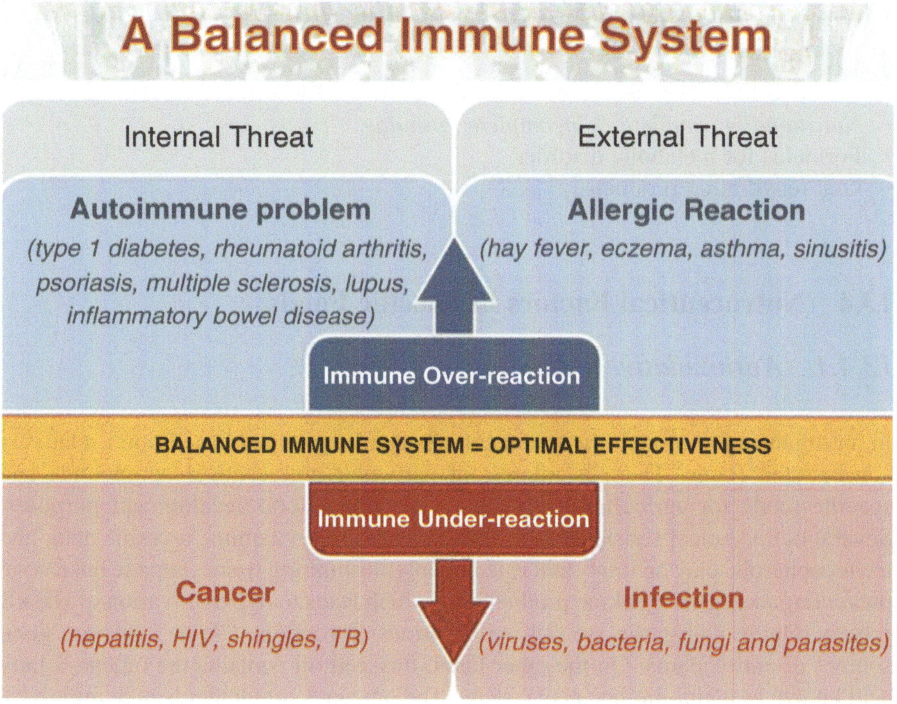

Fig. 13.3 Balanced immune system

13.3 Functional Food

13.3.1 Medical Foods as Nutritional Supplements

Medical foods are foods that are specially formulated and intended for the dietary management of a disease that has distinctive nutritional needs that cannot be met by normal diet alone. They were defined in the Food and Drug Administration's 1988 Orphan Drug Act Amendments and are subject to the general food and safety labeling requirements of the Federal Food, Drug, and Cosmetic Act. Medical foods are distinct from the broader category of foods for special dietary use and from traditional foods that bear a health claim. In order to be considered a medical food the product must, at a minimum:

- Be a food for oral ingestion or tube feeding (nasogastric tube),
- Be labeled for the dietary management of a specific medical disorder, disease or condition for which there are distinctive nutritional requirements, and
- Be intended to be used under medical supervision.

13.3.2 Medical Foods Can Be Classified into the Following Categories

- *Nutritionally complete & incomplete formulas.*
- Formulas for metabolic disorder.
- Oral rehydration products.

13.4 Nutraceutical Factors in Specific Foods

13.4.1 Antioxidative Vitamins

In an organization model, the nutraceuticals can be grouped based upon relatively concentrated foods. There is interest in particular nutraceutical compounds and specific foods for agricultural reasons or functional food-development purposes. Several nutraceutical substances, such as antioxidative vitamins, ascorbic acid, and the tocopherols, play an important role not only for limiting tissue damage but also in preventing increased cytokine productions, which leads to hyperactivation of NF-κB protein. Onion and garlic are rich with various flavonoids. Citrus foods are good sources of antioxidants. On the other hand, flaxseed oil contains the omega-3 fatty acid known as alpha-linoleic acid (ALA). The omega-3 supplements may improve a number of skin disorders, including dermatitis, psoriasis, and skin damage attributed to ultraviolet (UV) exposure. There are no guarantees that closely related foods

Table 13.1 Nutraceutical substances grouped by food sources (Holban and Grumezescu 2018)

S. no	Name of nutrients	Name of the elements
1	Essential microbes	Prebiotic
2	Microbial flora	Probiotic
3	Minerals	Nitrogen
4	Marine algae	Phosphorous
5	Sea weed	Potassium
6	Sea weed	Molybdenum
7	Milk	Calcium
8	Marine products	Magnesium
9	Green leafy vegetables	Sulfur
10	Healthy microbes	Zinc
11	Green leafy veggies	Copper
12	Fruits	Iron
13	Microelements	Manganese
14	Microelements	Boron
15	Microelements	Selenium
16	Essential fatty acids	Omega-3 fatty acids
17	Essential phytochemicals	Flavonoids
18	Phyto constituents	Tannin
19	Primary metabolites	Lignin
20	Primary metabolites	Glycosides
21	Primary metabolites	Serpentines
22	Primary metabolites	Terpenoids
23	Primary metabolites	Saponins
24	Phytoconstituents	Alkaloids
25	Phyto bioactives	Isoprenoid
26	Fiber sources	Oats, peas, beans, apples, citrus fruits, carrots, barley and psyllium
27	Antioxidative vitamins	Tocopherol (vit E)
28	Antioxidative vitamins	Ascorbic acid (vit C)
29	Micronutrients	Vit. A, B1, B2, B6, B12 (vit. B complex), C, D.
30	Macronutrients	Folic acid, biotin & carotenoids

contain the same nutraceutical compounds, for example, onions and garlic are in the same lily family; however, onions are loaded with quercetin and garlic is not (Gutiérrez et al. 2019) (Table 13.1).

13.4.2 Macronutrients

For a healthy person both micro and macronutrients are important, and as its name implies, macronutrients are required in large quantity for normal human daily life.

Proteins are the building blocks of the human body and are also known as functional and physiological units of life. They are also a source of energy after carbohydrate and fat consumption. Macronutrients play a unique role in repairing mechanisms and homeostasis. Our proteinaceous diets provide essential amino acids. Our bodies are like recycling geniuses that can take an old pallet (plant and animal protein), break it down (into amino acids), and make a bench from the parts (new protein).

13.4.3 Micronutrients

Vitamins and minerals are types of micronutrients required in small amounts for the function of the human body. They do not provide calories but are required to modulate the production of enzymes, hormones, and other substances that affect the capacity to defend against pathogens and are vital to development, prevention of many diseases, and boost our immunity. Micronutrients are commonly referred to as vitamins and minerals, and it is adequate intake of these micronutrients that help reduce the risk of chronic disease, promote a longer life, and improve overall wellbeing. Some research even indicates that higher intake of micronutrients is associated with improved mood, energy levels, and appetite control. Micronutrients such as essential proteins, antioxidants, essential amino acids, vitamins (A, B1, B2, B6, B12, C, D, E, and folic acid), omega-3 fatty acids, minerals (iron, calcium, selenium, zinc, and copper), and other certain phytochemicals are important for a healthy immune system. However, over reaction of the immune system leads to hypersensitivity reactions such as autoimmunity, which ultimately causes auto immune diseases and allergic reactions. The key role of dietary factors in immunotolerance promotion and skin disorders and allergic diseases prevention has become an important tool in management of several diseases. The aim of the chapter is the analysis of the impact of immunomodulatory dietary components, consumed by many people, on the development of skin disorder, pollen allergy, and hypersensitivity in their offspring. The antioxidative vitamins, ascorbic acid and the tocopherols, play an important role not only for limiting tissue damage but also in preventing increased cytokine production, which is a consequence of excessive activation of NF-κB. Glutathione is a major endogenous antioxidant and is important for lymphocyte replication. Two vitamins, vitamin B6 and riboflavin, participate in the maintenance of glutathione status. The former vitamin acts as a cofactor in the synthesis of cysteine and the latter vitamin is a cofactor for glutathione reductase. Deficiencies in tocopherol, vitamin B6, and riboflavin reduce cell numbers in lymphoid tissues of experimental animals and produce functional abnormalities in the cell mediated immune response. Ascorbic acid and tocopherols have shown anti-inflammatory effects in human and animal studies. In humans, dietary supplementation with ascorbic acid, tocopherols, and vitamin B6 enhances a number of aspects of lymphocyte function.

13.4.4 Omega-3 Essential Fatty Acids

Omega-3 fatty acids (ω-3 FA) are essential fats that every dietician recommends. They provide numerous health benefits, such as a reduced risk of heart disease, inflammation, and improved mood. Fish and flaxseed oils are good sources of omega-3 fatty acids, and the two main types found in them are eicosapentaenoic acid (EPA) and docosahexaenoic acid (DHA). A typical fish oil supplement contains 180 mg of EPA and 120 mg of DHA (Freeman 2000).

13.4.4.1 Good Source of Omega-3 Essential Fatty Acid Foods

Flaxseed Oil The flax plant (*Linum usitatissimum*) is an ancient crop that is rich in omega-3 fatty acid. The flax plant contains nutritious seeds commonly known as flax seeds. Flaxseed oil is obtained by cold-pressing ripened and dried flax seeds. The oil is also commonly known as linseed oil. Flaxseed oil can be used in a variety of ways and is available commercially in both liquid and capsule form. Flaxseed oil provides powerful health benefits, likely related to its high content of omega-3 fatty acids (ω-3 FA), which have shown their efficacy in the treatment of chronic and acute diseases due to their pleiotropic effects on cell signaling pathways linked to inflammation, angiogenesis, and cell cycle progression (Mirshekar et al. 2015).

Fish Oil Fish oil is one of the most popular dietary supplements on the market. It is made by extracting oil from fish tissue that is particularly rich in omega-3 fatty acids and provides heart health benefits. It may improve some skin disorders, including dermatitis, psoriasis, and skin damage attributed to sun exposure. Fish oil supplements can help you consume an adequate amount of omega-3 fatty acids, especially if you are not much of a seafood fan. Typical fish oil supplements contain 1000 mg of omega-3 fatty acids. In fact, certain fish oil supplements are often prescribed by healthcare providers to lower blood triglyceride levels, boost your immunity, and boost your memory (Boberg 1990).

13.4.4.2 Examples of Different Forms of Omega-3 Fatty Acid/Oil Products

I. Triacylglycerol (TAG) or TAG concentrate
II. Ethyl ester (EE) or EE concentrate of eicosapentaenoic acid
III. (EPA) and/or docosahexaenoic acid (DHA)
IV. Phospholipid 5. Chromium (III) – DHA complex
V. Phytosterol-DHA ester
VI. Epigallocatechin gallate (EGCG) – DHA ester (Fig. 13.4)

Fig. 13.4 Cod liver oil (Guy 1923)

13.5 Omega-3 Essential Fatty Acids

13.5.1 Cod Liver Oil

Cod is the common name for fish from the genus Gadus. The most known species are the Atlantic cod (*Gadus morhua*) and the Pacific cod (*Gadus macrocephalus*). Throughout the world, the cod fish and its liver are well known. Cod liver oil is exactly what it sounds like: the oil extracted from the liver of the cod fish. Cod liver oil is a nutrient-dense oil made from the livers of several species of the cod fish, and its powerful composition may help to reduce inflammation, promote brain function, improve eyesight, and boost the immune system. The oil is known in traditional folklore as a remedy for a wide range of different health issues. Research has found it to be one of the richest sources of vitamins A and D, as well as omega-3 fatty acids, including eicosapentaenoic acid (EPA) and docosahexaenoic acid (DHA), which manage the production of prostaglandins that regulate blood pressure, blood clotting, inflammation and allergic responses, and gastrointestinal issues. It has the highest concentrations in amounts of vitamins A and D, omega-3 fatty acids, structured fat, and has been used for centuries to boost immune system health and prevent many diseases such as type 1 diabetes and rickets. Rickets is a bone condition in children caused by a lack of vitamin D. However, the health benefits of cod liver oil may not end there. Eating fresh cod fish livers may provide many important health-boosting benefits during pregnancy for the developing baby, especially brain and eye development.

13.6 Health Benefits of Cod Liver Oils as Nutraceutical Supplements

13.6.1 Prevention of Rickets

In the past, rickets was a common disorder of the bones caused by a severe deficiency of vitamin D. In rickets, the bones fail to mineralize, which leads to soft bones and skeletal deformities in children, including:

- Bowed legs,
- Thick wrists and ankles,
- Projected breast bone.

The best source of vitamin D is sunlight, but people who live in northern latitudes often do not get a lot of sun during the winter months. Before the discovery of cod liver oil, many children suffered from deformed bones. Once mothers began including cod liver oil in their child's daily routine, incidence of rickets dropped dramatically (Rajakumar 2003). Vitamin D drops for children are also widely available. Along with the use of cod liver oil, these changes have made rickets a rare disease in the United States, but a few cases are seen today. Rickets remains a major public health concern in many developing countries.

13.6.2 Lowering the Risk of Type 1 Diabetes

Type 1 diabetes is an autoimmune disease that usually occurs in children, but its exact cause is unknown. A meta-analysis of 11 different studies found that children who took vitamin D supplements during their first year of life, including cod liver oil or a supplement with vitamin D, had a significantly lower risk of type 1 diabetes. The effect may be attributed to cod liver oil's high vitamin D content. Another point of view indicates the mother's vitamin D deficiency as a culprit in type 1 diabetes. In one article, researchers found that the odds of type 1 diabetes were more than two times higher in children whose mothers had the lowest levels of vitamin D, compared to children of mothers with high levels of vitamin D. At present, there is not yet enough research evidence to show that vitamin D deficiency is definitely linked to type 1 diabetes or that cod liver oil can reduce the risk.

13.6.3 Preventing Infections

Cod liver oil could mean fewer bouts of the cold and flu for your child, and fewer trips to the doctor. It is theorized that an immune system boost comes from the oil's high content of vitamin D, though research has not yet shown this. In research published in the Journal of the American College of Nutrition, cod liver oil

Fig. 13.5 Omega-3 fatty acid rich foods

supplements decreased trips to the doctor for upper respiratory illnesses by 36–58% (Cannell et al. 2008) (Fig. 13.5).

13.6.4 Protecting Eye Sight

Cod liver oil is rich in vitamins A and D. Both of these vitamins are essential for maintaining healthy eyesight over the long term. Vitamin A is especially important for preserving normal vision. It is also an antioxidant and could prevent the damage that leads to glaucoma. Glaucoma is an eye disease that can damage the optic nerve, leading to vision loss or even blindness. Scientists are exploring the relationship between cod liver oil supplementation and glaucoma. It is thought that the high omega-3 fatty acid content of cod liver oil could help improve blood flow to the eyes, keeping children's eyesight strong and healthy for a long time (Huang et al. 2011).

13.6.5 Reducing Depression

The vital component of cod liver oil is omega-3 fatty acids, which have been shown to reduce depressive symptoms in people suffering from major depression. A research study with over 20,000 people showed that those who regularly took cod liver oil were roughly 30% less likely to have symptoms of depression than those who did not taken cod liver oil. Research also suggests that omega-3 fatty acids may improve overall mood and brain functions (Raeder et al. 2007).

Table 13.2 Non-essential AA and essential AA

Non-essential AA	Essential AA
Alanine	Histidine
Arginine	Isoleucine
Asparagine	Leucine
Aspartate	Lysine
Cysteine	Methionine
Glutamate	Phenylalanine
Glutamine	Threonine
Glycine	Tryptophan
Proline	Valine
Serine	
Tyrosine	

13.6.6 Amino Acids (AA)

There are two types: non-essential amino acids and essential amino acids. They are used to boost immunity and increase energy levels (Table 13.2).

13.7 Essential Phytochemicals

Phytochemicals is a broad term meaning plant (phyto) chemicals referring to a wide variety of compounds that occur naturally in plants. Essential phytochemicals help in the prevention of chronic diseases and are bioactive, but their intensities are reduced during processing and handling. First described and discovered in the 1950s by Julius Axelrod, cytochrome P450 (CYP450) are a set of enzymes capable of promoting excretion of xenobiotics through hydroxylation reactions using a heme molecule as a cofactor (Elseehy and El-Shehawi 2020) (Fig. 13.6).

13.7.1 Food and Non-Food Sources of Nutraceuticals

One of the broader models of organization for nutraceuticals is based upon their potential as a food source for humans. Many nutraceuticals are found in plants and animals and sometimes in microbes (i.e., bacteria and yeast) groups. An example is conjugated linoleic acid (CLA), which is part of the human diet, mostly as a component of beef and dairy foods. Food chains or symbiotic combination may occur for some.

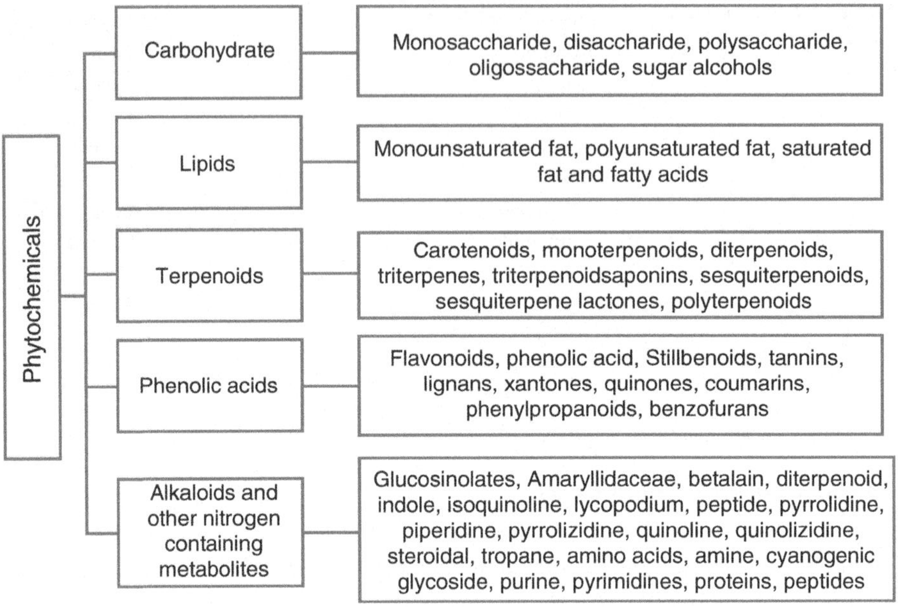

Fig. 13.6 Types of essential phytochemicals (Alamgir 2017)

13.8 Classified by their Proven Physiological Properties

The nutraceutical ingredients with fish oil and flaxseed oil contains the hearty dose of protein and fiber.

Dietary fiber is defined as a carbohydrate with three or more monomeric units, which are neither digested nor absorbed in the small intestine, and soluble fiber is found in oats, peas, beans, apples, citrus fruits, carrots, barley, and psyllium (Dwyer et al. 2018) (Table 13.3).

13.9 Good Source of Balanced Dietary Supplements

13.9.1 Insoluble Fiber

This type of fiber promotes the movement of material through your digestive system and increases stool bulk; therefore, it can be of benefit to those who struggle with constipation or irregular stools. Whole-wheat flour, wheat bran, nuts, beans and vegetables, such as cauliflower, green beans and potatoes, are good sources of insoluble fiber. The amount of soluble and insoluble fiber varies in different plant foods. To receive the greatest health benefits, eat a wide variety of high-fiber foods.

Table 13.3 Nutraceuticals grouped by mechanisms of action (Khalaf et al. 2021)

S. no.	Anticancer	Positive influence on blood lipid profile	Antioxidant activity	Anti-inflammatory	Osteogenetic or bone protective
1	Capsaicin	α-Glucan	CLA	Linolenic acid	CLA
2	Daidzein	δ-Tocotrienol	β-Carotene	DHA	Genestein
3	Genestein	γ-Tocotrienol	Ascorbic acid	EPA	Soy protein
4	Sphingolipids	Tannins	α-Tocopherol	Quercetin	Inulin
6	Limonene	β-Sitosterol	Ellagic acid	Curcumin	Cholcalciferol
7	α-Tocopherol	Pectin	Glutathione	–	Phosphopeptides
8	Diallyl sulfide	Saponins	Lycopene	–	–
9	Ajoene	Guar	Lutein	–	–
10	*Lactobacillus acidophilus*	Resveratrol	Indole-3-carbonol	Capsaicin	–
11	Ellagic acid	–	Gingerol	–	–
12	Lutein	–	Chlorogenic acid	–	–

13.9.2 Benefits of a High-Fiber Diet

A high-fiber diet may lower your risk of developing hemorrhoids and small pouches in your colon (diverticular disease). Some fibrous foods are fermented in the small intestine (colon). Increasing your dietary fiber foods intake, especially cereal fiber, is associated with a reduced risk of dying from CVD, liver dysfunction, and cancer (Chhabra 2018). Elements of a high-fiber diet are described in the following sections.

13.9.2.1 Normalizes Bowel Movements

Dietary fiber increases the weight and size of your stool and softens it. A bulky stool is easier to pass, decreasing your chance of constipation. If you have loose, watery stools, fiber may help to solidify the stool because it absorbs water and adds bulk to stool.

13.9.2.2 Lowers Cholesterol Levels

Soluble fiber found in beans, oats, flaxseed, and oat bran may help lower total blood cholesterol levels by lowering low-density lipoprotein, or "bad," cholesterol levels. Studies have also shown that high-fiber foods may have other heart-health benefits, such as reducing blood pressure and inflammation.

13.9.2.3 Control Blood Sugar Levels

In people with diabetes, fiber, particularly soluble fiber, can slow the absorption of sugar and help improve blood sugar levels. A healthy diet that includes insoluble fiber may also reduce the risk of developing type 2 diabetes.

13.9.2.4 Aids in Achieving Healthy Weight

High-fiber foods tend to be more filling than low-fiber foods, so you are likely to eat less and stay satisfied longer. In addition, high-fiber foods tend to take longer to eat and to be less "energy dense," which means they have fewer calories for the same volume of food.

13.9.2.5 Helps to Live Longer Life

Increasing your dietary fiber intake, especially cereal fiber, is associated with a reduced risk of dying from cardiovascular disease and all cancers.

13.10 Conclusion

This chapter discussed the role of nutraceuticals and functional foods in immunomodulation and the effects of nutritional foods intake for good health. Smart nutrition and food choices can help to prevent many diseases. Eating the right foods can help your body cope more successfully with an ongoing illness. Understanding good nutrition and paying attention to what you eat can help you maintain or improve your health. Modeling new eating habits using the existing knowledge is needed for the eventual ideal of "health for all" vision.

Competing Interest The authors declare that there are no competing interests.

References

Alamgir ANM (2017) Therapeutic use of medicinal plants and their extracts, progress in drug research, progress in drug research. Springer International Publishing, Cham

Barham JB, Edens MB, Fonteh AN, Johnson MM, Easter L, Chilton FH (2000) Addition of eicosapentaenoic acid to γ-linolenic acid-supplemented diets prevents serum arachidonic acid accumulation in humans. J Nutr 130:1925–1931

Boberg M (1990) Clinical effects of fish oil. Näringsforskning 34:133–134

Cannell JJ, Vieth R, Willett W, Zasloff M, Hathcock JN, White JH, Tanumihardjo SA, Larson-Meyer DE, Bischoff-Ferrari HA, Lamberg-Allardt CJ, Lappe JM, Norman AW, Zittermann A, Whiting SJ, Grant WB, Hollis BW, Giovannucci E (2008) Cod liver oil, vitamin a toxicity,

frequent respiratory infections, and the vitamin D deficiency epidemic. Ann Otol Rhinol Laryngol 117(11):864–870
Chhabra S (2018) Dietary fibre-nutrition and health benefits. In: Rani V, Yadav U (eds) Functional food and human health. Springer, Singapore, pp 15–25
Dwyer J, Coates P, Smith M (2018) Dietary supplements: regulatory challenges and research resources. Nutrients 10:41
Elseehy MM, El-Shehawi AM (2020) Methylation of exogenous promoters regulates soybean isoflavone synthase (GmIFS) transgene in T0 transgenic wheat (Triticum aestivum). Cytol Genet 54:271–282
Freeman MP (2000) Omega-3 fatty acids in psychiatry: a review. Ann Clin Psychiatry 12:159–165
Galindo A, Garrido D, Monroig PJA, Betancor MB, Acosta NG, Kabeya N, Marrero MA, Bolaños A, Rodríguez C (2021) Polyunsaturated fatty acid metabolism in three fish species with different trophic level. Aquaculture 530:735761
Gupta RC, Doss RB, Banerjee A, Lall R, Srivastava A, Sinha A (2021) Nutraceuticals in gastro-intestinal disorders, In: Nutraceuticals. Elsevier, 141–155
Gutiérrez S, Svahn SL, Johansson ME (2019) Effects of Omega-3 fatty acids on immune cells. Int J Mol Sci 20:5028
Guy RA (1923) The history of cod liver oil as a remedy. Arch Pediatr Adolesc Med 26:112
Holban AM, Grumezescu AM (2018) Advances in biotechnology for food industry. Adv Biotechnol Food Ind 14:1–451
Huang WB, Fan Q, Zhang XL (2011) Cod liver oil: a potential protective supplement for human glaucoma. Int J Ophthalmol 4(6):648
Joyce P, Meola TR, Schultz HB, Prestidge CA (2020) Biomaterials that regulate fat digestion for the treatment of obesity. Trends Food Sci Technol 100:235–245
Keservani RK, Kesharwani RK, Vyas N, Jain S, Raghuvanshi R, Sharma AK (2010) Nutraceutical and functional food as future food: a review. Pharm Lett 2(1):106–116
Keservani RK, Sharma AK, Kesharwani RK (2016) Nutraceutical and functional foods for cardio-vascular health. In: Food process engineering. Apple Academic Press, pp 291–312. (pp. 257-278). ISBN: 9781771884020
Khalaf AT, Wei Y, Alneamah SJA, Al-Shawi SG, Kadir SYA, Zainol J, Liu X (2021) What is new in the preventive and therapeutic role of dairy products as nutraceuticals and functional foods? Biomed Res Int 2011:8823222
Martin C, Li J (2017) Medicine is not health care, food is health care: plant metabolic engineering, diet and human health. New Phytol 216:699–719
Mirshekar R, Boldaji F, Dastar B, Yamchi A, Pashaei S (2015) Longer consumption of flaxseed oil enhances n-3 fatty acid content of chicken meat and expression of FADS2 gene. Eur J Lipid Sci Technol 117:810–819
Moynihan KD, Irvine DJ (2017) Roles for innate immunity in combination immunotherapies. Cancer Res 77(19):5215–5221
Nijhawan P, Behl T (2020) Nutraceuticals in the management of obesity. Obes Med 17:100168
Raeder MB, Steen VM, Vollset SE, Bjelland I (2007) Associations between cod liver oil use and symptoms of depression: the Hordaland health study. J Affect Disord 101:245–249
Rajakumar K (2003) Vitamin D, cod-liver oil, sunlight, and rickets: a historical perspective. Pediatrics 112:e132–e135

Siscovick DS, Barringer TA, Fretts AM, Wu JHY, Lichtenstein AH, Costello RB, Kris-Etherton PM, Jacobson TA, Engler MB, Alger HM, Appel LJ, Mozaffarian D (2017) Omega-3 polyunsaturated fatty acid (fish oil) supplementation and the prevention of clinical cardiovascular disease: a science advisory from the American Heart Association. Circulation 135:e867–e884

Springmann M, Wiebe K, Mason-D'Croz D, Sulser TB, Rayner M, Scarborough P (2018) Health and nutritional aspects of sustainable diet strategies and their association with environmental impacts: a global modelling analysis with country-level detail. Lancet Planet Health 2(10): e451–e461

Williamson EM, Liu X, Izzo AA (2020) Trends in use, pharmacology, and clinical applications of emerging herbal nutraceuticals. Br J Pharmacol 177(6):1227–1240

Yu Y, Shen M, Song Q, Xie J (2018) Biological activities and pharmaceutical applications of polysaccharide from natural resources: a review. Carbohydr Polym 183:99–101

Chapter 14
Therapeutic Effects of *Withania somnifera*: An Overview with Special Focus on Alzheimer's Disease and Infertility among Youth

Deepika Saini, Mansi Srivastava, Sandhya Vaid, and Vipin Kesharwani

Abstract Immunomodulators are biomolecules that are recognized for their characteristics to stimulate or suppress the adaptive or innate immunity. These agents modify the immune response by either enhancing (immunostimulators) or suppressing (immunosuppressants) the expression of inflammatory mediators such as cytokines and chemokines that activate immune cells. *Withania somnifera* (WS) is an ancient ayurvedic herb popularly known as "Ashwagandha" in India in Sanskrit and "winter cherry" outside India in English that has been used in indigenous medicines for over 3500 years or more, owing to its inexhaustible therapeutic value. Studies have shown a wide range of therapeutic applications associated with WS such as antibacterial, anti-inflammatory, antioxidant, anti-stress, antitumor, and anxiolytic, along with its protective role in numerous other ailments like cardiovascular, epilepsy, arthritis, diabetes, and depression. WS has also been recognized for its neuroprotective role in a number of degenerative conditions like Parkinson's, Huntington's, and most importantly Alzheimer's disease. In today's global scenario, the problem of infertility in males and disturbed sexual behavior in females is commonly observed. *Withania* has been shown to cure disorders related to reproductive system across countries with acceptable results. Currently, there is a drastic universal demand for development of herbal therapies effective for these ailments to minimize the side effects arising from chemotherapies. An overview of introduction and immunomodulatory role of *Withania somnifera* will be presented here with

D. Saini (✉)
Department of Zoology, Chaman Lal Mahavidyalaya, Landhaura Haridwar, Uttarakhand, India

M. Srivastava
Department of Bio Health Informatics, Indiana University & Purdue University Indianapolis (IUPUI), Indianapolis, IN, USA

S. Vaid
Department of Zoology, Chinmaya Degree College, Haridwar, Uttarakhand, India

V. Kesharwani
Maharishi University of Information Technology, School of Pharmaceutical Sciences, Lucknow, India

R. K. Kesharwani et al. (eds.), *Nutraceuticals and Functional Foods as Immunomodulators*, https://doi.org/10.1007/978-981-19-2507-8_14

special focus on the mechanistic insights into the therapeutic effect of WS in treating Alzheimer's and infertility problem. This chapter will also describe the latest attempts in biotechnological production of withanolides.

Keywords Immunomodulator · Alzheimer's · Parkinson's · Therapeutic · Herbal therapies

14.1 Introduction

Ayurveda is one of the world's oldest traditional system of treating uncountable human ailments. Since the life originated on the earth or from the origin of the human civilization, man is continuously using plants/herbs in one or other way, either as a food or as a medicine. India is the country purely known for its culture and traditions.

Ayurveda is originated in India and is the world's oldest and successfully accepted medicinal system known to combat uncountable ailments using the therapeutic powers of herbs and plants. In the current scenario, we are living in highly polluted, industrialized environment where we are consuming bulk of chemicals and escalating the deadly ratio of various neurodegenerative, cardiovascular diseases and cancer. Nowadays, the herb in demand is *Withania somnifera*, also known by the name Ashwagandha, winter cherry, and Indian ginseng possessing an extremely noticeable capacity as an immunomodulator. *Ashwagandha* name was basically given due to horse like smell from the plants root, in addition when consumed gives horse like power. *WS* belongs to family Solanaceae and is mostly found in tropical and subtropical zones. *Withania* is known globally for its various pharmacological potentials; in Ayurveda, it is known by the word Rasayana, which means potent rejuvenator; it is known to enhance hemoglobin count, treat melanin-based pigmentation (Singham et al. 2011), and general health. *Withania* is known for its extreme potentiality of treating anti-aging effects, body revitalization, and anti-stress (Changhadi 1938; Mahima et al. 2012). Though Ashwagandha is nowadays also available in capsule form, traditionally it is available in powder form which can be consumed with water, honey, and clarified butter as per comfort (Singham et al. 2011). All the parts of *Withania* are well-known to treat various body ailments, e.g., roots are widely used to treat rheumatic swelling as well as show health restorative properties on old age people and pregnant woman (used in powder as well as paste form). Few other ailments including insomnia, nervous breakdown, leukoderma, constipation, and many more are also treated successfully using *Withania somnifera stem extract* (Bhandari 1970). On the other hand, leaves of *Withania* are also in high demand due to its antipyretic, antihelminthic, anticancer, and antituberculosis properties as they are rich in secondary metabolites. Seeds are not lesser in potentiality of combatting ailments than other parts. They have any amazing capacity of curing white cornea spots, when mixed with rock salt. Despite this, they also possess an amazing power of treating incurable mental health issues like insomnia, memory loss, stress, and anxiety disorders with acceptable results. Ashwagandha is also recognized nowadays due to its remarkable potentiality of improving sperm count in males to combat infertility problems among youth (Govardhan 1938).

14.2 Immunomodulatory Role of *Withania somnifera* (WS)

WS is widely appreciated over the centuries for its distinguished potential as an effective medicinal herb for the treatment of broad range of diseases and for improving health and longevity. Besides its notable Ayurvedic properties, several studies have reported immunomodulatory, cardioprotective, neuroprotective, anti-aging, and anticancer functions of WS (Mishra et al. 2000a, b; Ng et al. 2020). The most vital immunomodulatory function of WS is reported to enhance the immune system to improve body's defense mechanism against infections and diseases (Kaur et al. 2017; Singham et al. 2011). Furthermore, the antioxidant nature of WS provides protection against free radicals that damage the cells (Singham et al. 2011). In addition to the antioxidant properties, it also includes other healing properties such as anti-inflammatory and herbal supplement for curing various ailments (Mandlik Ingawale and Namdeo 2020). Many glycowithanolides isolated from *W. somnifera* were investigated for their immunomodulatory properties and have shown to be effective in the activation of lysosomal enzymes, macrophage-mediated phagocytosis, and anti-stress effects on the central nervous system (Kuboyama et al. 2002; Pratte et al. 2014). Similar studies that have tested immunomodulatory efficacy of *W. somnifera* in azoxymethane induced colon cancer in mice, showing that it dramatically regulated the proportions of active immune cells including neutrophils, lymphocytes, leukocytes, and immunoglobulins (IgG, M and A) (Muralikrishnan et al. 2010). Furthermore, the immune regulatory properties in the leaf extract components of *W. somnifera* are investigated in previous study and have found to activate the immune system (Khan et al. 2009). Moreover, the immunomodulatory properties of *W. somnifera* have been attributed to the anticancer effects, such as in the prostate cancer cells in a previous finding (Palliyaguru et al. 2016; Mulabagal et al. 2009). Another study conducted on myelosuppression mice models demonstrated a significant upregulation of red blood cells, white blood cells, platelets, and higher body weight in *W. somnifera*-treated mice compared to control mice suggesting immunostimulatory functions of *W. somnifera*. A previous study evaluated the effect of *W. somnifera* on the function of macrophages in carcinogen ochratoxin A (OTA)-treated mice, showing that it is an inhibited form of OTA-induced suppression of chemotaxis and production of critical cytokines such as TNF-alpha and IL-1 in *W. somnifera*-treated mice compared to control mice (Dhuley 1997). Another study on tumor-bearing mice model investigated the effect of *W. somnifera* on cellular immune response. The valid research study revealed that *W. somnifera* significantly enhanced the proliferation of immune cells such as bone marrow cells, lymphocytes, thymocytes, and splenocytes in response to mitogens compared to control mice (Davis and Kuttan 2002). Furthermore, a study revealed that administration of *W. somnifera* in mouse leads to a substantial increase in the production of nitric oxide (NO) in macrophages in a concentration-dependent manner, as a result of increased production of enzyme nitric oxide synthase and other inflammatory mediators (Iuvone et al. 2003). In another report investigating the aqueous suspension of *W. somnifera* root powder for their immunomodulatory properties, the authors found immunosuppressive effects observed by inhibition of lymphocyte proliferation, complement activation, and delayed-type hypersensitivity reaction, suggesting a potential for expanding the applications of *W. somnifera* as an immunosuppressive drug. For the past few year, number of scientific researches has been carried out to prove the incredible therapeutic potential of *W. somnifera*. Previous study to explore the efficacy of aqueous extracts of *W. somnifera*-treated animals immunized with DPT (diphtheria, pertussis, tetanus) vaccine has shown a dramatic upregulation in the antibody titers as compared to untreated animals after challenge (Gautam et al. 2004). Overall, these results are encouraging and establish the role of *W. somnifera* beyond a traditional Ayurvedic medicinal herb. Owing to the extraordinary potential of *W. somnifera*, additional studies must be conducted to assess and validate its therapeutic efficacy in various clinical settings. The major ingredients of *W. somnifera* include alkaloids and steroidal lactones that contribute to its immunomodulatory and other medicinal properties. An interesting study examined the key signaling mechanism in the prostate cancer cells using genomic microarray analysis, indicating that *W. somnifera* treatment significantly downregulated the pro-inflammatory cytokines IL-6 and IL-1β and transcription factor STAT2 while at the same time upregulated crucial signaling molecules including PI3K, p38 MAPK, cyclin D, and c-myc. These observations suggested that *W. somnifera* modulated the expression of genes associated with immune response, cellular signaling, and transcriptional regulation, thus indicating its potential effectiveness in prostate cancer (Aalinkeel et al. 2010). Oxidative stress due to the rapid production of reactive oxygen species (ROS) and nitric oxide (NO) is the primary cause of cell damage during chronic inflammation. A published study established the inhibitory role of the withanolides during oxidative stress and inflammatory disorders. In this study, the authors examined withaferin A and withanolide A on mice BV-2 microglial cells; the results showed that *W. somnifera* effectively inhibited the LPS-induced ROS and NO production and activated the nuclear factor (erythroid-derived 2)-like 2 (Nrf2) pathway, leading to induction of heme oxygenase-1 (HO-1). Among the two active constituents of *W. somnifera*, withaferin A was found tenfold more effective compared to withanolide A. In this context, another scientific evidence shows that *W. somnifera* inhibited apoptosis via Akt-mediated inhibition of oxidative stress during cardiac ischemia reperfusion injury illustrating its immunomodulatory potential for treatment of cardiovascular diseases in the future (Mohanty et al. 2008; Gupta and Kaur 2019). Further, *W. somnifera* root powder has also shown potent inhibitory effect on mitogen-induced lymphocyte proliferation, complement activation, and delayed-type hypersensitivity reaction, making it a promising candidate as anti-inflammatory drug. Taken together, *W. somnifera* extracts possess immense potential as effective immunomodulatory agent by suppression of oxidative stress during inflammatory outburst by targeting the Nrf2 pathway in the microglial cells. Immunomodulatory role of *W. somnifera* is also described by its anti-inflammatory activity of its component "withaferin A" (WFA). Several studies have been conducted to assess the properties of steroidal lactone withaferin A extracted from *W. somnifera*, showing its potential for immunosuppression, anti-inflammatory, neuroprotective, hepatoprotective, and antimicrobial activity (Dutta et al. 2019). These extracts are reported to regulate immune response by diminishing the production of pro-inflammatory cytokines such as IL1-β and IL-18 and transcription factors such as NF-kB and AP1 in several in vitro models (Dubey et al. 2018; Martorana et al. 2015). Furthermore, the discovery of methanol extract of *Withania somnifera* was a great success provided a novel compound "withanolide sulfoxide" was reported with high selectivity for inhibiting cyclooxygenase-2 (COX-2) enzyme that is rapidly produced during inflammation and suppression of TNF-induced NF-kappa B activation in multiple cancer cell lines including human gastric (AGS), breast (MCF-7), central nervous system (SF-268), and colon (HCT-116) cancer cells (Mulabagal et al. 2009; Ichikaw et al. 2006). More specific studies have been conducted on withaferin A and withanolide E (steroidal lactone extracts from *W. somnifera*), for uncovering their immunosuppressive potential on immune cells such as B and T lymphocytes and mice thymocytes. The study demonstrated that withaferin A inhibited the formation of E-rosetting (erythrocyte rosetting, a phenomenon where red blood cells (erythrocytes) form a cluster around by normal T and B lymphocytes giving an appearance of flower) (Plunkett et al. 1987). Several other studies have illustrated that withaferin A is implicated in inhibition of T-cell responses by exerting its mechanism via thiol-dependent inhibition of NF-kB pathway. In addition to targeting NF-kB, withaferin A has also exhibited immunosuppression by targeting signal transducer and activator of transcription 3 (STAT3) in human breast cancer cells. These observations indicate therapeutic potential of withaferin A by inhibiting the proliferation and migration of breast cancer cells by suppressing IL-6-mediated STAT3 activation (Gambhir et al. 2015). More evidences show that administration of *W. somnifera* extract reduced leucopenia induced by cyclophosphamide (CTX) treatment in mice, delineating its potential for anticancer therapy (Davis and Kuttan 2000). The immunomodulatory and pharmacological properties of *W. somnifera* described by multiple studies over the years show a promising potential for its medicinal uses and therapeutic interventions. Overall, these observations further provide important insights into the immunomodulatory functions of *W. somnifera* in various pathological conditions and cancers. Thus, *W. somnifera* has emerged as a plant with enormous ability to modulate the immune system, and several of its extracts have proven their usefulness for biomedical applications. Thus, there is a need to further investigate and explore the constituents of *W. somnifera* extracts and establish their therapeutic efficacy in specific disease models.

14.3 Therapeutic Role of *W. somnifera* with Special Insights on Alzheimer's Disease (AD)

Alzheimer's disease (AD) is basically a neurodegenerative disorder, commonly associated with old age. The most common symptom of the disease is loss of memory as well as cognitive functions. AD is characterized by behavioral changes such as anxiety, inability to learn, irritability, stress, and depression (Apostolova 2016; Fan et al. 2019). The molecular mechanism of AD has been attributed to the accumulation of (a) neuritic plaques consisting of amyloid β peptide (Aβ) and loss of synapses, (b) neurofibrillary tangles in the brain composed of intracellular hyper phosphorylated tau proteins (Apostolova 2016), and (c) oxidative stress due to Aβ accumulation leading to neuronal death. Currently, the drugs available for treatment of AD show very limited effectiveness against the disease pathology and are only useful for suppressing the symptoms (Holtzman et al. 2012). In the recent years, Ayurvedic plants and their phytochemical constituents are widely appreciated for their efficacy in the treatment of neurodegenerative diseases including AD (Ma et al. 2019; Akram and Nawaz 2017; Eckert 2010). Active constituents of medicinal plants such as polyphenols, sterols, alkaloids, flavonoids, tannins, and lignans have shown promising anti-inflammatory, antioxidant and anti-amyloidogenic properties that are utilized to improve the treatment efficacy of neurodegenerative diseases (Rao et al. 2012). Owing to its enormous calming effect on the central nervous system, *W. somnifera* has shown a promising potential for AD patients (Singham et al. 2011; Dimpfel et al. 2020). Previous studies on double-blind, randomized, and placebo show that *W. somnifera* significantly reduced stress and memory loss in a dose-dependent manner (Chandrasekhar et al. 2012). Several interesting studies have documented the ability of *W. somnifera* for their cholinergic activity and memory enhancement by increasing the production of acetylcholine in the brain (Gautam et al. 2016; Kuboyama et al. 2005). *W. somnifera* has also demonstrated an immense potential for regeneration of dendrites and axons in human neuroblastoma cells (Kuboyama et al. 2005; Keuboyama et al. 2014) and, therefore, has appeared to be of great interest to researchers. In-depth studies to understand the molecular mechanism exerted by *W. somnifera* have been conducted by several research groups. The same study has shown that two dendritic markers PSD-95 and MAP2 were significantly upregulated in cells exposed to *W. somnifera* that lead to dendrite formation (Kuboyama et al. 2005; Keuboyama et al. 2014). The same study also exhibited that rat cortical neurons having amyloid peptides that caused axon and dendritic atrophy regenerated both axons and dendrites upon treatment with *W. somnifera* extracts. These observations present valuable insights into the effectiveness of *W. somnifera* against the neurodegenerative mechanism underlying AD. Similar study on aqueous extract of *W. somnifera* roots showed its ability and potential of inhibiting the formation of mature amyloid β fibrils in the brain (Kumar et al. 2012; Jayaprakasham et al. 2010). More such studies have investigated the methanol extracts of *W. somnifera* dried roots using LC-MS analysis showing that the extract is rich in alkaloid and steroidal lactone such as withanolides, and more importantly it showed neuroprotective properties against β-amyloid-induced cytotoxicity (Kurapati et al. 2013). These observations indicate towards the neuroprotective role of *W. somnifera* in AD and related neurodegenerative disorders. Another interesting study provided evidence that semi-purified extracts of *W. somnifera* roots effectively reversed the plaque pathology, β-amyloid peptide accumulation, and behavioral deficits in the brains of middle-aged and old AD transgenic mice that was mediated by enhancing the low-density lipoprotein receptor-related protein (LRP) in liver. In a recent study, an "i-Extract" of *W. somnifera* was observed to restore the memory loss in scopolamine (SC)-induced mice by targeting the muscarinic receptors, thus providing a mechanistic evidence for the applications of *W. somnifera* in neurodegenerative disorders (Konar et al. 2019). Another study has illustrated that the leaf water extracts of *W. somnifera* effectively inhibited LPS-induced impairment of synaptic plasticity, neuroinflammation, and cognitive dysfunction with a gain of memory and learning ability (Gupta and Kaur 2019). In the same study, the possible signaling mechanisms that are targeted by *W. somnifera* are PI3K-Akt and BDNF-TrkB and PLCγ-IP3 known for regulating the survival and maintenance of neurons. Thus, this study suggests *W. somnifera* leaves extract as a potential therapeutic candidate for neuroinflammation. Collectively, these studies indicate that there could be multiple mechanisms that are implicated in the neuroprotective ability of *W. somnifera*, and the synergistic action of its constituents such as withaferin A and withanolides could contribute to its therapeutic value as a neuroprotective agent. While the medicinal uses of *W. somnifera* are well established in Ayurveda for centuries, there is still a need to evaluate the therapeutic efficacy and underlying mechanism of various unexplored components of this multipurpose herb. The current drug discovery process for neurodegenerative disorders in the pharmaceutical industry is inefficient, time taking, and expensive. In such a scenario, Ayurvedic medicines based on plant extracts such as *W. somnifera* offer more rapid path towards treatment possibilities with least toxicity and have therefore gained enormous attention in the recent years. Although more rigorous efforts are required to describe the molecular mechanism and enhance the applicability of herbal extracts for age-related neurodegenerative therapies. Therefore, a combination of high-throughput screening methods for all possible active components of *W. somnifera* along with mechanistic insights will create hope for improved treatment approaches for neurological diseases.

14.4 Role of Withania Somnifera in Infertility Treatment

Worldwide herbs are used as alternative medicines to cure infertility. Herbal therapy uses WS, senna (*Cassia angustifolia*), Gokshura (*Tribulus terrestris*), Anantmool (Indian Sarsaparilla, *Hemidesmus indicus*), Shatavari, Swet musli, wild yarn saw palmetto, and rhodiola for men to cure infertility. For women, dandelion, aka chasteberry, song quaint, black cohosh, maca, and schisandra, is used to cure infertility. Whereas *Withania Somnifera* (WS) is the flagship herb in Ayurveda to cure infertility in both men and women. WS (Ashwagandha, winter cherry, or Indian ginseng) is a major source of phytochemicals and physiochemicals. WS is a leading therapeutic plant which finds place in many alternative medicine (Meis et al. 2003). Every part of WS plant has medicinal value. Shoots of WS plant contain calcium, phosphorus, protein, and scopoletin. They are helpful in curing impotency. The stem structure consists of condensed tannins free amino acids and flavonoids. Green leaves are chewed to control diabetes, high blood pressure, and bad cholesterol. Leaves which are bitter in taste are also recommended by practitioners in painful swellings and fever. Extract of flowers is diuretic, depurative, aphrodisiac, and astringent. The seeds in extract or powder form are treated as astringent and anthelmintic. Among all parts of WS, roots are most valuable and have innumerable uses. Therapeutic claims of WS roots are mentioned in the coming paragraphs (Bonilla et al. 2021).

14.5 Phytochemical and Physiochemical Aspects

Previous topic is dealt with biochemical structure of WS. Biochemical structure outlines chemical processes which occur in living organisms. All the chemical reactions which undergo in human body are biochemical reactions. Every part of WS (shoots, stem, leaves, flowers, seeds, or roots) is an important alternative medical curing agent. Extract or powder of WS roots contains steroidal and non-steroidal alkaloids, neurotransmitters, ergostane, steroidal lactones, fatty acids (non-essential and essential), flavonoids, amino acids, and withanolides. Withanolides are further divided into withanosides, withaferin A, and beta-sitosterol and sitoindosides. Among these constituents, various amino acids, withanolides, and alkaloids play vital role in curing infertility (Mishra et al. 2000a, b). WS in pure herbaceous powder form, apart from above, also contains saponins, namely, anaferon, anahygrine, cuscohygrine, tyrosine, withanine, visamine, sitoindoside, somnine, and amino acids like aspartic acid, glutamic acid, butyric acid, and chlorogenic acid. WS is a rich source of iron, calcium, phosphorous, starch, glycosides volatile oil, and chymase enzyme. All these components of WS deliver countless therapeutic functions including infertility remover. WS has large historical background of over 3500 years of its medical use in India. Abundance of steroidal and non-steroidal alkaloids, withanolides, amino acids, withaferin A, starch, glycosides, saponins, calcium, phosphorus, and iron make it herbal pick to perform broad spectrum of biomedical functions. The broad spectrum exhibited by WS includes anti-glycemic, anti-arthritic, anticancer, lactogenic, anti-inflammatory, antimicrobial, immunomodulatory, adaptogenic, and aphrodisiac. Classically, WS is treated as stimulant, antioxidant, narcotic, aphrodisiac, diuretic, astringent, anthelmintic, and thermogenic. However, most dependable application of WS is curing infertility in both men and women. Classically, WS is used as Rasayan (tonic or rejuvenator) in Ayurveda. Generally, root of WS is dried and granulated as powder in a grinder. WS root powder taken in doses of 450 mg to 2 g twice a day certainly cures infertility (Ilayperuma et al. 2002; Patil et al. 2012).

14.5.1 Infertility at a Glance

Infertility is a complex jargon of problems arising out of physiological, biochemical, psychological, or economical variants. According to the WHO report, worldwide about 15% of married couples suffer from infertility. WHO estimate shows worldwide 70–80 million couples have infertility problem. Infertility is a failure of couple to conceive after having regular unprotected sex. Often, in simple words, infertility can be defined as non-conceiving situation in couples after years of regular sexual intercourse without introduction of birth control measures. Causes of infertility in both men and women may be different. Low sperm count, poor sperm quality, less motility, abnormal sperm, ejaculation disorder, stress, and hormonal imbalance may be the few causes for men infertility. Etiology of men infertility classified on the basis of testicular disorder is as follows:

- Pretesticular: Etiologies prior to testes, like hypothalamus-pituitary-gonadal axis disorders, psychopathic, sexual disorder, or syntonic ailments
- Testicular: Causes like varicocele (low sperm production and poor sperm quality), cryptorchidism (dislocation of testes), semenal abnormality, and infection
- Post-testicular: Causes responsible for unsmooth sperm transfer like hypospadias (anatomical abnormalities, disorder of penis), dysfunction, or obstruction of epididymis
- Idiopathic: Unexplained or unidentified causes

In women, various reproductive dysfunctions along with physio-psycho stresses are reasons of infertility. Failure to ovulate, menstrual disorder, egg immaturity, and reproductive system structural problems are the common causes of women infertility. Other etiologies of women infertility include dysfunction of hypothalamus-pituitary-ovarian axis, endometriosis, ovarian ailments, fallopian tubal disorders, or uterine pathologies. Confirm studies have proved that regular intake of WS (especially root powder) in doses specified by medical experts can eliminate problem of infertility in both men and women. WS has potential to regulate hormonal imbalance and can strengthen and rejuvenate reproductive system of both male and females (Krauss 2011; Nasimi et al. 2018; Kelgane et al. 2020).

14.5.2 Ability of Withania somnifera *to Cure Infertility*

Studies reveal that WS roots have gonadotropic functions which increase seminiferous tubular cell layers in males and general weight in females by developing follicles size. WS enhances spermatogenic activity due to hypothalmic-hypophysical gonadal hormonal axis supporting and balance of testosterone in testes. WS affects semen parameters of infertile men. The presence of high amount of alkaloids, amino acids (non-essential and essential), and ergostane steroids increases vitamins A, C, and E and improves semen quality thereby increasing fertility. These constituents of WS are supposed to restore testosterone secretion, improve detoxification due to scavenging activity, and reduce stress. Treatment by aqueous extract of WS into infertile married couple has shown significant statistical increased sexual function index (SFI) and retarding sexual distress index (SDI) in both male and female. Endocrinologists prescribe medications such as human menopausal gonadotropin (hMG), human chorionic gonadotropin (hCG), gonadotropin releasing hormone (GnRH), recombinant FSH, or clomiphene citrate (estrogen receptor) for treatment of reproductive dysfunctions of infertile couples. However, Indian Ayurvedacharyas recommend use of WS to stimulate gonadotropin release (Lopresti et al. 2019).

14.5.3 Mechanism of Function in Men and Women

Mechanism of effect of WS on endocrine glands/reproductive system is interesting to study. Mechanism is linked with both oxidative and non-oxidative features and the capability to expedite detoxification process and enhancement of the hormonal balance of follicle stimulating hormone (FSH), luteinizing hormone (LH), and testosterone. WS induces gonadotropin hormone secretion release. It also maintains hormonal balance. WS provides metal ions which modify oxidative stress, prevent cell apoptosis, and facilitate enzyme activities in men. Alanine in seminal fluid increases due to alanine transaminase activity inducement by WS which results in less oxidative stress index (OSI) and improved semen parameters. Another salient feature of WS is to balance citrate glutamine, histidine, lactate, and phenyl amine in seminal fluid thereby increasing enzymatic activity in fatty acid metabolism and tricarboxylic acid (TCA) cycle. WS improves sperm concentration and increases motility, release of mature ejaculation, and normal spermatozoa to counter stress-related damages. Spermatogenesis function streamlines by WS. Stress also contributes towards the etiology of infertility. WS stimulate hypothalamus-pituitary-adrenal (HPA) axis to nullify stress. Cortisol is the lead stress hormone. WS reduces the generation of cortisol and reactive oxygen species (ROS). WS also controls GABA-mimetic action. WS is known to be the best adaptive to secure fertility by balancing homeostasis thereby protecting human body from the toxicity of stress. Regular intake of WS decreases production and release of cortisol, epinephrine, and norepinephrine to overcome stress and increase fertility.

Mechanism of function of effect of WS on women endocrine glands needs to be investigated. Researchers have vast untamed field to explore. Stress-related impacts of WS on women infertility are similar to that of men as already explained above. Infertility cases may occur due to men, women, or both. Studies show that each factor represents one third of the total infertility cases. Infertility in women may primarily be due to endometriosis, abnormality in fallopian tubes, evolution problem, or polycystic ovary syndrome. Women could have hypoactive sexual desire disorder (HSDD), female sexual dysfunction (FSD), female orgasmic disorder (FOD), or female sexual arousal disorder (FSAD). These disorders give rise to reduced arousal, dryness in vagina, reduced libido, problems to achieve orgasm, reduced genital perception, and pain during intercourse. WS due to its phytochemical and physiological characteristics, as already explained in the previous paragraphs, has potential to address sexual disorders and their effects on women. Symptoms of polycystic ovarian syndrome (PCOS) are significantly suppressed by WS. Intake of WS by women strengthens endocrine system and regulates adrenal and thyroid glands which are responsible for balancing the reproductive hormones. Lust, appeal for intercourse, readiness for courtship, and reaching orgasm exacerbate after regular intake of WS by women. Vigor, vitality, and sexual satisfaction are increased in women by prolonged WS intake. Lactation and ova production (Oogenesis) increase on WS administration. Endocrine/exocrine glands get activated and ovarian ducts nourished by WS due to increased secretion of prolactin and progesterone hormones, resulting in improved fertility in women having conceiving disorders. Thus, WS plays a vital role in female infertility and is a panacea for females to infertility dysfunctions (Singh et al. 2011; Sharma et al. 2018).

14.5.4 Benefits of Withania somnifera

WS in pure aqueous extract or powder form have many benefits over other substitute herbs:

- WS as an adaptive, aphrodisiac, narcotic, or astringent is deemed to be the best available herbal medicine to cure infertility.
- WS is well trusted by infertile couples being flagship of Ayurveda for curing infertility of both men and women.
- WS is panacea for all women sexual disorders.
- WS is lactogenic and oogenesis promoter, helpful for infertile women to conceive.
- WS is anti-hyperglycemic and can be taken by diabetic patients.
- WS as rejuvenator can be taken by all age groups in small doses to boost immunity.
- WS easily grows in almost every region of India irrespective of topography.
- Every part of WS plant has medicinal utility and produces no herbal waste.
- Medical advantages of WS are well explained and reported in Ayurveda; infertile couples are well aware about WS use.
- Agricultural productivity cost of WS is fairly low.
- Manufacturing cost of WS formulations is minimum, easily affordable by common citizen.
- WS being a purely herbal pack has no side effects.

- WS can be easily consumed along with other medicines.
- Export of WS formulations contributes in countries GDP.

14.6 Biotechnological Outlook/Technologies for the Production of Withanolides

Due to its immense medicinal values and therapeutic and pharmacological approach towards various diseases, *Withania somnifera* is getting extreme importance from the researchers and pharma industries. Nowadays, more focus is paid to WS to develop biotechnological techniques and approaches to enhance the withanolide production. The estimated production rate WS is 1500 tons/annum(approx.), whereas its annual demand crosses 7000/annum thus creating a huge gap. The major fact that renders the path to meet the current global demand is long cultivation process of *Withania*, which hampers all criteria to meet global consumption demand. Nowadays, biotechnologists and researchers are emphasizing more on outstretching the approach towards in vitro withanolide production in order to come up with the shortage faced by pharma market globally. The main reason of drifting the outlook from in-field to in vitro cultivation of WS is unavoidable factors which limit the cultivation strategies, ultimately challenging the global market demands and hampering the way to meet the input and output supply chain of medicine world. Poor seed viability and germination rate, downwards seedling survival graph, water, and climatic conditions are adverse yet need to be considered factors for moving towards biotechnology for in vitro cultivation and production of secondary metabolites to get immunomodulatory advantages. In *Withania somnifera*, withanolide production occurs in two distinctive ways: first is mevalonate and the second is non-mevalonate pathway. Squalene, cholesterol, and mevalonic acid act as three immediate precursors in metabolic pathway during withanolide production.

14.6.1 Enhancement of Withanolide Production in WS by Using Hairy Root Culture Technique

Currently, hairy root culture technique is emerging as one of the efficient techniques giving revolutionary changes in the production strategies of secondary metabolites. In this technique, susceptible explants are infected with *Agrobacterium rhizogenes* for a particular time period and are observed by certain distinguished factors like negative geotropism, positive growth rate symptoms, accurate genetic stability, and remarkable profuse branching (Cai et al. 2012). The potentiality of hairy root culture is directly proportional to any alterations in nutrient medium. A series of research work has been carried out by Praveen and Murthy (2010) on WS hairy root culture technique for the withanolide A production by simply optimizing the growth parameters. As per their findings, they investigated that any changes in the type and concentration of carbon source may show different growth and increased withanolide production under optimal circumstances. It was concluded by their series of experimental setup that sucrose as a carbon source (11.92 g/l DW and 11.96 mg/g DW of withanolide A) is best for maximum withanolide production and bioaccumulation. Biomass accumulation was observed to be maximum (11.92 g/l DW) at 3% sucrose concentration; on the other hand, production of withanolide A was observed to have a favorable growth approach at 4% concentration of sucrose. Elicitor treatment has also been introduced and giving new heights to the biotechnological approaches for increasing secondary metabolite production in case of *Withania somnifera.* Hairy root when treated with methyl jasmonate (Mej) and salicylic acid (SA) enhances the bioaccumulation and production rate of withanolide A and of course of other metabolites possessing global pharmaceutical importance. As per new findings by the team of researchers, it was revealed that ammonia nitrate (NH_4^+ /NO_3^-) ratio imposes critical viability in the metabolite production rate. When the concentration of NO_3^- is higher than that of NH_4^+ (0.00/18.80 mM), the withanolide A production rate increases gradually. They concluded the series of experimental setup by plotting a graph showing the maximum biomass accumulation rate (148.75 g/l FW and 14.79 g/l DW) at 14.58/37.60 Mm (NH_4^+/ NO_3^- ratio) concentration (Praveen and Murthy 2012).

14.6.2 Use of Cell Suspension Culture for Enhancing the Production of Withanolides

Nowadays, cell suspension culture technique is well known for its phenomenal property of enhancing the ability to yield maximum metabolite production and contribute maximum in global pharma market. According to research outcomes, production of withaferin A in suspension cell culture was supported by proven evidences of adding salicin at 750 μm, which resulted in an outstanding production rate. WithaferinA, 25 ± 2.9 mg/l, which was comparatively very much more than the yield observed under controls, 0.47 ± 0.03 mg/l (Ciddi 2006; Nagella and Murthy 2010), used shake flask for setting up cell suspension culture technique for metabolite production in *Withania somnifera*. Nagella et al. studied the basic co-relation between the effect of various factors like P^H, media, growth regulators (like auxins and cytokine), and source of carbon and their strength and of P^H for withanolide A production. As per their findings, for productive and maximum biomass accumulation and withanolide A production (2.95 mg/g DW), the optimal condition was observed in Murashige and Skoog (MS) medium, 10 g/l(FW) inoculum and 3% w/v sucrose, and pH of the medium was adjusted to 5.8 along with culture period of 4 weeks (Table 14.1). Production of withanolide A is also observed to be enhanced by the action of an endophytic fungus *Piriformospora indica.* A comparative study was carried out by Ahlawat et al. (2015) to show how enrollment of fungus can help

Table 14.1 Biotechnological techniques for withanolide production using various plant tissue culture techniques

Procedure applied	Elicitor used	Time taken	Metabolite produced	Reference
Hairy root culture	NH_4^+/NO_3^-	4–5 weeks	**Withanolide A** 15.27 mg/g	Praveen and Murthy (2010)
Hairy root culture	Salicylic acid	4–5 weeks	**Withaferin A** 70.72 mg/g DW	
Adventitious root	Aluminium chloride 10 mg/L	4 hours, 6 weeks	**Withanone** 84.35 mg/g DW	Sivanandhan et al. (2012)
Adventitious root	Chitoan-100 mg/L	4 hours, 6 weeks	**Withanolide A** 323.85 mg/g **Withanolode B** 0.275 mg/g **Withaferin A** 4.347 mg/g	Sivanandhan et al. (2012)
Adventitious root culture	Salicylic acid 150 μm	Approx.10 days elicitation period	**Withanolide A** 64.65 mg/g **Withanolide B** 33.74 mg/g **Withaferin-A** 17.47 mg/g	Sivanandhan et al. (2012)
Cell Suspension culture	NH_4^+/ NO_3^-	–	**Withanolide A** 3.96 mg/g	Nagella and Murthy (2010)
Cell suspension Culture	Seaweed extracts (*Gracilaria edulis*)	–	**Withanolide A 7.21 mg/g Withanolide-B 4.23 mg/g Withaferin A 3.88 mg/g**	
Cell Suspension	Homogenate suspension *of Priformospora indica*	15 days elicitation time	Withaferin A 4.9 ± 0.23 mg/L	Ahlawat et al. (2015)

in increasing the production of secondary metabolites in *Withania somnifera*. In the repeated experimental setups, cell suspension and callus of WS were inoculated with various disc of *P. indica* at different time intervals and obviously with varying concentrations of cell homogenate and culture filtrate. Elicitation capacity of *P. indica* when compared with controls was observed to be increased by 2.04 times, sequentially followed by 1.78 times and 1.46 times after the addition of 3% (v/v) cell homogenate, 3% (v/v) culture filtrate, and culture disc. From the research it has been found that after inoculation of *P. indica* cell homogenate, withaferin A production was 4.9 ± 0.23 mg/l with an exposure period of 7 hours 9 (Table 14.1). Productive and maximum biomass accumulation and withanolide production (2.95 mg/g DW) were observed in MS medium, 10 g/l (FW) inoculum, and 3%W/V sucrose, and pH of the medium was adjusted to 5.8 along with culture period of 4 weeks (Table 14.1). Addition of polyamines like spermidine, spermine, and putrescine in nodal cultures of WS is known to increase the withanolide production. Increased production of withanolide was remarkably noticed in the in vitro regenerated plant parts (leaves, stem, root) when compared to the in-field grown samples. Comparative production analysis of in vitro and in-field leave extract has shown that in vitro regenerated leaves sample has shown 1.14- and 1.20-fold in withanone and withaferin A production. On the other hand, in vitro regenerated root has shown a remarkable increase of 1.10 times in withanolide A production than in the in-field-grown roots (Table 14.1). No significant change was observed in withanolide production of in vitro and in-field regenerated plant (stem).

14.7 Conclusion

Withania somnifera has been used for centuries as an efficient immunomodulator and is well known in pharmaceutical world especially in Ayurvedic medicines for curing many human ailments like inflammation, cancer, and neurogenerative diseases as well. The presence of few extremely important secondary metabolites like withanolides, sitoindosides, and few alkaloids has been proved by valid scientific research analysis, hence making it widely acceptable in pharma sector for treating acute to chronic ailments. Acceptable and remarkable results of treating Alzheimer's patients with *Withania somnifera* are observed worldwide as it is known to improve the cognitive declines and, on the other hand, increases the level of two major factors responsible for synaptic plasticity and amyloid beta elimination, i.e., PSD-95 and low-density LRP1. Somnifera has shown significant results in infertility treatment. Valid experimental researchers have proved that *Withania's* rich chemical composition helps in reducing stress level which is anyhow known as root cause for infertility among youth. In addition, it improves the semen quality and antioxidant level. Nowadays, biotechnology and biotechnologists are giving new heights and direction to the in vitro production of withanolides and withaferin, in order to meet the global demand with maximum yield and higher concentrations of metabolites.

Competing Interest The author declares that there are no competing interests.

References

Aalinkeel R, Hu Z, Nair BB, Sykes DE, Reynolds JL, Mahajan SD, Schwartz SA (2010) Genomic analysis highlights the role of the JAK-STAT signaling in the anti-proliferative effects of dietary flavonoid-'Ashwagandha' in prostate cancer cells. Evid Based Complement Alternat Med 7: 177–187

Ahlawat S, Saxena P, Ali A, Abdin MZ (2015) *Piriformospora indica* elicitation of withaferin a biosynthesis and biomass accumulation in cell suspension cultures of *Withania Somnifera*. Symbioisis 69:37–46

Akram M, Nawaz A (2017) Effects of medicinal plants on Alzheimer's disease and memory deficits. Neural Regen Res 12:660–670

Apostolova LG (2016) Alzheimer disease. Continuum (Minneap Minn) 22:419–434

Bhandari C (1970) Ashwagandha(*Withania Somnifera*) ' Vanaoshadhi Chandroday" (An encyclopedia of Indian Herbs) Vol. 1 series of Varanasi Vidyavilas Press, Varanasi, pp. 96–97

Bonilla DA, Moreno Y, Gho C, Petro JL, Odriozola-Martínez A, Kreider RB (2021) Effects of ashwagandha (Withania somnifera) on physical performance: systematic review and Bayesian meta-analysis. J Funct Morphol Kinesiol 6:20

Cai Z, Kastell A, Knorr D, Smetanska I (2012) EXUDATION: an expanding technique for continuous production and release of secondary metabolites from plant cell suspension and hairy root cultures. Plant Cell Rep 31(3):461–477

Chandrasekhar K, Kapoor J, Anishetty S (2012) A prospective, randomized double-blind, placebo-controlled study of safety and efficacy of a high-concentration full-spectrum extract of ashwagandha root in reducing stress and anxiety in adults. Indian J Psychol Med 34:255–262

Changhadi GS (1938) Ashwagandharishta- Ras Tantra Saar and Sidhyaprayog Sangrah. Krishna-Gopal Ayurveda Bhawan (Dharmarth Trust), Nagpur, 743–744

Ciddi V (2006) Withaferin a from cell cultures of Withania Somnifera. Indian J Pharma Sci 68:490–492

Davis L, Kuttan G (2000) Immunomodulatory activity of Withania somnifera. J Ethnopharmacol 71:193–200

Davis L, Kuttan G (2002) Effect of Withania somnifera on cell mediated immune responses in mice. J Exp Clin Cancer Res 21:585–590

Dhuley JN (1997) Effect of some Indian herbs on macrophage functions in ochratoxin A treated mice. J Ethnopharmacol 58:15–20

Dimpfel W, Schombert L, Keolinger-Dimpfel IK & Pannosian A (2020) Effects of an adaptogenic extract on electrical activity of the brain in elderly subjects with mild cognitive impairment: a randomized, double-blind, placebo-controlled, two-armed cross-over study. *Pharmaceuticals (Basel),* 13

Dubey S, Yoon H, Cohen MS, Nagarkatti P, NagarkattI M, Karan D (2018) Withaferin A associated differential regulation of inflammatory cytokines. Front Immunol 9:195

Dutta R, Khalil R, Green R, Mohapatra SS, Mohapatra S (2019) Withania Somnifera (Ashwagandha) and withaferin a: potential in integrative oncology. Int J Mol Sci 20:5310

Eckert GP (2010) Traditional used plants against cognitive decline and Alzheimer disease. Front Pharmacol 1:138

Fan L, Mao C, Hu X, Zhang S, Yang Z, Hu Z, Sun H, Fan Y, Dong Y, Yang J, Shi C, Xu Y (2019) New insights into the pathogenesis of Alzheimer's disease. Front Neurol 10:1312

Gambhir L, Checker R, Sharma D, Thoh M, Patil A, Degani M, Gota V, Sandur SK (2015) Thiol dependent NF-kappaB suppression and inhibition of T-cell mediated adaptive immune responses by a naturally occurring steroidal lactone Withaferin A. Toxicol Appl Pharmacol 289:297–312

Gautam A, Wadhwa R, Thakur MK (2016) Assessment of cholinergic properties of Ashwagandha leaf-extract in the amnesic mouse brain. Ann Neurosci 23:68–75

Gautam M, Diwanay SS, Gairola S, Shinde YS, Jadhav SS, Patwardhan BK (2004) Immune response modulation to DPT vaccine by aqueous extract of Withania somnifera in experimental system. Int J Immunopharmacol 4:841–849

Govardhan S (1938) Ashwagandharishta-Rastantra Sar Evam Sidhyaprayog Sangrah. Krishna-Gopal Ayurveda Bhawan (Dharmarth Trust), Nagpur, pp 743–744

Gupta M, Kaur G (2019) Withania somnifera (L.) Dunal ameliorates neurodegeneration and cognitive impairments associated with systemic inflammation. BMC Complement Altern Med 19:217

Holtzman DM, Mandelkow E, Selkoe DJ (2012) Alzheimer disease in 2020, vol 2. Cold Spring Harb Perspect Biol Med, p a011585

Ichikaw AH, Takada Y, Shishodia S, Jayaprakasham B, Nair MG, Aggarwal BB (2006) Withanolides potentiate apoptosis, inhibit invasion, and abolish osteoclastogenesis through suppression of nuclear factor-kappaB (NF-kappaB) activation and NF-kappaB-regulated gene expression. Mol Cancer Ther 5:1434–1445

Ilayperuma I, Ramasooriya WD, Weerasooriya TR (2002) Effect of Withania somnifera root extract on the s sexual behavior of male rats. Asian J Androl 4(4):295–298

Iuvone T, Esposito G, Capasso F, Izzo AA (2003) Induction of nitric oxide synthase expression by Withania somnifera in macrophages. Life Sci 72:1617–1625

Jayaprakasham B, Padmanabhan K, Nair MG (2010) Withanamides in Withania somnifera fruit protect PC-12 cells from beta-amyloid responsible for Alzheimer's disease. Phytother Res 24: 859–863

Kaur T, Singh H, Mishra R, Manchanda S, Gupta M, SainiI V, Sharma A, Kaur G (2017) Withania somnifera as a potential anxiolytic and immunomodulatory agent in acute sleep deprived female Wistar rats. Mol Cell Biochem 427:91–101

Kelgane SB, Salve J, Sampara P, Debnath K (2020) Efficacy and tolerability of ashwagandha root extract in the elderly for improvement of general Well-being and sleep: a prospective, randomized, double-blind, placebo-controlled study. Cureus 12:e7083

Keuboyama T, Tohda C, Komatsu K (2014) Effects of Ashwagandha (roots of Withania somnifera) on neurodegenerative diseases. Biol Pharm Bull 37:892–897

Khan S, Malik F, Suri KA, Singh J (2009) Molecular insight into the immune up-regulatory properties of the leaf extract of Ashwagandha and identification of Th1 immunostimulatory chemical entity. Vaccine 27:6080–6087

Konar A, Gupta R, Shukla RK, Maloney B, Khanna VK, Wadhwa R, Lahiri DK, Thakur MK (2019) M1 muscarinic receptor is a key target of neuroprotection, neuroregeneration and memory recovery by i-Extract from Withania Somnifera. Sci Rep 9:13990

Krauss C (2011) Male infertility: pathogenesis and clinical diagnosis. Best Pract Res Clin Endocrinol Metab 25(2):271–278

Kuboyama T, Tohda C, Komatsu K (2005) Neuritic regeneration and synaptic reconstruction induced by withanolide A. Br J Pharmacol 144:961–971

Kuboyama T, Tohda C, Zhao J, Nakamura N, Hattori M, Komatsu K (2002) Axon- or dendrite-predominant outgrowth induced by constituents from Ashwagandha. Neuroreport 13:1715–1720

Kumar S, Harris RJ, Seal CJ, Okello EJ (2012) An aqueous extract of Withania Somnifera root inhibits amyloid beta fibril formation in vitro. Phytother Res 26:113–117

Kurapati KR, AtluriI VS, Sammikannu T, Nair MP (2013) Ashwagandha (Withania somnifera) reverses beta-amyloid1-42 induced toxicity in human neuronal cells: implications in HIV-associated neurocognitive disorders (HAND). PLoS One 8:e77624

Lopresti AL, Smith SJ, Malvi H, Kodgule R (2019) An investigation into the stress-relieving and pharmacological actions of an ashwagandha (Withania somnifera) extract: a randomized, double-blind, placebo-controlled study. Medicine (Baltimore) 98:17186

Ma Y, Yang MW, Li XW, Yue JW, Chen JZ, Yang MW, Huang X, Zhu LL, Hong FF, Yang SL (2019) Therapeutic effects of natural drugs on Alzheimer's disease. Front Pharmacol 10:1355

Mahima RA, Deb R, Latheef SK, Abdul Samad H, Tiwari R, Verma AK, Kumar A, Dhama K (2012) Immunomodulatory and therapeutic potentials og herbal traditional/indigenous and ethno veterinary medicines. Pak J Biol Sci 15:754–774

Mandlik Ingawale DS, Namdeo AG (2020) Pharmacological evaluation of Ashwagandha highlighting its healthcare claims, safety, and toxicity aspects. J Diet Suppl:1–44

Martorana F, Guidotti G, Brambilla L, Rossi D (2015) Withaferin A inhibits nuclear factor-kappaB-dependent pro-inflammatory and stress response pathways in the astrocytes. Neural Plast 2015: 381964

Meis PJ, Klebanoff M, Thom E, Dombrowski MP, Sibai B, Moawad AH et al (2003) Prevention of recurrent preterm delivery by 17 alpha-hydroxyprogesterone caproate. N Engl J Med 348:2379–2385

Mishra L, Singh BB, Dagenias S (2000a) Scientific basis for the therapeutic use of *Withania Somnifera* (Ashwagandha): a review. Altern Med Rev 5(4):334–336

Mishra LC, Singh BB, Dagenais S (2000b) Scientific basis for the therapeutic use of Withania somnifera (ashwagandha): a review. Altern Med Rev 5:334–346

Mohanty IR, Arya DS, Gupta SK (2008) Withania Somnifera provides cardioprotection and attenuates ischemia-reperfusion induced apoptosis. Clin Nutr 27:635–642

Mulabagal V, Subbaraju GV, Rao CV, Sivaramakrishna C, Dewitt DL, Holmes D, Sung G, Aggarawal BB, Tsay HS, Nair MG (2009) Withanolide sulfoxide from Aswagandha roots inhibits nuclear transcription factor-kappa-B, cyclooxygenase and tumor cell proliferation. Phytother Res 23:987–992

Muralikrishnan G, Dinda AK, Shakeel F (2010) Immunomodulatory effects of Withania somnifera on azoxymethane induced experimental colon cancer in mice. Immunol Invest 39:688–698

Nagella P, Murthy HN (2010) Effect of macro elements and nitrogen source on biomass accumulation of and Withanolide A production from cell suspension cultures of Withania Somnifera(L.) Dunal. Plant Cell Tissue Cult Organ Cult 104:119–124

Nasimi R, Zomossdi A et al, (2018) Effects of Withania Somnifera on reproductive system: a systematic review of the available Evidence. ID 4076430, 17 pages

Ng QX, Loke W, Foo NX, Tan WJ, Chan HW, Lim DY, Yeo WS (2020) A systematic review of the clinical use of Withania somnifera (Ashwagandha) to ameliorate cognitive dysfunction. Phytother Res 34:583–590

Palliyaguru DL, Singh SV, Kensler TW (2016) Withania somnifera: from prevention to treatment of cancer. Mol Nutr Food Res 60:1342–1353

Patil RR, Vora SR, Pillai MM (2012) Protective effect of spermatogenic activity of Withania somnifera (Ashwagandha) in galactus stressed mice. Anim Biol Res 3(8):4169–4165

Plunkett ML, Sanders ME, Selvaraj P, Dustin ML, Springer TA (1987) Rosetting of activated human T lymphocytes with autologous erythrocytes. Definition of the receptor and ligand molecules as CD2 and lymphocyte function-associated antigen 3 (LFA-3). J Exp Med 165: 664–676

Pratte MA, NanavatiI KB, Young V, Morley CP (2014) An alternative treatment for anxiety: a systematic review of human trial results reported for the ayurvedic herb ashwagandha (Withania somnifera). J Altern Complement Med 20:901–908

Praveen N, Murthy HN (2010) Production of Withanolide-A form adventitious root culture *of Withania Somnifera*. Acta Physiologiae Plantarum 32(5):1017–1022

Praveen N, Murthy HN (2012) Withanolide A production from *Withania Somnifera* hairy root cultures with improved growth by altering the concentrations of macro elements and nitrogen sources in the medium. Acta Physioogiae Plantarum 35:811–816

Rao RV, Descamps O, John V, Bredesen DE (2012) Ayurvedic medicinal plants for Alzheimer's disease: a review. Alzheimers Res Ther 4:22

Sharma AK, Basu I, Singh S (2018) Efficacy and safety of Ashwagandha root extract in subclinical hypothyroid patients: a double-blind, randomized placebo-controlled trial. J Altern Complement Med 24:243–248

Singh N, Bhalla M, de Jager P, Gilca M (2011) An overview on ashwagandha: a Rasayana (rejuvenator) of Ayurveda. Afr J Tradit Complement Altern Med 8(5 Suppl):208–213

Singham N, Bhalla P, Jager MG (2011) An overview on ashwagandha: a Rasayan (rejuvenator) of ayurveda. Afr J Tradit Complement Altern Med 8(5):208–213

Sivanandhan G, Arun M, Mayavan S, Rajesh M, Mariashibu TS, Manickavasagam M, Selvaraj N, Ganapathi A (2012) Chitosan enhances withanolides production in adventitious root cultures of Withania somnifera (L.) Dunal. Ind Crop Prod 37(1)

The manufacturer's authorised representative in the EU is Springer Nature Customer Service Centre GmbH, Europaplatz 3, 69115 Heidelberg, Germany. If you have any concerns regarding our products, please contact ProductSafety@springernature.com

Printed and bound by CPI Group (UK) Ltd, Croydon, CR0 4YY
30/09/2026
02225245-0010